Physics and Chemistry in Space   Vol. 20
Space and Solar Physics

Series Editors: M.C.E. Huber · L.J. Lanzerotti · D.Stöffler

R. Schwenn E. Marsch (Eds.)

Physics of the Inner Heliosphere I

Large-Scale Phenomena

With 103 Figures

Springer-Verlag
Berlin Heidelberg New York
London Paris Tokyo
Hong Kong Barcelona

Dr. RAINER SCHWENN
Dr. ECKART MARSCH
Max-Planck-Institut für Aeronomie, Postfach 20,
D-3411 Katlenburg-Lindau, Fed. Rep. of Germany

Series Editors:

Professor Dr. M. C. E. HUBER
European Space Research and Technology Centre,
Keplerlaan 1, NL-2200 AG Noordwijk, The Netherlands

Dr. L. J. LANZEROTTI
AT&T Bell Laboratories, 600 Mountain Avenue,
Murray Hill, NJ 07974-2070, USA

Professor Dr. D. STOEFFLER
Institut für Planetologie, Universität Münster,
Wilhelm-Klemm-Str. 10, D-4400 Münster, Fed. Rep. of Germany

ISBN-13: 978-3-642-75363-3   e-ISBN-13: 978-3-642-75361-9
DOI: 10.1007/978-3-642-75361-9

Library of Congress Cataloging-in-Publication Data. Physics of the inner heliosphere I: large-scale phenomena / R. Schwenn, E. Marsch (eds.). p. cm. – (Physics
and chemistry in space; vol. 20) Includes bibliographical references and index.

                              1. Heliosphere. 2. Sun – Corona. 3. Inter-
stellar matter. 4. Interplanetary magnetic fields. I. Schwenn, R. (Rainer), 1941–
II. Marsch, E. (Eckart), 1947–. III. Series: Physics and chemistry in space; v. 20.
QC801.P46 vol. 20 [QB520]   530'.0919 s – dc20   [523.7'5]   90-9975

© Springer-Verlag Berlin Heidelberg 1990
Softcover reprint of the hardcover 1st edition 1990

2156/3150-543210 – Printed on acid-free paper

Preface

The idea of producing a book on the inner heliosphere dates back to 1982. At that time, shortly after the activity maximum of solar cycle 21, the scientific evaluation of the data returned by the *Helios* solar-probe mission had also reached peak intensity. It was a common feeling of the scientists involved that this was an appropriate time to summarize in a tutorial fashion the results from *Helios*. The basic idea was put into action by the designated editors F.M. Neubauer and L.J. Lanzerotti, and it has not changed much since then, except in one crucial respect: the mission went on, surprisingly enough, for another four years, and the data evaluation has also continued to produce new results at a fast pace. A substantial expansion of the book project in terms of both page allocation and delivery deadlines was therefore inevitable.

After F.M. Neubauer had withdrawn as an editor, L.J. Lanzerotti appointed us as new co-editors of the book. We were fortunate in that we found Springer a very cooperative and patient partner in the project. They agreed with our proposal to spread the material over two volumes, and they accepted an extension of the time schedule. The authors turned out to be highly motivated and productive as well, though fairly slow in some cases. However, they all made proper use of the time and put a lot of effort into maturing their work.

All articles were scrutinized by well-known experts in their fields: B. Bavassano, W.C. Feldman, S.P. Gary, J.T. Gosling, J.T. Hoeksema, R.B. McKibben, M. Neugebauer, D.A. Roberts, T.R. Sanderson, S.J. Schwartz, N.R. Sheeley, Jr., S.T. Suess, B.T. Tsurutani, R. Woo, and H.A. Zook. It is a particular pleasure for us to thank here these referees for their immense efforts, which were highly appreciated and carefully taken into consideration by the authors and were, we think, to the benefit of the readers.

It is certainly appropriate to express here our gratitude to the many individuals, organizations, and companies who have rendered possible such ambitious space missions as the *Helios* solar probes. Thus, they have laid the basis for the many new scientific achievements that finally led to the compilation of this book. As representatives for all them, let us just mention the names of the *Helios* project scientists H. Porsche and J. Trainor. Further thanks are due to various persons who promoted the final production of the book: F.M. Neubauer and L.J. Lanzerotti for inaugurating it, W. Engel and H.U. Daniel at Springer, M.K. Bird and P.W. Daly for linguistic advice, and Mrs. G. Bierwirth and Mrs. U. Spilker for their secretarial services.

H. Porsche deserves special credit, since we were particularly inspired and motivated in editing this book by the impressive booklet he had edited at the tenth anniversary of the *Helios 1* launch. We realize that his work has gained the considerable attention of the public and has served as a major source of attraction to our field of research for many students. We hope that the present book will, in turn, continue to inspire students and scientists to further expand our views on the physics of the inner heliosphere.

Katlenburg-Lindau
April 1990

Eckart Marsch
Rainer Schwenn

Contents

Index of Contributors

Bird, Michael K.
Radioastronomisches Institut, Universität Bonn, Auf dem Hügel 71,
D-5300 Bonn, Fed. Rep. of Germany

Edenhofer, Peter
Institut für Hoch- und Höchstfrequenztechnik, Ruhr-Universität Bochum,
Universitätsstrasse 150, D-4630 Bochum 1, Fed. Rep. of Germany

Grün, Eberhard
Max-Planck-Institut für Kernphysik, Postfach 10 39 80,
D-6900 Heidelberg, Fed. Rep. of Germany

Leinert, Christoph
Max-Planck-Institut für Astronomie, Königstuhl 17,
D-6900 Heidelberg, Fed. Rep. of Germany

Mariani, Franco
Universitá Degli Studi di Roma, Departimento di Fisica,
Via Orazio Raimondo, I-00173 (La Romanina) Roma, Italy

Marsch, Eckart
Max-Planck-Institut für Aeronomie, Postfach 20,
D-3411 Katlenburg-Lindau, Fed. Rep. of Germany

Neubauer, Fritz M.
Institut für Geophysik und Meteorologie, Universität zu Köln,
Albert-Magnus-Platz, D-5000 Köln 41, Fed. Rep. of Germany

Schwenn, Rainer
Max-Planck-Institut für Aeronomie, Postfach 20,
D-3411 Katlenburg-Lindau, Fed. Rep. of Germany

1. Introduction

Eckart Marsch and Rainer Schwenn

1.1 Why the "Inner Heliosphere"?

The central region of our solar system contains a commonplace G2 main-sequence star in the prime of life: the sun. It is the source of the light that illuminates and warms the earth, and it thus sustains life on our home planet. Humans have, from prehistoric beginnings to the present time, watched the sun with both bold admiration and scientific curiosity. They have continually refined their observational means, moving from the naked eye to the present sophisticated telescopes on the ground or in space. Modern science has unraveled mysteries of the sun, which even our contemporaries would not have dreamed of a few decades ago.

In common language the sun is usually identified with the photosphere, the luminous disk apparent to observers on earth. However, both ancient and modern man have been occasionally diverted by the splendor and mystery of the sun's atmosphere, the corona. We may never have guessed that the corona existed had not the brilliance of the photosphere been extinguished every so often by the earth's fortuitously positioned moon. Its angular extent on the sky happens to match accurately that of the sun. It is through these eclipses of the solar disk that the dim outer atmosphere of the sun is naturally revealed in the guise of the chromosphere and the corona.

At least a feeble impression of what has fascinated eclipse observers ever anew may be conveyed by Fig. 1.1: a contrast-enhanced photograph of a typical solar minimum corona, taken during the solar eclipse of 30 June 1973 [1.6]. The central part is an almost simultaneous H-alpha picture of the solar disk inserted for comparison. This amazing picture shows striking details and structures on various scales of the distribution in space of electrons revealed by Thomson scattering of photospheric light. The hot coronal plasma has a temperature in excess of one million degrees. As such, it produces a complex though weak emission pattern of its own, particularly in X-rays and the extreme ultraviolet, that is closely associated with the underlying structures of the sun's outer magnetic field. The corona is the source region of the solar wind, the continually outward-streaming supersonic flow that ultimately fills the whole solar system with its tenuous plasma.

The "heliosphere" is the spatial cavity carved by the solar wind ram pressure out of the local interstellar medium. It is the habitat of the sun and its planetary companions, also encompassing their satellites, the asteroids, and comets, each of which has its own particular gas and plasma environment. The solar system is pervaded by electromagnetic radiation, extending from gamma rays to radio

Physics and Chemistry in Space - Space and Solar Physics, Vol. 20
Physics of the Inner Heliosphere I   Editors: R. Schwenn · E. Marsch
© Springer-Verlag Berlin Heidelberg 1990

Fig. 1.1. The solar corona observed during the total eclipse on 30 June 1973. Photographic processing was performed to enhance small-scale intensity gradients. The central part is an almost simultaneous H-alpha picture of the solar disk inserted for comparison. From [1.6]

waves, of predominantly solar origin. There is also "corpuscular radiation" in the form of solar wind ions and electrons and more energetic particles emanating from the solar atmosphere or entering from outside as cosmic rays of galactic origin. Dust particles and micrometeorites sparsely populate the inner heliosphere, manifesting themselves in the form of the zodiacal light produced by scattering of white sunlight.

The prominent role played by the sun in shaping the interplanetary medium and influencing the magnetospheres and atmospheres of the planets is now widely recognized. Within only a few decades, solar system plasma physics [1.7] has in the age of space exploration emerged as a broad and mature scientific discipline concerned with the investigation and theoretical interpretation of the properties of ionized gases and charged particles in the solar system. The causal relations between physical processes in the interior and atmosphere of the sun and in the terrestrial environment appear to be of paramount importance for mankind's ability to comprehend and eventually even predict the impact of interplanetary "weather" on spaceship earth.

Further progress in the research field of solar–terrestrial relations crucially hinges on whether we can improve our understanding of the solar "input" into the heliosphere. In retrospect, it is true that many satellites in earth orbit had already gathered a wealth of scientific data in the early years of solar wind research. In parallel, scientists had been active in developing appropriate theoretical models for the solar wind and other interplanetary phenomena. However, owing to a lack of reliable information about the plasma state in the extended spatial domain between the corona and 1 AU, the modeling efforts were plagued by too many free parameters. Most models were able to produce more or less successfully the "right" parameters at 1 AU (e.g. the review by Hollweg [1.4]). Nevertheless, essential differences in their basic assumptions clearly showed up in the radial profiles they predicted for various plasma parameters. Any measurements of these radial gradients, expected to help constrain the then existing models, were therefore highly desirable.

Such reasoning, driven by scientific curiosity, led almost 25 years ago to the concept of a solar-probe mission designed to investigate and explore the interplanetary medium by approaching the sun as close as possible, using all the technical means available. The ambitious space mission emerging from this development was named, appropriately enough, *Helios*. After several years of careful scientific and technical preparation, two nearly identical space probes were finally launched toward the sun. They carried a well-selected set of scientific instruments designed to acquire new information about the interplanetary medium by means of *in situ* measurements. The name *Helios* was chosen to remind us of the Greek sun god who was believed to give light and warmth to human beings and was often pictured as racing across the heavens on fiery stallions.

The desired mission goals, in fact, have been fully reached. The present two-volume book provides ample evidence on how the numerous scientific achievements of the *Helios* mission have changed our view of the heliosphere. The success of the mission is due, in part, to the unplanned and unexpectedly long survival of the spacecraft and essential parts of the scientific payload. Many investigations covered nearly a complete solar activity cycle! This long mission duration came as a real surprise, since the technological problems involved had been huge and unprecedented. The interested reader is referred, e.g., to the publication celebrating the tenth anniversary of the launch of *Helios 1* [1.10], which also illustrates and summarizes in simple terms many of the results by then obtained. These will be treated in this book in considerably more depth and completeness.

The reader might wonder about the title of this book: what is it that physically discriminates the "inner" as opposed to the "outer" heliosphere? Admittedly, this separation seems somewhat artificial, and indeed it primarily reflects our vantage point, viewing from earth out into the solar system. Yet, there are also good physical reasons for this division in that the solar wind is a dynamically evolving magnetofluid. Whereas the state of the interplanetary medium inside the earth's orbit is still largely determined by the solar initial and boundary conditions, this no longer holds true beyond 1 AU. With increasing distance from the sun, the streams collide and interact more intensely with each other owing to the strongly bent spiral magnetic field. Near 1 AU the average inclination of the solar magnetic field to the radial direction already amounts to about 45 degrees. Therefore, at this distance range even the largest-scale streams undergo compression and deflection by their neighboring flows. The primordial coronal imprints on the plasma get

3

increasingly lost as the wind expands further. For this reason, we tend to draw the borderline of the inner heliosphere near 1 AU. Obviously, in the outer heliosphere dynamic phenomena of intrinsically interplanetary origins prevail, which modify and reprocess the original streams such that they can hardly be recognized. The comprehensive review by Burlaga [1.1] is devoted to these interesting processes.

1.2 Scientific Objectives and General Aspects of the Helios Mission

The *Helios* mission was aimed at performing *in situ* measurements of the inter-planetary medium in the near-solar environment. Two almost identical spacecraft were launched into highly elliptical, small inclination orbits with a perihelion distance of 0.31 AU for *Helios 1* (launched on 10 December 1974) and 0.29 AU for *Helios 2* (launched on 15 January 1976). Since the rotation axis of the sun is tilted by 7.25 degrees, the *Helios* orbits covered a heliographic latitude range between +7.25 and −7.25 degrees. In Fig. 1.2 the orbits of both *Helios* probes and the earth are shown. The tick marks referring to the calendar days in 1976 clearly demonstrate the variable orbital speeds along the highly eccentric orbits of the probes. The fast swing-by through perihelion caused *Helios 1* to traverse a solar latitudinal range from −6 to +6 degrees in about 20 days, i.e., in less than a solar rotation. (Some other aspects of the spacecraft and the mission strategy are given in [1.9] and [1.10]).

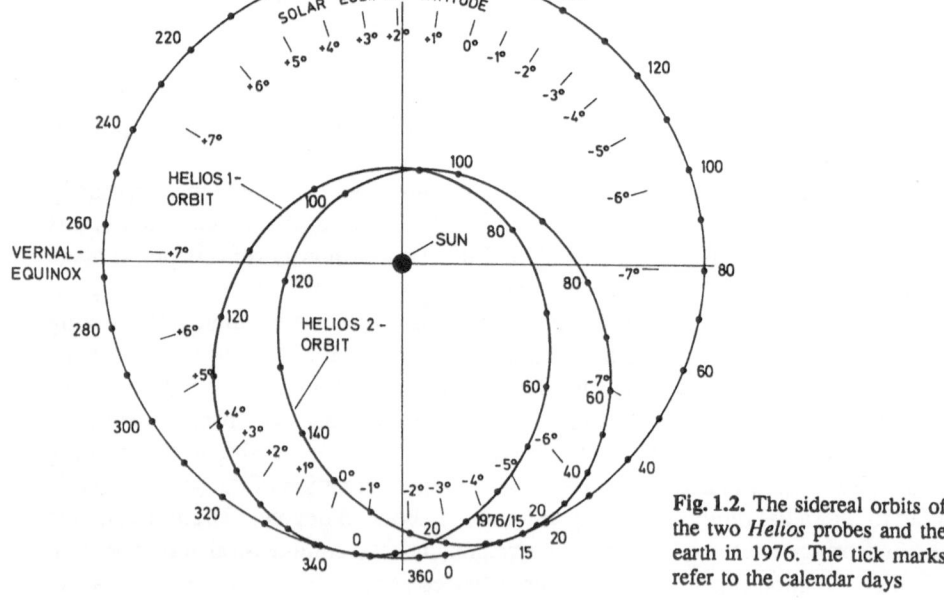

Fig. 1.2. The sidereal orbits of the two *Helios* probes and the earth in 1976. The tick marks refer to the calendar days

These orbital characteristics were crucial for performing studies of the near-sun latitudinal variations of the solar wind plasma and magnetic field properties. Another fortuitous outcome of the orbit configuration and the twin-probe concept was the occasionally occurring "line-up" constellations between the two probes. In these cases (e.g. on days 70 and 123 in 1976, see Fig. 1.2) the same plasma fluid element of the solar wind could be investigated from two space probes located at different heliocentric distances but equipped with sets of identical and cross-calibrated instruments. Similar to these "plasma line-ups" of the probes, constellations occurred where both spacecraft were connected magnetically along the Parker spiral, thus enabling energetic particle detectors to "see" the same particle populations at two locations on the same field line. Finally, it should be mentioned that many studies of the radial variations of solar wind properties were rendered possible by combining observations on *Helios* in the inner heliosphere with corresponding measurements made at earth orbit by the IMP satellites or in the realms of the outer planets by the *Voyager* space probes.

The scientific payload comprised twelve independent experiments described in detail by Porsche [1.9] (and in the several papers subsequent to his article). Table 1.1 lists these experiments, their principal investigators, and the affiliated institutions. The number of scientists engaged in the *Helios* project steadily increased with time over the number of those researchers who initiated it, built the instruments, and carried out the primary part of the mission in the mid-seventies.

Table 1.1. HELIOS experiments

Exp. 1	Plasma Particles	H. Rosenbauer	MPI für Physik und Astrophysik,
		R. Schwenn	Inst. für extraterr. Physik Garching
Exp. 2	Flux Gate Magnetometer	G. Musmann	Inst. für Geophysik und Meteorologie,
		F.M. Neubauer	TU Braunschweig
Exp. 3	Flux Gate Magnetometer	F. Mariani	Universita de l'Aquila, Italy
		N.F. Ness	NASA-GSFC Greenbelt, Md.
Exp. 4	Search Coil Magnetometer	G. Dehmel	Inst. für Nachrichtentechnik
		F.M. Neubauer	Inst. für Geophysik und Meteorologie,
			TU Braunschweig
Exp. 5a	Plasma Waves	D.A. Gurnett	Univ. of Iowa, Dep. of Physics and
			Astronomy, Iowa City
Exp. 5b	Plasma Waves	P.J. Kellogg	Univ. of Minnesota, School of
			Physics and Astronomy, Minneapolis
Exp. 5c	Radio Waves	R.R. Weber	NASA-GSFC Greenbelt, Md.
Exp. 6	Cosmic Rays	H. Kunow	Inst. für Reine und Angewandte Kernphysik,
			Univ. Kiel
Exp. 7	Cosmic Rays	J.H. Trainor	NASA-GSFC Greenbelt, Md.
Exp. 8	Low Energy Cosmic Rays	E. Keppler	MPI für Aeronomie, Inst. für
			Stratosphärenphysik, Katlenburg-Lindau
Exp. 9	Zodiacallight Photometer	C. Leinert	MPI für Astronomie Heidelberg
Exp. 10	Micrometeorite Analyzer	E. Grün	MPI für Kernphysik Heidelberg
Exp. 11	Celestial Mechanics	W. Kundt	Inst. für Theoretische Physik I Univ. Hamburg
		W.G. Melbourne	JPL Pasadena, Cal.
Exp. 12	Coronal Sounding	G.S. Levy	JPL Pasadena, Cal.
		H. Volland	Radioastronomisches Institut Univ. Bonn
		P. Edenhofer	DFVLR Oberpfaffenhofen

As is apparent from the list of authors of this book, several of the early pioneers are still actively taking part in the data analysis. Most of the data are now available through international data centers to a wider community of interested researchers. Numerous *Helios* investigators of the second and third generation have used these data and produced a wealth of new results, an activity continuing to the present day.

The payload of both *Helios* probes was composed of three different groups of experiments (the reader may forgive us that we use the term "experiment" for what is actually an apparatus for performing measurements – the use of this misnomer has become common within the community of space researchers):

1. The largest group of seven instruments was devoted to the interplanetary plasma and magnetic fields. The plasma analyzer provided energy spectra and two- or three-dimensional velocity distributions of solar wind electrons, protons, and helium ions. One search-coil and two flux-gate magnetometers and two plasma-wave experiments obtained measurements of the electromagnetic-field fluctuations over a wide range of frequencies. The three vector components of the magnetic field were recorded from DC up to 5 Hz, fast magnetic-field oscillations up to 2 kHz. Electrostatic waves were monitored up to 3 MHz. A radio-astronomy experiment was designed to track bursts of radio waves traveling between the sun and the earth.

2. In the second group, there were three experiments, whose aim was to measure the mass and energy spectra and directional distributions of solar and galactic cosmic "rays" (i.e., energetic particles). They were also assigned the task of monitoring the solar X-ray activity. On *Helios 2* a supplementary device was installed to register gamma-ray bursts that had then just been detected and were suspected to be of galactic origin.

3. In the third group, two investigations focused on establishing the physical characteristics of micrometeorites in the inner solar system. Three photometers registered intensity and polarization of the zodiacal light in three spatial directions. Two dust-particle analyzers directly measured the particle fluxes in order to determine the mass distribution and chemical composition of micrometeorites.

In addition to these "active" experiments a group of "passive" experiments was performed, making use of the probes' telemetry signal. A coronal sounding experiment was rendered possible by *Helios*' orbital characteristics. At times near superior conjunction the *Helios* radio signal had to pass through the solar corona to be received on earth. By analyzing the polarization of the signal and exploiting the Faraday rotation effect the plasma state of the solar corona could be probed. Other radio-science measurements using Doppler techniques provided further information on coronal and interplanetary properties. A celestial-mechanics experiment, intended to corroborate the validity of Einstein's theory of gravitation, was unfortunately unsuccessful because of various difficulties in achieving the required measurement accuracy.

The officially agreed duration, the minimum essential for designating the mission a "success", ran from the date of launch to the first perihelion passage about ninety days later, with a desired mission extension of 18 months. As the actual mission history shows, these objectives were more than accomplished: *Helios 1* lasted for more than eleven years until February 1986; *Helios 2* died on 3 March 1980. The main phase of the *Helios* mission fell into solar cycle 20 and was fortuitously close to the activity minimum. During that same phase, although a few months earlier, the famous *Skylab* mission carrying the *Apollo* Telescope Mount was active. This unique set of solar telescopes dramatically changed our view of the sun. *Skylab* provided invaluable solar and coronal observations in the visible, the extreme ultraviolet, and soft X-rays, which greatly helped to put the *Helios* interplanetary measurements into the proper solar-physics perspective, and thus to improve our understanding of the full complexity of the solar–terrestrial environment.

Complementary as these two missions were, they gave us many new insights into the structure of the solar atmosphere and the inner heliosphere and its variability over the solar activity cycle. For instance, during the *Skylab* era coronal holes were identified and rigorously established as the source regions of solar wind high-speed streams (see, e.g., the several articles in [1.16]). The same recurrent streams were subsequently seen again by both *Helios* spacecraft during their primary missions. The *Skylab* coronagraph detected spectacular coronal mass ejections [1.3], which apparently occurred in close association with erupting prominences, optical flares, and radio bursts. They were found to leave their imprints on the interplanetary medium by a variety of phenomena like shock waves and other disturbances [1.14] that, in turn, have a strong bearing on the earth's magnetosphere and ionosphere.

Skylab and *Helios*, in particular, have shaped our current perception of the sun and the inner heliosphere, thereby setting the stage for possible subsequent missions. We name only the ongoing *Ulysses* project and the future ambitious *SOHO* and *Cluster* missions, all of which are dedicated to studying the "wonders" of the inner solar system.

Of course, we should also mention the accomplishments of the *Voyager* and *Pioneer* missions, which were designed for the investigation of the outer heliosphere and the giant planets. It is merely the editorial constraint to concentrate on the physics of the inner heliosphere that here we cannot fully delineate the abundance of results obtained by these spacecraft. Nevertheless, many cross-references to the relevant literature given in the various chapters will hopefully enable the reader to obtain a coherent picture of the physical processes in the entire heliosphere.

Finally, it should be recognized in retrospect that the *Helios* mission at the time of its conception represented an unprecedented challenge to the management and space-industry structures then existing in Germany and to the network facilities available for national and international collaboration, and ground operations (for details see [1.10]). The *Helios* project offered the opportunity to develop the professional skills needed in an expanding space industry and in the nascent field of space science. The technological expertise and the scientific abilities of German industry and research institutes were thereby greatly enhanced.

The achievements resulting from the *Helios* enterprise thus had and still have a lasting impact on the space-science and space-technology community.

1.3 Scientific Highlights and Summary

In this section we shall have a quick look at the subsequent chapters and some of the main results described there. The whole book is divided into two volumes. Whereas the first volume deals primarily with large-scale phenomena in the inner heliosphere, the second volume addresses kinetic and microscopic aspects of the interplanetary medium and is concerned with particles, waves, and turbulence.

"Remote Sensing Observations of the Solar Corona" is the title of Chap. 2, by M. Bird and P. Edenhofer. They present a comprehensive account of the general structure and phenomenology of the corona itself and the corona as the source region of the solar wind, including brief discussions of coronal emission and rotation. Variations in the plasma-density distribution and the magnetic-field configuration as derived from radio-sounding techniques and similar means are analyzed. Particular attention is paid to the problem of solar wind acceleration and coronal heating in relation to the spectrum of observed turbulence. Radio-sounding techniques also make possible the detection of coronal mass ejections and the tracing of their evolution.

Chapter 3, by R. Schwenn, extensively describes the "Large-Scale Structure of the Interplanetary Medium". Among the outstanding accomplishments of the *Helios* mission is the observational establishment of sharp and distinct boundaries of solar wind streams in heliographic coordinates. Problems related to the spatial evolution of stream fronts and interfaces are elaborated in detail. Scenarios of the longitudinal and latitudinal stream structures are developed from the plasma observations and comprehended within the framework of the "ballerina" model of the heliospheric magnetic field. Emphasis is put on the distinction between several "types" of solar wind: the high-speed wind from coronal holes, the interstream-type slow wind emerging above coronal streamers close to the current sheet, the maximum-type slow wind emerging from all around the sun at times of high solar activity, and the coronal mass ejecta in association with solar transients.

"The Interplanetary Magnetic Field" is then investigated in depth by F. Mariani and F.M. Neubauer in Chap. 4. The three-dimensional structure of the magnetic field and its relation to the Parker spiral are elucidated, and radial gradients for the components are presented. The complex configuration of the current sheet during the sun's magnetic activity cycle and its fine structure is analyzed. Observations reveal that sector boundaries are complex structures in which the field direction alternates between smooth and abrupt changes while the large-scale sector polarity transition is achieved.

Chapter 5 on "Interplanetary Dust", by C. Leinert and E. Grün, completes the first volume. The spatial distribution of dust particles in the inner heliosphere was studied in unprecedented detail by exploiting the *Helios* orbits to scan through the near-sun dust environment, thus enabling us to probe the dust cloud *in situ*.

Its luminosity and consequently its density were clearly found to increase with proximity to the sun. Overall, the zodiacal light intensity was found to be surprisingly constant in time to within a few percent. Only temporary stray-light enhancements were observed in connection with solar activity or the passage of a comet. The mass and velocity distributions and chemical composition of the dust were established. The observed particles could be divided into chondritic and heavier iron-rich particles.

The second volume deals with the more dynamic phenomena in the interplanetary medium. It starts with two fascinating aspects of large-scale transients. In Chap. 6, L.F. Burlaga reports on "Magnetic Clouds". The existence and topology of such "plasma clouds" propagating outward from the sun was proposed and disputed even before the existence of the solar wind itself was contemplated. The final proof of their existence in 1981 is a unique example of the new quality and power that open-minded scientific collaborations have recently achieved: data from six spacecraft cruising through different parts of the solar system were assembled by scientists based at different parts of the earth. In this way, they solved a puzzle which none of them could have ever solved on his own! To this day, the origin and topology of magnetic clouds have remained at the focus of intense research activity.

There are strict associations with coronal transients on the one hand and with geomagnetic effects on the other. Coronal mass ejecta often drive large-scale shock waves, which, in turn, may literally shake large parts of the heliosphere, including the earth's system. The *Helios* twin mission has contributed much to exploring the properties of shock waves. However, we decided to treat this whole issue in this book only in passing (e.g. in Chap. 2, 6, 7, and 11), since there already exists a series of comprehensive reviews in the literature (see [1.5, 11, 13] and further referencess therein).

One further special aspect is dealt with in Chap. 7 by A.K. Richter. "Interplanetary Slow Shocks" have surprised the specialists mainly by their striking deficiency. In fact, there are severe dynamical constraints found for the evolution and steepening of slow mode waves and the propagation of slow shocks. These constraints appear to be most effective at solar distances beyond 0.4 AU. Inside that distance, slow shocks should occur fairly frequently, and some were actually observed by *Helios*. To bring this issue to the attention of a wider community, we therefore chose to include in this book Richter's detailed review, which is a revised and updated version of a paper given at the Sixth Solar Wind Conference in 1987 [1.8].

"Kinetic Physics of the Solar Wind Plasma" is discussed in Chap. 8 by E. Marsch. It addresses in detail the microscopic aspects of the plasma, the characteristics of ion and electron velocity distributions, and their radial evolution with heliocentric distance from the sun. Coulomb collisions, but even more so plasma microinstabilities, are demonstrably established as the major agents in shaping velocity distributions and in providing the dissipation that determines the development of internal energy of the solar wind plasma. Nonequilibrium thermodynamics and nonclassical transport theory are thoroughly discussed, also in connection with the dissipation of the observed magnetohydrodynamic fluctuations in the wind and its effect on ion heating. Finally, the global energy, mass, and momentum balances of the wind are evaluated and the constraints imposed by *Helios* observations on the acceleration of the solar wind in the corona are analyzed.

Chapter 9 is a review by D.A. Gurnett on "Waves and Instabilities". Observations of high-frequency waves and related plasma instabilities are presented. An overview of the various types of waves that can exist in the solar wind is given and experimental results on electron-plasma oscillations, ion acoustic waves, electromagnetic ion cyclotron waves, and whistler mode noise are discussed. The intensity of all these waves is found to increase upon approaching the sun. The reason for this radial variation is examined, and the effects of the waves on the solar wind's internal-energy state are evaluated.

"MHD Turbulence in the Solar Wind" is reviewed in Chap. 10 by E. Marsch. The observed MHD structures and fluctuations are described on various scales. Then the set of basic equations and the theoretical concepts needed to diagnose and comprehend the turbulence are established, and the relevant spectral densities are defined. The observed fluctuations show a distinct correlation with the stream structure and clearly evolve radially, whereby a state akin to fully developed fluid turbulence is ultimately reached. The origin and spectral properties of the fluctuations are investigated in detail and subsequently explained within the context of numerical simulations and recent models for turbulence in inhomogenious media. The role of the compressible turbulence component is analzyed and found to need further study.

The final chapter, Chap. 11, by H. Kunow, G. Wibberenz, G. Green, R. Müller-Mellin, and M.B. Kallenrode is devoted to a study of "Energetic Particles in the Inner Solar System". Origin, acceleration, and propagation of solar and galactic cosmic rays are analyzed. The structure of the interplanetary medium, in particular the magnetic field and its fluctuations, is shown to modulate strongly the propagation of energetic particles. Solar transients and flares are observationally established as the strongest sources of particle acceleration on the sun. Strong interplanetary shock waves, driven by coronal transients or created by large-scale stream interactions, may also accelerate particles to the observed high energies. Prominent features of acceleration and the long-term variability of cosmic rays in relation to the solar cycle are discussed in detail. Finally, the view is turned to distant astrophysical objects, and their role in generating cosmic rays is examined.

1.4 Heliospheric Physics and Some Astrophysical Connections

The sun is and will certainly remain the only star that, because of its proximity, can be scrutinized in considerable detail, and whose sphere of influence in the form of the heliosphere is accessible to *in situ* investigations by space probes.

We expect solar and heliospheric research to become increasingly important for a better understanding of many basic fields of astronomy and astrophysics such as the formation of stellar systems, stellar atmospheres, in particular coronae and winds, and the associated stellar magnetic activity cycles. The three-volume monograph *Physics of the Sun* [1.15] clearly demonstrates the invaluable example solar system physics generally sets to other fields of astrophysics.

Many of the plasma processes discussed in the following chapters may similarly take place in other astrospheres of sun-type stars even though they will remain elusive to detailed observations because they are so remote. The stellar wind is an example. How are we ever going to understand wave-driven stellar winds if we do not properly comprehend the impact that the observed coronal and solar wind turbulence has on the coronal expansion and acceleration of the solar wind? Likewise, the propagation of cosmic rays can be studied in detail only in the heliosphere. This is used to verify the validity of basic theoretical concepts, which may then be safely applied to the interstellar or intergalactic medium. The observed solar X-ray corona reveals the leading role played by the magnetic field in the spatial structuring of the emission regions in close relation to magnetic loops of various scale sizes. Of course, we are entitled to regard these revelations as relevant to stellar coronae in general.

The solar magnetic activity cycle provides an eminent paradigm of stellar magnetic activity originating from a dynamo believed to act in the convection zone. Theories of chromospheric and coronal heating developed for the sun have still not withstood the experimental tests and need to be advanced further before the very existence of the solar corona and the generation of the solar wind are fully comprehended. An understanding of these subjects would also advance our overall knowledge of stellar atmospheric and coronal activity. Although the energy flux in the solar wind is six orders of magnitude smaller than the sun's luminosity in visible light, it is comparable to the flux in the extreme ultraviolet and soft X-rays, emissions that come from the transition region and corona.

Therefore, solar mass and angular momentum carried away by the solar wind are intimately linked to the physical properties of the sun's atmosphere and magnetic envelope and to the emission in Lyman alpha and in multiple lines of coronal minor ions. Only in the case of the sun are we able to diagnose the coronal plasma properties by remote, yet highly resolving, optical observations from space, together with dedicated particle measurements that allow us to infer stellar atmospheric features in detail impossible to obtain even for nearby stars. Future solar studies are bound to provide more valuable information for comparative astrophysics. Without intending completeness, we mention the fairly new scientific field of helioseismology [1.12], which, once established, was immediately extended to other stars and evolved into the promising research branch of stellar seismology.

Following *Helios* and the many other missions we have mentioned, the forthcoming *SOHO* mission represents a new concerted effort to study the sun and its heliosphere and to understand more deeply solar–terrestrial relationships and their astrophysical implications. Among the main objectives of *SOHO* are the reassessment of the key issues of how the corona is heated and how the solar wind is generated, and the study of the solar interior and its outer atmospheric layers from the photosphere through the chromosphere into the corona. These goals will be achieved by remote sensing with spectrometers, telescopes, and *in situ* measurements of solar wind and energetic particles [1.2].

Before this ambitious project is embarked on in the mid-nineties, we felt it was the right time to summarize the accomplishments of earlier missions, to determine where we stand, and to collect our current knowledge and ideas about the physics of the inner heliosphere. We are glad to present here to the scientific community overviews of essential components of heliospheric and solar–terrestrial physics provided by several distinguished researchers in the field, many of whom are also actively engaged in the preparation of *SOHO* and in other ongoing heliospheric studies.

References

1.1   Burlaga, L.F., MHD processes in the outer heliosphere, Space Sci. Rev., **39**, 255–316, 1984.
1.2   Domingo, V. (ed.), The SOHO mission, scientific and technical aspects of the instruments, ESA SP-1104, 1989.
1.3   Gosling, J.T., E. Hildner, R.M. MacQueen, R.H. Munro, A.I. Poland, C.L. Ross, Mass ejections from the sun: a view from Skylab, J. Geophys. Res., **79**, 4581–4587, 1974.
1.4   Hollweg, J.V., Some physical processes in the solar wind, Rev. Geophys. Space Phys., **16**, 689, 1978.
1.5   Hundhausen, A.J., The origin and propagation of coronal mass ejections, in *Proceedings of the Sixth International Solar Wind Conference*, V.J. Pizzo, T.E. Holzer, D.G. Sime (eds.), NCAR/TN 306+Proc, Boulder, Colorado, 181–214, 1987.
1.6   Koutchmy, S., P. Lamy, G. Stellmacher, O. Koutchmy, N.I. Dzubenko, V.I. Ivanchuk, O.S. Popov, G.A. Rubo, S.K. Vsekhsvjatsky, Photometrical analysis of the June 30, 1973 solar corona, Astron. Astrophys., **69**, 35–42, 1978.
1.7   Parker, E.N., C.F. Kennel, L.J. Lanzerotti (eds.), *Solar System Plasma Physics*, Vols. I–III, North-Holland, Amsterdam, 1979.
1.8   Pizzo, V.J., T.E. Holzer, D.G. Sime (eds.), *Proceedings of the Sixth Solar Wind Conference*, NCAR/TN 306+Proc, Boulder, Colorado, 1987.
1.9   Porsche, H., General aspects of the mission Helios 1 and 2, introduction to a special issue on initial scientific results of the Helios mission, J. Geophys., **42**, 551–559, 1977.
1.10 Porsche, H. (ed.), *10 Jahre HELIOS, Festschrift aus Anlaß des 10. Jahrestages des Starts der Sonnensonde Helios am 10. Dezember 1974*, DFVLR Oberpfaffenhofen, 1984.
1.11 Richter, A.K., K.C. Hsieh, A.H. Luttrell, E. Marsch, R. Schwenn, Review of interplanetary shock phenomena near and within 1 AU, in *Collisionless Shocks in the Heliosphere: Reviews of Current Research*, Geophysical Monograph, **35**, 33–50, 1985.
1.12 Rolfe, E.J. (ed.), *Seismology of the Sun & Sunlike Stars*, Proceedings of a Symposium held in Puerto de la Cruz, Tenerife, Spain, ESA SP-286, 1988.
1.13 Schwenn, R., Relationship of coronal transients to interplanetary shocks: 3D aspects, Space Sci. Rev., **44**, 139–168, 1986.
1.14 Sheeley, N.R., Jr., R.A. Howard, M.J. Koomen, D.J. Michels, R. Schwenn, K.-H. Mühlhäuser, H. Rosenbauer, Coronal mass ejections and interplanetary shocks, J. Geophys. Res., **90**, 163–175, 1985.
1.15 Sturrock, P.A., T.E. Holzer, D.M. Mihalas, R.K. Ulrich (eds.), *Physics of the Sun*, Vols. I–III, D. Reidel Publishing Company, Dordrecht, Holland, 1986.
1.16 Zirker, J. (ed.), *Coronal Holes and High Speed Streams*, Colorado Associated University Press, Boulder, 1977.

2. Remote Sensing Observations of the Solar Corona

Michael K. Bird and Peter Edenhofer

2.1 Introduction

In situ measurements of the outer solar atmosphere have been conducted only down to a distance of 0.3 AU ($62R_\odot$) by the *Helios* spacecraft. Follow-on missions deeper into the corona such as the *Solar Probe* [2.162] are not likely to be realized for many years. As a result, the only reliable tools presently available for investigations inside of 0.3 AU are remote sensing of emissions which either (a) are generated naturally within the region, or (b) travel through the medium as a probe signal.

Observation and interpretation of type (a) emissions were the scientific objectives of the *Solar Maximum Mission* (*SMM*) launched in February 1980 [2.37]. The *SMM* satellite, which had ceased operating after only nine months in orbit, was repaired during a Space Shuttle mission in April 1984. Most of the investigations have since completed an impromptu "solar minimum mission" and were able to continue their surveillance almost up to the maximum of sunspot cycle no. 22 [2.143]. The next coordinated venture in solar observations from space is scheduled to be the *Solar and Heliospheric Observatory (SOHO)*, which will be launched in 1995 into a halo orbit at the Lagrangian point L1 in the undisturbed solar wind about $2R_\odot$ upstream of the earth [2.203]. *SOHO* would be the first solar observatory with a continually unobstructed view of the sun. It was also originally conceived to be its own interplanetary monitoring station with an assortment of particles and fields instrumentation. Unfortunately, budgetary constraints resulted in the cancellation of many of the important *in situ* investigations.

For the most part, the quieter region of the solar corona suspected of harboring the nascent solar wind are inert, and thus devoid of intrinsic radiation. In this case, it is possible to derive the characteristic properties of the medium by examining its effects upon an external probe signal, i.e. emission type (b) mentioned above. Radio-sounding investigations of the corona with both natural and spacecraft signals, which include but also extend well beyond the various techniques referred to as interplanetary scintillations (IPS), have been a productive source of information about this hitherto inaccessible region of interplanetary space. Of special significance in this regard were the two *Helios* spacecraft, whose linearly polarized S-band carrier signals were exploited during the frequent solar occultations to derive estimates of the electron density, magnetic field, turbulence spectrum, and bulk velocity of the solar corona [2.15].

Physics and Chemistry in Space - Space and Solar Physics, Vol. 20
Physics of the Inner Heliosphere I   Editors: R. Schwenn · E. Marsch
© Springer-Verlag Berlin Heidelberg 1990

It is the purpose of this chapter to describe the solar corona, the source region of the solar wind, based on observations of the above two types of emission. A somewhat arbitrary lower bound on the region of interest of $2R_\odot$, which is violated only when absolutely necessary, is imposed in order to limit the otherwise overwhelming scope of the subject matter. Radio-sounding investigations, which are feasible only in the source regions of the solar wind at distances $R > 2R_\odot$, will be treated in some detail. On the other hand, solar radio emission, which originates primarily in active regions presumably decoupled from the coronal material that flows into the inner heliosphere, will not be covered except when directly applicable to solar wind origins. A contemporary review on solar radio-physics was written by Dulk [2.51]. Specialized topics are covered in the recent compendium by McLean and Labrum [2.149]. No attempt will be made to review rigorously the morphology or the physical processes of the solar corona. Excellent overviews directed at these subjects have appeared recently [2.179, 189]. More specialized summaries of topics in coronal physics were published by Withbroe [2.256] and Zirker [2.276].

The general structure and phenomology of the solar-wind source region, including brief discussions of coronal emission and coronal rotation, are described in the following section. The space/time variations in the plasma density distribution in the solar corona, as derived from radio-sounding and other techniques, are presented in Sect. 2.3. The strength and configuration of the coronal magnetic field, which is known to be the dominant component of solar wind energetics throughout most of the region of interest, is discussed in Sect. 2.4. Section 2.5 addresses the problem of the solar wind acceleration, the measurement of coronal plasma velocities and the determination of the spatial spectrum of inhomogeneities (turbulence) with its implications for coronal MHD-wave activity and heating. Section 2.6 is devoted to the coronal mass ejection event with emphasis on its detection by radio-sounding techniques. The final section is an appendix containing a brief description of the measurement techniques used in coronal radio-sounding experiments.

2.2 Morphology of the Outer Corona: A Brief Survey

Sources of the solar wind would probably be obvious if all the coronal plasma had free access to interplanetary space. This would be the case, for example, if only radial dependences were important for the transfer of mass, momentum, and energy from the coronal base outwards. All solar magnetic field lines would be radial, with half the total magnetic flux pointing out and half pointing in. The continual interplay between magnetic and thermal forces in the corona, however, results in an extremely complex, sometimes nonstationary lower boundary condition, which can often mask the true origins of the coronal plasma that eventually forms the solar wind. The search for these origins becomes considerably more problematic as it is continued down to lower coronal heights.

There are three proven sources of solar wind: coronal holes, coronal streamers, and coronal mass ejections [2.257]. The remaining solar wind, comprising about 50% of the total mass flow, originates in difficult-to-define coronal regions referred to simply as "quiet". Each of these will be briefly described here in turn.

Coronal Holes are permanently open magnetic field regions that expand outward from the photosphere into the outer corona at a rate much greater than purely radial. The solid angle subtended by a coronal hole at the sonic point $(2-10R_\odot)$ is typically a factor of 5 to 7 greater than its value at the coronal base [2.258]. Plasma at the base of coronal holes is funneled by the gradient in magnetic field pressure and easily escapes into interplanetary space, leaving behind a density-depleted "hole" in the corona. Coronal holes can be recognized on the solar disk as regions of relatively low X-ray, XUV, or radio brightness. They can be observed only on the limb in the visible using a coronagraph to occult the photosphere. The contrast between holes and their neighboring regions decreases at lower coronal heights (e.g. upon observing at higher radio frequencies).

Coronal Streamers are sources of solar wind that is generally slower, cooler, and denser than average. The streamers are associated with the location of the "neutral line" in synoptic H-alpha photospheric maps. They are often recognized in interplanetary space by their containment of the interplanetary current sheet that delineates the neutral boundary between magnetic sectors [2.75]. Although the standard model implies a strictly open-field configuration for the source of solar wind in streamers, their magnetic foot points in the photosphere are necessarily very close to active regions with closed magnetic structure. It is conceivable that some field lines could intermingle and form open flux tubes in the immediate vicinity of closed-field regions.

Coronal Mass Ejections, which are discussed in more detail in Sect. 2.6, are observed to travel at speeds exceeding the sun's escape velocity and are therefore bona fide sources of solar wind. They contribute about 5% of the total solar mass loss [2.96]. Since they are observed to occur almost exclusively in and around complex coronal structures (streamers, loops, etc.), the plasma contained in coronal mass ejections may have started from a region of closed magnetic field.

About half of the coronal base area is neither streamer nor hole. For lack of identifying structure the remaining regions are usually just called the *quiet corona*. In view of the dynamic evolution of observable coronal structure, however, it should not be assumed that the structureless part is truly "quiet". In fact, the magnetic configuration of the quiet corona may well be transiently open, allowing coronal plasma to escape intermittently into interplanetary space [2.9]. The lifetime of a given open-field structure would presumably be short enough to prevent the permanent establishment of a high-speed solar wind expansion from the momentarily hole-like magnetic configuration.

The role of fine-scale structure in the source regions of the solar wind is still unknown [2.257]. It is quite possible that some of the corona's fine structure could be preserved in the solar wind to produce characteristic signatures in interplanetary space. Such signatures have been identified [2.237], presumably confined to individual magnetic flux tubes with typical angular widths of 5°. This size, perhaps allowing for moderate nonradial expansion, suggests a direct connection to individual supergranular cells of the photospheric network with their scale sizes of ca. 30 000 km (2.5°). Other coronal features that have attracted attention in this regard are the upward moving spicules, macrospicules, and high-speed jets. The ubiquitous spicules have velocities of some 20–30 km/s, typical diameters of about 1000 km, and cover about 1% of the total solar surface area. The larger specimens (macrospicules) can travel at speeds exceeding 100 km/s and have been followed even up to $1.1 R_\odot$. Taken together, the upward mass flux of the spicules is about 100 times greater than that of the solar wind. Most of the spicule material thus undoubtedly falls back to the photosphere, but it is not known whether some small fraction remains in the corona to escape eventually as solar wind. The high-speed jets, observed to rise in the transition region at velocities of the order of 400 km/s, are also conjectured to be the ultimate sources of the solar wind [2.28]. The mass and energy fluxes of the high-speed jets are estimated to be comparable with those of the solar wind. A comprehensive description of coronal fine-scale phenomenology may be found in the book by Priest [2.180].

2.2.1 The White-Light Corona

The best opportunities for ground-based photometry or polarimetry of the white-light corona continue to be the few minutes of totality available approximately once a year during solar eclipse. The evolution of coronal structure in response to the level of solar activity has been recently described in a pictorial survey of 19 solar eclipses [2.132]. Detailed reports of the observational campaigns and results are also available for the eclipses of 16 February 1980 [2.54, 193, 198], 31 July 1981 [2.64, 120, 195], and 11 June 1983 [2.197, 221].

The invention of the coronagraph over 50 years ago made coronal viewing possible without requiring the moon's intervention. Nevertheless, synoptic observations of the white-light corona from space have now been recognized to be an equally important advance in the field. Whereas light scattered in the earth's atmosphere essentially limits ground-based observations to coronal heights less than $2 R_\odot$, present-day concepts such as that envisioned for the *SOHO* mission [2.203] utilize complementary instruments with different optical parameters that can cover the dynamic range necessary for measurements from $1.1 R_\odot$ out to $30 R_\odot$. The initial step in this direction was taken on *OSO-7* [2.116], but the great potential of this observing technique was first widely publicized by the ATM images from *Skylab* [2.140]. Two instruments flown since *Skylab* include the "Solwind" coronagraph on the *P78-1* satellite [2.150, 213, 214] and the coronagraph/polarimeter (C/P) on the *Solarmax (SMM)* satellite [2.139]. The *SMM* C/P, having been once revived for extended duty in 1984, was forced to cease operations somewhat earlier than originally hoped. The ironic reason for this was the greatly increased atmospheric drag on the *SMM* satellite due to the exceptionally steep rise in solar activity during the ascending phase of solar cycle 22. The highly successful "Solwind" coronagraph took its last solar image on 13 September 1985, when the *P78-1* satellite was destroyed in a demonstration of antisatellite weaponry [2.145]. The observational characteristics of these various space-borne coronagraphs are given together with those of one of the better ground-based systems [2.65] in Table 2.1.

The corona's white-light appearance at two different phases of the solar cycle, recorded in this case by the *SMM* C/P, is demonstrated in Fig. 2.1. These two full-sun views were constructed by a superposition of four images of each solar quadrant recorded near solar maximum (1980: upper picture) and near solar minimum (1985: lower picture). The field of view extends from the edge of the occulter at $1.6R_\odot$ out to ca. $4.6R_\odot$. A 60-degree sector roughly centered on the south pole is obscured by the support pylon of the external occulting disk. These image mosaics are displayed in a heliographic (rather than ecliptic) coordinate system. The projection of the solar spin axis, inclined by 7.25° to the ecliptic, is aligned along the direction denoted N–S. The coordinates conventionally used to specify the position of features are apparent radial distance and azimuthal position angle PA, measured from north over east in the mathematically positive sense.

Table 2.1. Orbiting and ground-based coronagraphs

	ATM - *Skylab*	Solwind {*OSO-7*}	*SMM* - C/P	*MLSO - Mk III*
Observation Period	May 73–Feb 74	Mar 79–Sep 85 {Oct 71–Jun 74}	Feb 80–Sep 80 Apr 84–Nov 89	Aug 80–present
Field of view	1.6–$6R_\odot$ entire sun	2.6–$10R_\odot$ entire sun	1.6–$8R_\odot$ $6 \times 6R_\odot$ quadrant	1.2–$2.3R_\odot$ radial strip
Spatial Resolution	8 arcsec	1.25 arcmin	6.4 arcsec	20 arcsec
Image repetition time	ca. 90 min (orbit period)	10 min {44 min}	72 s (max rate)	90 s (limb circuit)
Detector	Eastman 026-02 emulsion	SEC Vidicon 256×256 pixel	SEC Vidicon 896×896 pixel	Reticon diode array
Passband	370–700 nm	395–650 nm	445–512 nm (6 other filters)	700–1080 nm (3 other filters)
Polarizers	polaroid wheel ($3 \times 60°$)	concentric rings	polaroid wheel ($3 \times 60°$)	rotating polaroid wave plate
References	[2.76, 140]	[2, 213, 214] {[2.116, 238]}	[2.94, 139, 246]	[2.65]

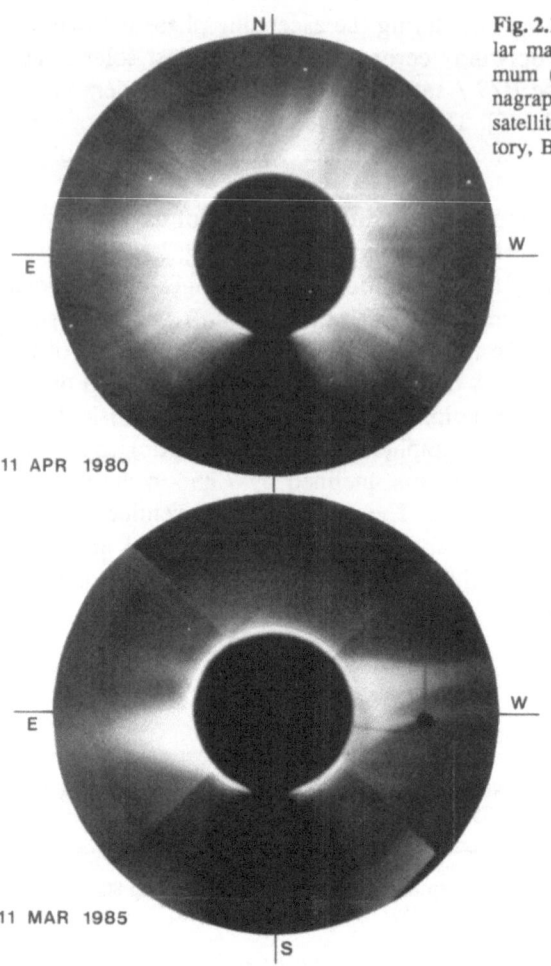

N

W

E

11 APR 1980

W

E

11 MAR 1985

S

Fig. 2.1. The sun's white-light corona at solar maximum (*upper image*) and solar minimum (*lower image*) as viewed by the coronagraph/polarimeter on the *Solarmax* (*SMM*) satellite (courtesy of High Altitude Observatory, Boulder, CO/USA)

The corona at solar maximum is characterized by a multitude of streamers radiating outward from virtually all solar latitudes, thereby yielding a quasi-symmetric distribution of brightness. The three-dimensional construction of a coronal streamer is generally thought to be an arcade, supported by a closed magnetic field configuration, that meanders more or less parallel to the solar surface and is gradually squeezed into a flat sheet as it extends up to greater coronal heights. The streamers are seen in projection onto the plane of the sky and thus appear to assume such typical descriptive forms as "helmets" and "rays" (viewed "edge-on") or "fans" and "plumes" (observed "flat" against the sky).

In contrast to the spherical symmetry of 1980, the brightness of the minimum corona of 1985 is concentrated in two large helmet streamers based at low solar latitude. These helmet streamers, located at PA ~ 100° (east limb) and PA ~ 290° (west limb) coincide with a region of magnetic field reversal, a warped band that encircles the sun and projects outward as the heliospheric current sheet.

Rather than the complex magnetic configuration at solar maximum, the coronal magnetic field at solar minimum is thought to be dominated by a dipolar component [2.87]. The polar regions of the minimum corona display essentially uniform magnetic polarity, and can therefore easily assume an open magnetic configuration that entices coronal material to escape into interplanetary space. The net result is the establishment of extended regions of relatively depleted plasma density that appear darker than neighboring regions at lower latitudes. These are the polar coronal holes that presumably develop during each solar minimum, occupying some 25% of the total solar surface. The fact that the bulk of the coronal material at solar minimum does not collect in the heliographic equator (see Fig. 2.1) has been offered as evidence that the solar dipole is tilted with respect to the solar rotation axis.

There can be little doubt that regions containing outward-flowing plasma that eventually becomes the solar wind are within the field of view of Fig. 2.1. It is perhaps somewhat counterintuitive, however, that the fastest solar winds originate from the darkest white-light features – the coronal holes. The association between high-speed streams in the solar wind and coronal holes has been well established for many years [2.275].

2.2.2 Sources and Properties of Coronal Emission

It has been known for some time that the spatial and spectral distributions of coronal emission are a direct consequence of the scattering properties of the various material components populating the corona. The radiance (or brightness) B_λ (SI unit: watts m^{-2} steradian$^{-1}\lambda^{-1}$) detected at wavelength λ is determined by an integral of the type

$$B_\lambda(R) = \int_0^\infty N(R, s)\sigma(R, s)b_\lambda(R, s)ds \;, \tag{2.1}$$

where the integral is taken along the line-of-sight (ray path) from the observer at earth to infinity. The proximate distance of the line-of-sight to the sun, also sometimes called the solar offset or solar elongation, is denoted by R. The three factors in the integrand of (2.1) are:

N = number density of source particles (m^{-3})

σ = differential cross section (m^2 steradian^{-1}) (2.2)

b_λ = brightness of an individual source particle.

The largest component (over 90%) of the total integrated coronal brightness, which is only about one millionth as bright as the mean solar disk brightness $B_\odot$, comes from the *K-corona* [2.117]. K-corona emission, resulting from Thomson scattering of photospheric radiation from coronal electrons, is a continuum (hence the designation "K" from the German *Kontinuum*) whose spectrum closely resembles that of the sun. The thermal velocities of the scattering electrons in the hot corona ($T \simeq 2 \times 10^6$ K) are so large that the solar Fraunhofer absorption lines are smeared out by the Doppler effect. The factors (2.2) to be used in the integral (2.1) for the K-corona become:

$$N = N_e, \qquad \text{the electron number density}$$
$$\sigma = \sigma_T f(\alpha), \qquad \text{cross section for Thomson scattering} \qquad (2.3)$$
$$b_\lambda = B_\odot [R_\odot / r]^2, \quad \text{solar brightness at radial distance } r$$

where

$$\sigma_T = (8\pi/3)[e^2/mc^2]^2 = \text{Thomson cross section}$$
$$f(\alpha) = (3/16\pi)[1 + \cos^2 \alpha] = \text{angular scattering function}$$
$$r^2 = R^2 + s^2 \; ; \quad \alpha = \text{scattering angle}.$$

The Thomson cross section ($\sigma_T = 6.6 \times 10^{-25} \text{cm}^2$) is essentially the effective area of a free coronal electron as seen by a typical photospheric photon. The optical depth of the corona, a measure of the Thomson scattering probabilty, can be estimated by the product $I_t \sigma_T$, where I_t is the total electron content from the photosphere out to the earth's orbit. As shown in the following section, a good estimate for I_t is about $5 \times 10^{18} \text{cm}^{-2}$, so that the optical depth is approximately $I_t \sigma_T = 3 \times 10^{-6}$.

Considering the factors involved in (2.3), it is apparent that the spatial distribution of the optically thin K-corona is controlled by the tenuous electron density N_e, whose distribution is known to vary over the course of a solar cycle. A more or less spherical symmetry is present at solar maximum and a pronounced polar flattening prevails during sunspot minimum. Mauna Loa K-coronameter data from 1965 to 1983 indicate that the total integrated brightness of the corona increases by a factor of two from minimum to maximum [2.66, 219]. This result was substantiated by a study of total brightness at many natural eclipses that yielded a mean max-to-min ratio of 2.6 [2.196]. Since the total integrated brightness is simply a measure of the total number of electrons in the corona, this result may indicate that the mean coronal scale height is larger during solar maximum (higher temperature). Only about 50% of this variation could be explained by the growth and decay of coronal holes, which occupy up to 25% of the total surface area at solar minimum. The remaining 50% variation was attributed to true brightness changes at equatorial latitudes. Up to an order of magnitude in brightness variation is possible between the average equatorial corona and within the darker coronal holes [2.200]. Coronal streamers are the brightest features of the quiescent corona and are thus clearly distinguished over neighboring regions.

The K-corona typically dominates the coronal emission out to about $2R_\odot$, but is overtaken at larger solar distances by the *F-corona* ("F" for Fraunhofer). The F-corona emission contains the Fraunhofer lines and is attributed to scattering from interplanetary dust [2.1]. The brightness integral factors (2.2) appropriate for the F-corona are:

$N = N_d$ = dust number density

$\sigma = \pi a^2 g(\alpha)$; a = dust particle radius   (2.4)

$b_\lambda = AB_\odot [R_\odot / r]^2$; A = particle albedo

where $g(\alpha) = J_1^2(ka\alpha)/\pi\alpha^2$; J_1 = first-order Bessel function. The angular scattering function $g(\alpha)$ is applicable for the case of Fraunhofer diffraction and is valid for $ka > 1$ ($k = 2\pi/\lambda$), i.e. when the wavelength of the incident radiation is smaller than the scattering dust particle. Only small angle scattering is relatively strong in this case, so that most of the F-corona radiance originates from regions between the sun and earth. This situation will also produce a reddening of the corona (i.e. an enhancement in radiance at longer wavelengths relative to the color of the sun). It is presently thought that the F-corona is indeed redder than the K-corona [2.117], and the difference may become greater at IR wavelengths [2.153]. It is unclear, however, if the IR enhancement stems from diffraction effects or perhaps an additional thermal component (see below). If $ka \ll 1$ (scattering dust particles smaller than the wavelength of the incident radiation), the process enters the regime of Rayleigh scattering and the angular scattering function becomes very similar to the $f(\alpha)$ of Thomson scattering. The F-corona would then be produced primarily by circumsolar grains and would display very little reddening [2.11]. The extension of the F-corona out to large solar elongations is the zodiacal light, which is described at length in Chap. 5 [2.123].

The controversy over the relative importance of the *thermal emission* of interplanetary dust close to the sun has not been resolved [2.117]. Since it also originates from dust, this component is sometimes inappropriately lumped together with the F-corona even though the nature of its emission is quite different. As impressively demonstrated by the *IRAS* observations, interplanetary dust is certainly detectable in the IR from its thermal radiation [2.252]. As the grains spiral in toward the sun under the influence of the Poynting–Robertson effect, their surface temperature increases and the particles begins to emit more profusely ($\sim T^4$) in accordance with Stefan's law for gray bodies. The thermal emission at a solar elongation R is obtained from (2.1) using the following integrand factors:

$N = N_d$ = dust number density

$\sigma = \pi a^2$; a = dust particle radius   (2.5)

$b_\lambda = (1 - A)\dfrac{2hc}{\lambda^5} \exp[-hc/\lambda k_B T]$

where

A = dust albedo ($A = 0$ for a perfect absorber)

h = Planck's constant

k_B = Boltzmann's constant

c = speed of light.

In this case the particle's brightness is given by the Planck Law for gray-body radiation in the Wien approximation ($k_B T \ll hc/\lambda$). The temperature is a function of solar distance and is known to increase rapidly after reaching the sublimation temperature of the grain material [2.10]. Peterson [2.174] was among the first to point out that the grain heating would quickly lead to vaporization. The result would be a "dust-free" cavity surrounding the sun, whose radius R_{min} would be directly related to the albedo (i.e. composition) of the least volatile interplanetary dust particles. Near-sun scans with the *Helios* photometer experiment found no evidence for this hypothetical dust-free zone down to a minimum solar distance of $19 R_\odot$ [2.124].

The sharp cutoff in interplanetary dust at R_{min} would be very difficult to detect from the earth at optical wavelengths if the F-corona were produced primarily by Fraunhofer diffraction [2.135]. This is because most of the coronal intensity in this case comes from small-angle scattering outside R_{min}. Although this is the preferred interpretation, the circumsolar component has been shown to be dominant in recent calculations [2.152]. At IR wavelenghts, however, a relative enhancement in coronal brightness due to thermal radiation would be expected to show up as a "bump" at that solar elongation corresponding to R_{min} [2.117, 174].

Searches for this thermal radiation have been inconclusive. Radial brightness scans at $\lambda = 2.2\,\mu m$ revealed 4 distinct "features" at different solar distances that were attributed to thermal emission [2.135]. A thermal feature was clearly detected in the $\lambda = 1.65\,\mu m$ band at $R = 4 R_\odot$ during the June 1983 eclipse which was attributed to olivine particles with radii $a = 100\,\mu m$ and surface temperatures at $1300\,K$ [2.153]. Ground-based IR observations both during eclipse [2.182] and outside of eclipse [2.142] have failed to detect the thermal emission. The ephemeral nature and multiplicity of features might be explained by a "ring around the sun" populated by hitherto undetected asteroids as large as $a = 10\,km$ [2.27].

The final contributor to coronal radiance is the *E-corona* ("E" for emission), which comes mostly from "forbidden" lines of multiply ionized ions. Iron is one of the more prolific sources of E-corona emission, two examples being the green line ($\lambda 5303$) of Fe XIV (13-times ionized) and the red line ($\lambda 6374$) of Fe X (9-times ionized). Although the vast majority of coronal material is a proton–electron plasma, neutral hydrogen is also observed in the Lyman-alpha ($\lambda 1215$) and H-alpha ($\lambda 6562$) emission lines. The results of using Lyman-alpha and other coronal emission lines for velocity–temperature diagnostics are described in Sect. 2.5.1. The E-corona radiance, which is only about 1% of that of the K-corona, can be neglected in studies of the spatial and temporal variations of the K- and F-coronae.

The brightness of the K- and F-coronae in units of the solar disk brightness $B_\odot$ (taken from the semiempirical model of Koutchmy and Lamy [2.117], is plotted as a function of solar elongation in Fig. 2.2 (upper panel). The mean background radiance of sky-scattered emission during eclipse is 1–$2 \times 10^{-9} B_\odot$. A clear blue sky at sea level outside eclipse is much brighter, typically around $10^{-6} B_\odot$. The K-corona is dominant close to the sun, but falls off more rapidly

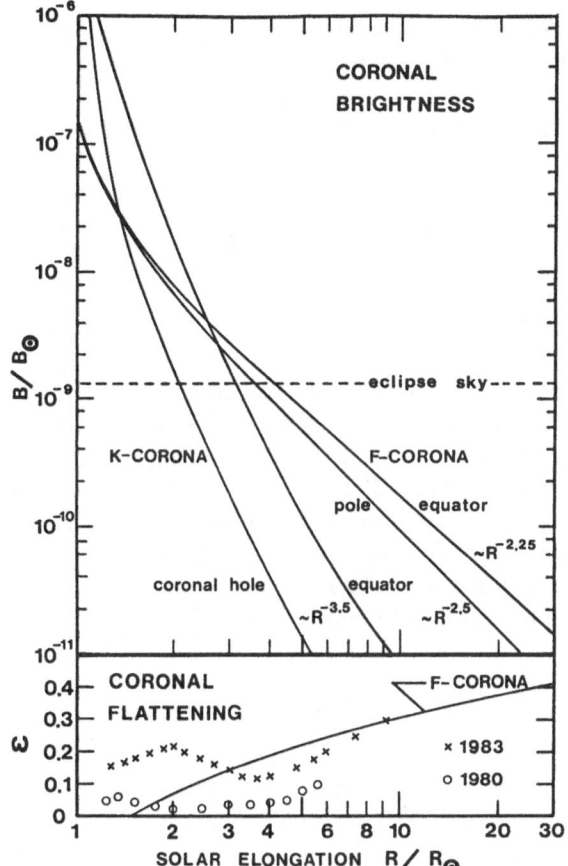

Fig. 2.2. Coronal brightness and flattening vs. solar elongation. *Upper panel:* Brightness of the K- and F-coronae at solar equator and pole in units of the mean photospheric brightness. The background sky level at total solar eclipse is indicated. *Lower panel:* Flattening parameter ε (see text) for the F-corona and for the K-corona at two different epochs of the solar cycle (adapted from Koutchmy and Lamy [2.117])

with solar distance than the F-corona because the electron density decreases faster than the dust number density. The asymptotic falloff rates for each contribution are shown. The K-corona brightness can vary by an order of magnitude from the mean equatorial profile to a coronal hole, including the polar coronal holes that appear at sunspot minimum. The F-corona brightness also decreases more rapidly at the poles than at equatorial latitudes.

The ellipticity of the solar corona is traditionally described by a flattening index ε defined by:

$$\varepsilon = \frac{R_{eq}}{R_{pol}} - 1 \tag{2.6}$$

where R_{eq}, R_{pol} are the radial distances of an isophote of coronal radiance at the equator and pole, respectively. The variation of ε with solar distance is shown in Fig. 2.2 (lower panel). Since the coronal radiance is dominated by the K-corona out to about $2R_\odot$, the flattening of the isophotes between $1-2R_\odot$ is determined by the electron spatial distribution. Beyond $4R_\odot$ the flattening index is controlled by the dominant F-corona. The solid line for $\varepsilon(R)$ is an empirical model for the flattening of the F-corona alone, which is observed to be constant at all phases of the solar cycle [2.117]. The individual points show two specific sets of measurements giving the total coronal flattening for times near solar maximum and during the descending phase of the solar cycle. The "X" points [2.117] were measured during the eclipse of 11 June 1983 and the "O" points were taken on 16 February 1980 [2.194]. Both sets tend toward a larger flattening index at large R because of the concentration of interplanetary dust near the ecliptic [2.123]. The value of the flattening index at $R = 2R_\odot$ is found from Fig. 2.2 to be $\varepsilon = 0.03$ in February 1980 and $\varepsilon = 0.22$ in June 1983. These are representative of the K-corona shape at those phases of the solar cycle [2.132].

The scattered emission from an isolated Thomson scattering event is highly polarized. This is because the scattering electron is confined to vibrate only in the direction of the incident photon's electric vector. The polarizing properties of Rayleigh scattering from dust grains much smaller than the wavelength ($ka \ll 1$) are virtually the same as for Thomson scattering. For outward-propagating light in the case of a point source, the electric vector is perpendicular to the radius vector from the sun to the scattering particle. The finite size of the sun and line-of-sight effects tend to decrease the degree of polarization of Thomson-scattered light, defined by

$$p_K = \frac{(B_{KP} - B_{KR})}{B_K} \tag{2.7}$$

where the K-corona brightness $B_K = B_{KP} + B_{KR}$ is composed of components B_{KP}, B_{KR} polarized perpendicular to and along the radial direction, respectively. Theoretical values for p_K range from $p_K = 0.2$ just above the solar disk to a value $p_K = 0.6$–0.7 far from the sun [2.241]. Defining corresponding quantities for the F-corona, the total coronal polarization can be expressed as:

$$p = \frac{p_K B_K + p_F B_F}{B_K + B_F} \, . \tag{2.8}$$

Under the assumption that the F-corona is unpolarized ($p_F = 0$), and using the theoretical value for $p_K(R)$, one can separate the F- and K-coronae using (2.8). If the coronal radiance from dust is due to Fraunhofer diffraction or thermal radiation, then $p_F = 0$ is a good approximation. It would not be applicable, however, if Rayleigh scattering controls the polarization of the outer corona, as implied by recent balloon-borne measurements that yielded $p = 0.17$ at about $4R_\odot$ [2.103]. Comprehensive polarization measurements of the corona in the IR, which do not presently exist [2.252], would provide a powerful diagnostic for determination of the characteristic size a, the minimum approach distance R_{min}, and the albedo A of the circumsolar dust [2.11].

2.2.3 Coronal Rotation

The solar rotation rate was first estimated by Galileo, who watched sunspots wander across the disk at a speed that would carry them approximately once around the sun per month. The systematic observations of sunspots by Carrington in the 1860s revealed the photospheric differential rotation, whereby the rotational speed of equatorial sunspots exceeds that of high latitude spots. It has also been known for quite some time from synoptic maps of the photospheric magnetic field that the differential rotation of large-scale unipolar regions is weaker than for the short-lived, small-scale features [2.30]. Since the solar corona is magnetically dominated by fields anchored in the sun, it might be expected that the corona would be subject to the differential rotation observed on the photosphere.

Nevertheless, optical observations of the rotational velocity of prominences, coronal holes, green line emission regions, and other features all indicated a substantially weaker differential rotation from equator to pole at coronal heights [2.118]. Since the corona presumably dictates the rotational characteristics of the entire heliosphere (corotation lag time, Parker spiral angle, etc.), knowledge of the latitude dependence of the rotation rate is not merely an academic novelty. Some remarkable recent developments arising from studies of coronal rotation, which may also provide vital clues about the nature of the solar magnetic dynamo, are briefly summarized here.

Fisher and Sime [2.67] have used the extensive Mauna Loa white-light coronameter observations to determine the latitudinal variation of the coronal rotation from 1965–1983. They derive an empirical formula for the time-averaged synodic rotation rate (rotation frequency as viewed from earth) given by

$$\omega_\odot = 13.22 - 0.57 \sin^2 \theta \quad [\text{deg/day}] \tag{2.9}$$

where θ is the heliographic latitude. The equatorial synodic rotation period corresponding to (2.9) is 27.23 days, increasing to 28.46 days at the poles. The latitude dependence of (2.9) is considerably weaker than that derived for the photospheric tracers, for which the polar rotational period is typically about seven days longer than the equatorial period. The most pronounced difference between the photospheric rate and the coronal rotation rate occurs at high solar latitudes. Persistent, large-scale features were found to adhere to a more rigid coronal rotation than smaller, relatively ephemeral structures. An indication for a cyclic variation in the rotation period of about one day was observed over the course of a solar cycle, the longest rotation periods occurring about 2 years before solar maximum.

Continuing studies of the changes in differential rotation with coronal height and phase of the solar cycle [2.170–172] were performed with a subset of the Mauna Loa coronameter data (solar cycle 20: 1964–1976). Differential rotation of the corona derived from autocorrelations of polarization brightness pB at $R = 1.5R_\odot$ was found to be much weaker than that from an identical analysis at $R = 1.125R_\odot$. The observed trend, if extrapolated to greater heights, would imply rigid coronal rotation above ca. $2.5R_\odot$. An asymmetry in the rotation rates of the solar hemisphere was also apparent in the results, the rotation period being roughly one day longer in the northern hemisphere (60° N) than at the same latitude in the southern hemisphere [2.171].

Hoeksema and Scherrer [2.88] studied coronal differential rotation by computing the autocorrelations and power spectra of their magnetic field strengths at $2.5R_\odot$ derived from potential field calculations (see also Sect. 2.4.1). The coronal fields were found to follow a differential rotation curve similar to (2.9). Power spectra of the magnetic field time series at all latitudes revealed two discrete synodic rotation periods at ca. 27 and 28 days, respectively, that could only be resolved for autocorrelation time lags of several solar rotations. These same two periods had also been derived earlier from the inferred IMF polarity in the ecliptic plane using geomagnetic records over the past six solar cycles [2.234]. The longer period was found to be more prevalent at southern latitudes during the time span covered by the data (1976–1985, essentially solar cycle 21). This rotational asymmetry is just opposite to that indicated from the analysis of data from solar cycle 20 [2.171, 172] and provides further evidence for a solar cycle dependence of the large-scale solar rotation.

Recent numerical simulations of the differential rotation for the large-scale photospheric field [2.216], the coronal magnetic field [2.248], and the boundaries of coronal holes [2.161] have provided a very credible description of the mechanism responsible for the quasi-steady rotation of these features. The evolution of the photospheric magnetic field is determined in these calculations from a magnetic flux-transport equation. Continuous new sources of flux erupt in bipolar magnetic regions and are subsequently transported across the solar surface via differential rotation, supergranular diffusion, and meridional flow. The meridional components of flux transport, when combined with differential rotation, are found to generate a striped alternating polarity pattern that rotates rigidly like a "barber pole" [2.216]. Based on this simulated photospheric field and assuming its current-free extension out into the corona, Wang et al. [2.248] used a spherical harmonic analysis to compute the potential field at $2.5R_\odot$ and to examine its rotational properties. Three principal features were singled out to describe the coronal rotation:

(1) The harmonic components of the curl-free magnetic field are "filtered" at coronal heights according to their different radial falloff rates. Only the lowest-order terms, usually just the dipole and quadrupole contributions, were necessary to represent the coronal field and its rotation adequately at $2.5R_\odot$.

(2) The dominant contribution of these low-order harmonics for the transfer of angular momentum to the corona generally comes from those solar regions containing the most freshly erupted magnetic flux. At solar minimum, for example, when most newly created flux is concentrated at the equator, the coronal magnetic field is seen to rotate at approximately the equatorial rate. Slower coronal rotation is expected near solar maximum, consistent with the cyclic variation found in the white-light data [2.67].

(3) New source eruptions are capable of inducing erratic phase shifts in the low-order modes, leading to sudden changes in the coronal rotation rate.

In summary, the results indicate a more or less rigid rotation of the corona and extended solar wind at all heliolatitudes [2.220]. There does exist some evidence to the contrary from a solar wind velocity correlation study using radio scintillations that appears to be more consistent with a strong differential rotation [2.21]. It would be difficult to understand why the solar wind velocity would be coupled to the photospheric rotation rather than that of the coronal magnetic field. More data are needed before further conclusions can be drawn. The rigidity and north–south symmetry of the heliospheric rotation are two of the many issues waiting to be addressed by the flight of the *Ulysses* spacecraft over both solar poles, now scheduled for the mid 1990s.

2.3 Coronal Plasma Density

Substantial contributions to our knowledge of the coronal plasma density distribution in the inner heliosphere have come from dispersion measurements using radio links from spacecraft. In fact, measurements of the plasma density in the inner corona are presently feasible only by remote sensing techniques such as white-light observations from earth-based or orbiting coronagraphs (see also Sect. 2.2.1) or from radio-sounding observations of natural [2.43] or artificial (spacecraft) radio sources during solar occultation [2.15, 273]. These techniques are sensitive to the electron density, assumed to be equal to the ion charge density in the neutral coronal plasma.

Spacecraft radio occultation experiments in the solar corona began in 1968 with S-band (carrier frequency: 2.3 GHz) observations using *Pioneer 6* [2.74, 127], an extremely durable spacecraft, by the way, which is still functional at this writing after more than 23 years in interplanetary space. The first comprehensive time-delay occultation measurements to exploit the dispersive plasma propagation effects associated with phase delay (Doppler frequency shift) and group delay (ranging) were performed with the spacecraft *Mariner 6* and *7* [2.157]. Dual-frequency differential time-delay measurements at S-band and X-band (8.4 GHz) were first recorded during the polar coronal occultations of the spacecraft *Mariner 10* and *Viking* [2.155, 239], and later with *Voyager 2* [2.3]. Similar experiments have been performed with the *Venera* spacecraft [2.192, 242]. Future coronal radio-sounding experiments are foreseen using the *Ulysses* and *Galileo* spacecraft during their respective cruise phases on the way to Jupiter.

A typical occultation geometry, using the arbitrary example of *Helios 1* in March 1975, is shown in Fig. 2.3. The radio propagation path from spacecraft to earth attains its minimum distance to the sun at point P. The heliocentric distance of this solar proximate point is often referred to as the ray impact parameter R or "solar offset". The point P is also characterized by its corresponding heliographic longitude ϕ and latitude θ (not labeled in the ecliptic plane view of Fig. 2.3). Radio propagation effects can be observed during both the entry (ingress) and exit (egress) phases of the solar occultation, but are excluded exactly at superior con-

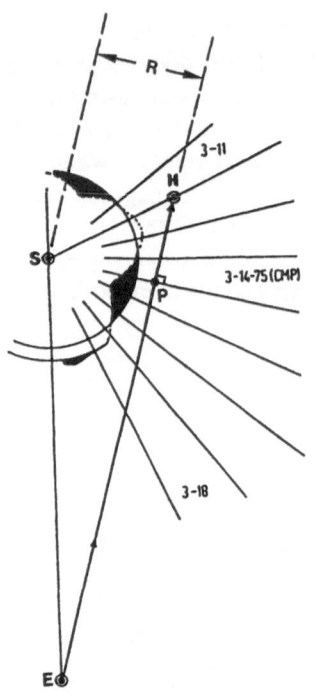

Fig. 2.3. Orbital geometry in ecliptic plane during solar occultation The relative positions of *Helios 1* (H), earth (E) and sun (S) on the date 21 March 1975 are indicated. The solar offset, or radial distance of the proximate point (*P*) along the *Helios*/earth ray path, is denoted by *R*. Eight angular segments of solar longitude are shown, defined by the central meridian passages (CMP) on the dates 11–18 March 1975. Enhancements and depletions in the equatorial K-corona brightness are indicated along the west limb. The point *P* is located over a distinct coronal hole extending over 30° in longitude [2.58]

junction. A brief survey of the various radio-sounding measurement techniques used for investigation of the corona is presented in an appendix (Sect. 2.7) to this chapter. The radio-sounding results to be reported in this section have provided information on (a) the large-scale spatial distribution of coronal plasma density, particularly the radial decrease and the solar cycle effect on the latitudinal dependence, and (b) temporal variations on shorter time scales which reflect the presence of turbulence and waves in the solar corona.

2.3.1 Spatial Distribution of Coronal Plasma Density

Several models have been used to represent the spatial distribution of coronal electrons. One of the first and perhaps most widely known empirical representations is the spherically symmetric Allen–Baumbach formula [2.1] given by

$$N_e(r) = \left(\frac{2.99}{r^{16}} + \frac{1.55}{r^6} \right) \times 10^{14} \quad [\text{el/m}^3] \tag{2.10}$$

with r in $R_\odot$. The Allen–Baumbach model (2.10) was derived from earth-based coronagraph observations of the K-corona (see Sect. 2.2). The first term in (2.10) is negligible for $r > 3 R_\odot$. For the outer corona and interplanetary space, (2.10) is often supplemented with an inverse-power-law term $\sim r^{-2}$, which accounts for conservation of mass assuming spherically symmetric outflow at a constant solar wind velocity. These models have been shown to be reasonably good for the quiet mean equatorial corona at solar minimum, and somewhat better at solar maximum. They are supported by ground-based polarization brightness (pB) observations [2.78] and by pulse delay-time measurements using occulted pulsars [2.253]. Further experimental verification is provided by an interferometric determination of the relative angular separation of the two radio sources 3C273 and 3C279 during solar occultation [2.156].

A modification of the coronal model (2.10) accounts for a radially dependent solar wind velocity profile $v_o(r) \sim r^\gamma$, for which

$$N_e(r) = \frac{N_A}{r^6} + \frac{N_B}{r^{2+\gamma}} , \tag{2.11}$$

where $\gamma = 0$ would represent a constant outflow velocity. The acceleration of the solar wind requires that $\gamma > 0$ in the corona out to that radial distance where the solar wind reaches its asymptotic velocity.

The observable time delay is proportional to the coronal columnar electron content $I_t(R)$ given by (see also Sect. 2.7)

$$I_t(R) = \int_{s_1}^{s_2} N_e(r) ds ; \quad r^2 = R^2 + s^2 , \tag{2.12}$$

an integral expression which must be inverted to obtain the coronal electron density N_e. One approach to this inversion problem is to perform a least-squares fit to a set (or subset) of coronal parameters in an appropriate density model such as (2.11). Other methods include the Abel transform to obtain an analytical solution [2.155, 175] and the numerical Kalman filtering technique [2.232]. The Abel transform is restricted by its assumptions of stationarity and spherical symmetry. Another disadvantage is that the data set must be differentiated prior to applying the transformation algorithm, thereby inevitably amplifying the noise in the inverted density profile. The Kalman filtering technique avoids this differentiation and noise amplification. Covariance matrices are introduced to account for the statistical properties of the input data measurements and the output data inversion (bias and variance for stochastic inversion). Most importantly, no assumptions about spherical symmetry are needed.

A rough estimate of the total electron content for a spherically symmetric coronal density (2.11) with $N_A = \gamma = 0$ can be obtained from (2.12) as

$$I_t(R) = N(R) \cdot R \cdot [ESP] \tag{2.13}$$

where [ESP] is the earth–sun–probe angle in radians. Near superior conjunction ([ESP] = π), one may derive the following analytical expression valid for any value of γ:

$$I_t(R) = f(\gamma) N(R) R = f(\gamma) \langle I_t \rangle R^{-1-\gamma} \tag{2.14}$$

where

$$f(\gamma) = \sqrt{\pi} \frac{\Gamma(\gamma/2 + 1/2)}{\Gamma(\gamma/2 + 1)} \quad \text{and} \quad \langle I_t \rangle = N_B R_\odot .$$

Table 2.2. Parameter values for coronal models derived from spacecraft time-delay experiments. (a) No measurements close enough to sun; (b) weak estimate; (c) fixed value (strong correlation between N_B, γ)

$N_A(10^{14}\text{el/m}^3)$	$N_B(10^{12}\text{el/m}^3)$	γ	Source	Ref.
1.3 ± 0.9	1.14 ± 0.7	0.3 ± 0.3	nominal (*a priori*)	[2.58]
0.69 ± 0.86	0.54 ± 0.56	0.047 ± 0.24	*Mariner 6*	[2.157]
(a)	0.66 ± 0.53	0.08 ± 0.24	*Mariner 7*	[2.157]
1.0 (b)	0.55 ± 0.1	0.1 (c)	*Helios 2*	[2.62]
(a)	3.72	0.6 (c)	*Voyager 2*, ingress	[2.3]
(a)	0.526	-0.06 (c)	*Voyager 2*, egress	[2.3]

The coronal density model (2.11) was adopted for the analysis of time-delay data collected during the solar occultations of *Mariner 6/7* [2.157], *Helios 1/2* [2.58, 62], and *Voyager 2* [2.3]. Table 2.2 shows the results for the parametric fits to the time-delay data from these three different spacecraft. The coronal parameters N_B and γ are highly correlated (correlation coefficient > 0.9), but less correlated with $N_A(< 0.4)$. The least-squares fit for the *Helios* data involved a set of up to 12 parameters to be estimated: in addition to the 3 coronal parameters N_A, N_B, γ, the 6 orbital elements of the spacecraft plus up to 3 parameters of the solar radiation pressure must be determined. Two types of independent observations were invoked to constrain the *Helios 2* analysis: (a) close to the sun ($R < 3R_\odot$), ground-based K-coronagraph measurements were used to define better the value of N_A [2.78], and (b) far away from the sun ($215R_\odot$), *in situ* plasma densities measured on the *IMP 7/8* earth-orbiting satellites provided *a priori* constraints on N_B and γ. The final parameter fit to the *Helios 2* time-delay measurements taken from 1 March to 15 June 1976 ($5R_\odot < R < 60R_\odot$) is shown in Fig. 2.4 [2.62]. The electron density distribution derived by this radio-sounding technique was found to be in reasonable agreement with the radial decrease determined on board *Helios 2* (from 65 to $215R_\odot$) over the same time interval [2.188, 205].

Following the 1976 solar occultation of *Helios* (S-band only), dual-frequency (S/X-band) radio-sounding measurements were performed later the same year with the *Viking* spacecraft [2.155, 239]. The downlinks were detected as close as $0.3R_\odot$ (X-band) and $0.8R_\odot$ (S-band), respectively, above the photosphere near the south solar pole. The radial electron content profiles were strongly asymmetric with respect to the entry and exit phases of the occultation (Fig. 2.5). This asymmetry was assumed to arise from a distinct latitudinal dependence of the coronal electron density, which was incorporated into a model for $N_e[\text{el/m}^3]$ given by the formula

$$N_e(r, \theta) = \left[\left(\frac{2.99}{r^{16}} + \frac{1.55}{r^6} \right) \times 10^{14} + \frac{3.44}{r^2} \times 10^{11} \right] F(\theta) , \qquad (2.15)$$

with

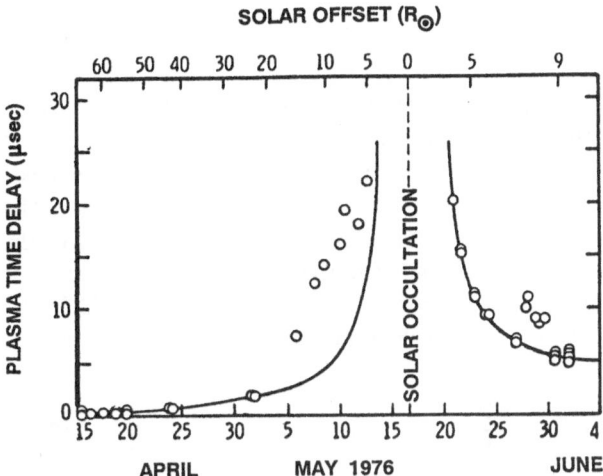

Fig. 2.4. Coronal group time delay (columnar electron content) from *Helios* 2 as a function of time and solar offset *R*. The solid curve corresponds to the electron density model of Table 2.2, deduced from 145 individual *Helios* 2 range measurements. The circles are time-delay residuals, the closest range point to the sun taken at a solar offset of $3R_\odot$. The systematic increases during ingress (5–14 May) and egress (28–30 May) are due to regions of enhanced electron content at the west and east solar limbs, respectively, as verified by ground-based K-coronagraph measurements [2.58, 62]

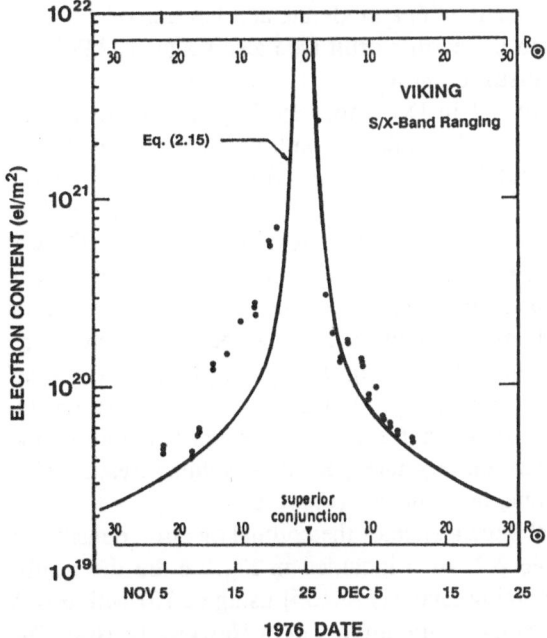

Fig. 2.5. Coronal electron content determined from *Viking* differential range data during solar occultation in 1976 (superior conjunction: 25 November). Altogether, 312 measurement points are included in the diagram at solar offsets $R < 30R_\odot$. The standard deviation associated with each measurement is of the order of 10^{18} el/m, i.e. considerably smaller than the size of the individual points. The solid curve is the content expected from an electron density model $N_e(r, \theta)$ given by (2.15) in text (from [2.239])

31

$$F(\theta) = \sqrt{\cos^2\theta + \frac{1}{64}\sin^2\theta} \; .$$

The model (2.15) assumes a constant solar wind speed ($\gamma = 0$) and yields a value of 7.5×10^6el/m^3 at the earth's orbit. The solar wind plasma could be detected for solar offsets as large as $215R_\odot$ (1 AU) due to the high sensitivity of the dual-frequency group time-delay measurements (standard error: 70×10^{16}el/m^2). The (heuristic) latitudinal factor represents an ellipse of axial ratio $8:1$ and is of the same order as latitudinal variations determined from white-light observations near solar minimum [2.78] and from dispersion measurements with pulsars [2.253].

A later analysis of the *Viking* data [2.155] found a better fit to the data with a model of the form

$$N_e(r,\theta) = \frac{1.32}{r^{2.7}} \times 10^{12}G(\theta) + \frac{2.3}{r^{2.04}} \times 10^{11} \quad [\text{el/m}^3] \; , \tag{2.16}$$

with

$$G(\theta) = \exp[-(\theta/8)^2] \; ,$$

for which the latitudinal dependence $G(\theta)$, with θ in degrees, is strong only near the sun. A solar wind radial falloff exponent slightly greater than 2 is also in good agreement with the $\sim r^{-2.1}$ dependence derived from a comparative study of *in situ* and solar radio-burst determinations of interplanetary electron density outside 0.3 AU [2.24]. This is consistent with a moderate acceleration of the solar wind continuing all the way out to the earth's orbit (see also Chap. 3 [2.208]).

The coronal electron density distribution was most recently investigated during the solar occultation of *Voyager 2* in December 1985, just prior to its encounter with the planet Uranus [2.3]. The radio ray path to earth remained very near the solar equator, thus minimizing propagation effects depending on solar latitude. Due to limited power in the dual-frequency downlink transmission, time-delay measurements were confined to solar offsets $R = 6$–$40R_\odot$. Only a radial variation of the coronal electron density was assumed, but different values of the model parameters were determined for the ingress and egress phases of the occultation. The electron columnar content $I_t(R)$ was found to decrease as $\sim 1/R$ in accordance with $\gamma = 0$ in (2.14) during the egress phase on the solar west limb. The ingress phase of the occultation, however, displayed a faster decrease $I_t(R) \sim R^{-1.6}$, implying a significant asymmetry in the structure of the equatorial solar corona. This is not entirely unexpected – asymmetries are also quite apparent in white-light coronagraph images (Fig. 2.1).

Figure 2.6 illustrates the radial dependence of the coronal electron density for each of the models given in Table 2.2. Also included in Fig. 2.6 are the results from the *Viking* dual-frequency ranging analysis [2.155] using (2.16) with $\theta = 0$, as well as the standard equatorial solar minimum model of Newkirk [2.164]. The *in situ* range of values measured by *Helios* [2.24] is indicated at the lower right of the diagram.

It is interesting to compare the radio-sounding measurements of various spacecraft in an attempt to detect the well-known latitudinal dependence of the

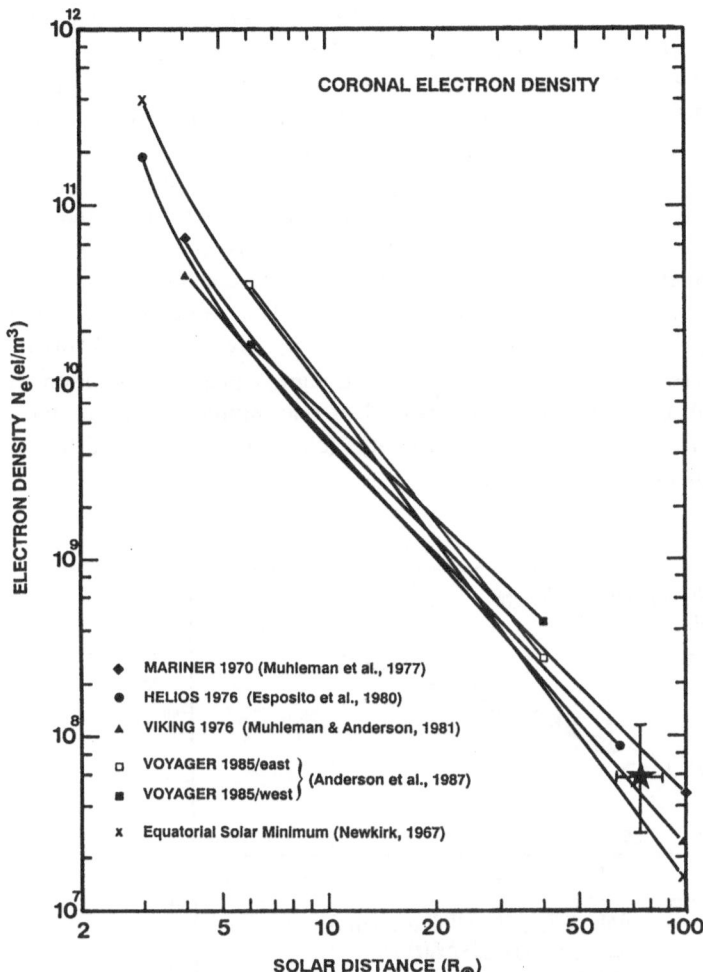

Fig. 2.6. Radial variation of coronal equatorial electron density models derived from radio science occultation experiments with four different spacecraft at various phases of the solar cycle. The standard solar minimum model of Newkirk [2.164] is included for comparison. The point plotted at the lower right indicates the range of *in situ* measurements from *Helios* [2.24]

coronal electron density at different phases of the solar cycle. Experiments were performed with *Mariner 6/7* at the solar maximum in 1970 [2.157], with *Helios* and *Viking* at the solar minimum in 1976 [2.58, 62, 155, 239], and with *Voyager 2* near solar minimum in 1985 [2.3]. The signal ray path from spacecraft to earth passed over the north and south poles of the sun during the occultations of *Mariner* and *Viking*, respectively. As a result, the ray path sampled a full range of heliolatitudes over only a few days around superior conjunction. Since *Helios* and *Voyager* were essentially in the ecliptic plane during solar occultation, however, latitudinal excursions of only about ±7° were incurred. As seen in Fig. 2.6, the electron densities inferred from the *Mariner 7* data are marginally higher than the *Viking* and *Helios* densities at solar offsets $R > 5R_\odot$. At smaller solar distances, however, the density from the solar minimum model actually exceeds that for solar maximum. The electron density distribution at solar maximum is known to be nearly isotropic (see Sect. 2.2), so that the northern polar region would presumably be very similar to the equatorial corona. As apparent in the deviations of the individual data points from the mean curves in Figs. 2.4 and 2.5, it is more likely that local enhancements (coronal streamers) or depletions (coronal holes) are responsible for the differences. Although the *Voyager 2* experiment was conducted close to solar minimum [2.3], the densities were found to be the highest observed at any occultation so far. The values for N_e from the derived model (Table 2.2) at the distance $R = 7R_\odot$ were $2.4 \times 10^{10}\text{el/m}^3$ for the ingress phase and $1.2 \times 10^{10}\text{el/m}^3$ for the egress phase, a variation of $2:1$ from east limb to west limb over a period of one month. The asymmetry is most likely due to large-scale plasma structures (coronal streamers and holes) that extend out into the equatorial corona.

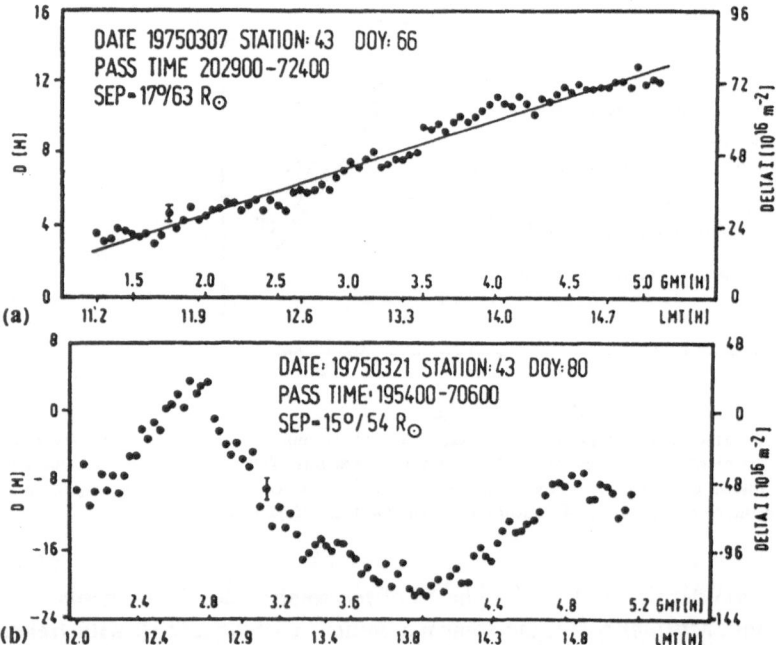

Fig. 2.7a,b. *Helios 1* electron content measurements close to the first perihelion passage. Two contrasting examples are shown:

    (a) "quiet" corona   ($R = 63R_\odot$,   7 March 1975), and
    (b) "disturbed" corona   ($R = 54R_\odot$,   21 March 1975).

The change in electron content $\Delta I_t(t)$ is given in units of 10^{16} el/m² on the right or in terms of residual range on the left scale (D = plasma time delay × velocity of light). The rms variation in $\Delta I_t(t)$ given by the error bars for each curve is ca. $\pm 2 \times 10^{16}$ el/m² (a), and $\pm 15 \times 10^{16}$ el/m² (b). The time scale LMT (local mean time) refers to the 64-m NASA ground station at Goldstone, CA/USA [2.58]

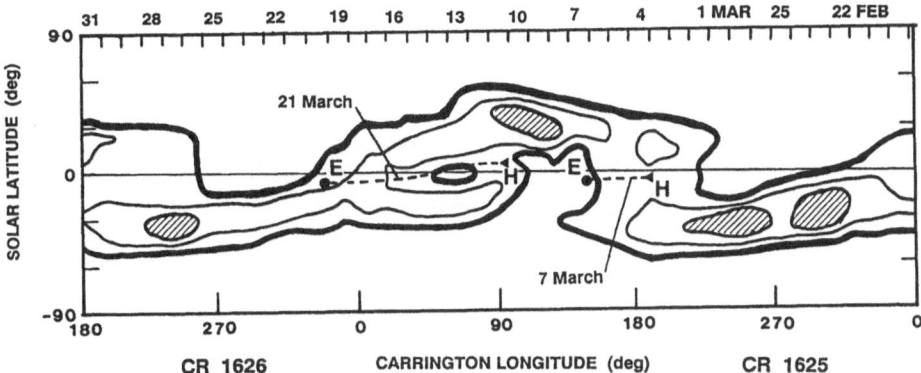

Fig. 2.8. Isophotal map of coronal white-light brightness at $1.5 R_\odot$ from west solar limb K-coronagraph measurements during Carrington rotations 1625 and 1626. Time runs from right to left; the central meridian passages (CMP) of each date are indicated in the upper abscissa scale, Carrington longitude in the lower scale. Coronal intensities are graded by contour levels equivalent to electron densities of $(5.5, 7.2, 8.9) \times 10^{12}$ el/m^3. The solar surface projections of the *Helios*/earth ray paths for the two 1975 dates 7 March and 21 March are indicated. Pronounced density enhancements (oblique hatching) associated with coronal streamers are distinguishable between solar longitudes 220° and 300° (southern hemisphere) and from 90° to 130° (northern hemisphere). Very little variation of electron density is expected along the relatively short ray path on 7 March. Larger variations occur on March 21 because (a) the electron content is about four times greater than on 7 March, and (b) the ray path traverses a region characterized by strong lateral gradients (after [2.58])

2.3.2 Temporal Variations

Typical temporal variations in the electron content on time scales of hours are shown in Fig. 2.7 – observations made with *Helios 1* near its first perihelion in 1975 (upper panel: 7 March; lower panel: 21 March). The geometry in the ecliptic plane on 21 March 1975, showing the longitudinal variation of electron density along the *Helios*/earth ray path, can be seen in Fig. 2.3. Ground-based K-coronagraph observations during this time, also showing the location of the *Helios* ray path with respect to the long-lived, large-scale global structure, are displayed in the synoptic map of Fig. 2.8 [2.58]. The ray path did not intersect either of the two largest brightness maxima (coronal streamers), located between heliolongitudes 220° and 300° in the southern hemisphere and from 90° to 130° in the northern hemisphere (oblique hatching). The (quiet) measurements on 7 March are most strongly affected by the coronal conditions right at the *Helios* location in front of the solar limb at solar longitude 180° (EPS angle $\simeq 40°$). The electron content during this pass is only 3×10^{18} el/m^2 and there is very little density variation in heliolongitude along the relatively short ray path. The (disturbed) measurements on 21 March, however, are dominated by rather chaotic coronal structure in the neighborhood of an equatorial coronal hole with a tongue running quasi-parallel to the radio ray path. Indeed, as seen clearly in the ecliptic plane view of Fig. 2.3, this coronal hole with its associated lateral density gradients, is closer to the *Helios* ray path than adjacent structures. The total variation

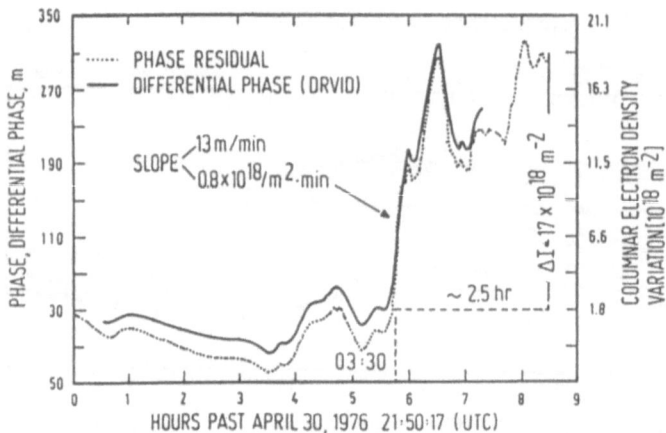

Fig. 2.9. Interplanetary transient detected in *Helios* 2 columnar content measurements on 30 April/1 May 1976 at a solar offset $R = 26.1 R_\odot$. The scaling is the same as Fig. 2.7. Both data types, phase residual (Doppler shift) and differential phase (DRVID), show the distinct signature of a sudden increase in electron content along the *Helios* S-band downlink at 0330 UT. Assuming the event is associated with a large solar flare that erupted about 6.7 hours earlier, one calculates a radial propagation speed of 1000 km/s for the leading edge of the disturbance [2.58]

measured over the pass duration of 3.2 hours is $1.4 \times 10^{18} \mathrm{el}/\mathrm{m}^2$, i.e. about 10% of the total electron content of $1.4 \times 10^{19} \mathrm{el}/\mathrm{m}^2$ given by the *Helios* 2 model in Table 2.2 and from (2.14). Considerable uncertainty, of course, is associated with the extrapolation of coronal structures at $1.5 R_\odot$ out to the radio ray paths applicable to Fig. 2.7 at distances of the order of $60 R_\odot$.

One unique event, shown in Fig. 2.9, was observed to cause an enormous variation in electron content of $1.7 \times 10^{19} \mathrm{el}/\mathrm{m}^2$ at a maximum rate of $1.3 \times 10^{16} \mathrm{m}^{-2} \mathrm{s}^{-1}$ [2.58]. An interplanetary traveling disturbance associated with a solar flare on 30 April 1976 (start 2048 UT, max. 2114 UT; position 9°S, 47°W) that also produced type II and III radio bursts presumably propagated radially outward and intercepted the *Helios* ray path at a solar distance of $34.7 R_\odot$. Taking the abrupt increase in electron content at 0330 UT on 1 May 1976 as the arrival time of the disturbance, one can derive a propagation speed of ca. 1000 km/s. The background electron content during the event was estimated to be ca. $2.5 \times 10^{19} \mathrm{el}/\mathrm{m}^2$, which is only about 50% more than the content in the interplanetary transient.

The sun is undoubtedly the source of most transient wavelike phenomena in the heliosphere, and considerable attention has been paid to coronal wave excitation and propagation [2.91]. Of special interest here are possible coupling effects between fast waves and those of Alfvén type [2.111]. Observational evidence verifying the existence of Alfvén waves, magnetoacoustic waves and fast waves in the solar wind has long been established.

The electron content (and Faraday rotation) measurements on *Helios* 1/2 were well suited for the investigation of coronal wave activity. The signal-to-noise ratio associated with the *Helios* radio subsystems and orbital geometry was

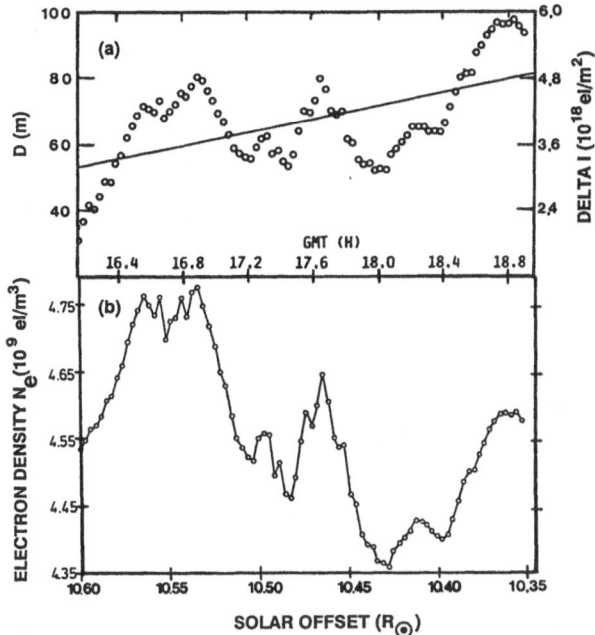

Fig. 2.10. Coronal plasma variations observed at $10R_\odot$ by *Helios* 2 on 8 May 1976. *Panel (a)*: S-band DRVID measurements (total of 84 points), taken at two-minute intervals with a mean signal-to-noise ratio of 7 dB [2.56, 57]. *Panel (b)*: Coronal electron density distribution derived from the electron content measurements of (a) using a recursive Kalman filter algorithm (after [2.232])

significantly better than for comparable space missions. Thus even small-scale structures of the *Helios* electron content data could be resolved. The majority of the *Helios* time-delay data (45 tracking passes from *Helios 1*; 60 passes from *Helios 2*) display structures of periodic nature superimposed on a steady increase/decrease (ingress/egress) with occasional abrupt transients. Subtle variations of electron density in the corona can be resolved using a Kalman filtering inversion scheme. As an example, panel (a) of Fig. 2.10 shows a set of columnar content measurements (DRVID technique) taken with *Helios 2* at a solar offset of $R \simeq 10R_\odot$ on 8 May 1976. Faraday rotation data also exist for this data pass, and have been compared with the DRVID data to determine the relative amplitudes of electron density and magnetic field fluctuations [2.57]. The data of Fig. 2.10 (a) were divided successively into arcs of up to 18 segments, each segment representing 10 minutes of data with a resolution in solar longitude of ca. 5°. The spatial and temporal resolution can be optimized for a given desired degree of numerical stability as controlled by the inversion condition number of the Kalman filter technique. Figure 2.10 (b) shows the result of such a stochastic inversion [2.232]. The *a priori* information is adjusted to the *in situ* plasma densities measured at *Helios 2* [2.188]. Starting with a properly chosen subset of the electron content data, the recursive algorithm inverts each of the successive data points up to the total number of 84 measurements. A comparison of panels

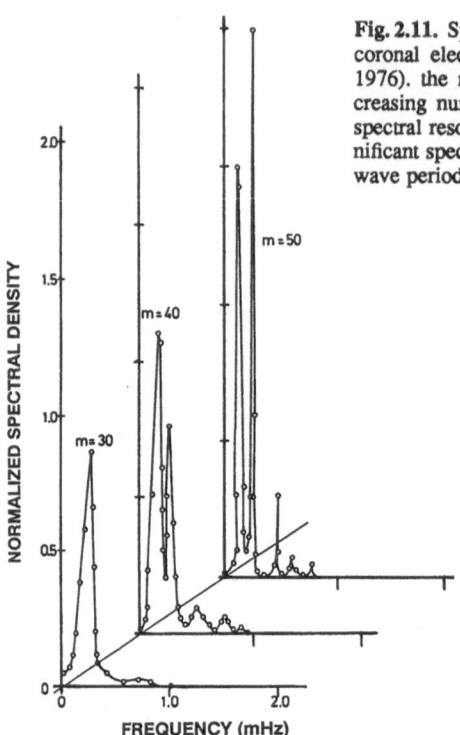

Fig. 2.11. Spectral power density calculated for the *Helios 2* coronal electron content measurements of Fig. 2.10 (8 May 1976). the maximum entropy method was used with an increasing number of filter coefficients (30 to 50) to improve spectral resolution. The Nyquist frequency is 4.2 mHz. A significant spectral peak appears at 0.24 mHz, corresponding to a wave period of 70 min [2.59]

(a) and (b) confirms local enhancements and depletions of the electron density in high correlation with the electron content measurements.

The spectral power density resulting from a maximum entropy spectral analysis [2.181] for the linear detrended data set of Fig. 2.10 (a) is shown in Fig. 2.11 [2.59]. The maximum entropy filter algorithm was combined with a computational procedure searching for individual spectral peaks. The spectra shown in Fig. 2.11 are plotted on a linear–linear scale in order to visualize more easily the spectral power distribution at the lowest frequencies. The total frequency interval up to the Nyquist frequency (4.2 mHz) was divided up into 400 cells in order to locate the position of peaks. The spectral calculations are controlled by subsequently proving the invariance of power between filter input and output according to Parseval's theorem. The number of prediction error coefficients of the spectral filter was varied by $m = 30, 40, 50$ (so that $m/M = 36\%, 48\%, 60\%$; where $M = 84 =$ total number of data points). The spectral analysis yields a lowest detectable frequency of about 0.24 mHz (period: 70 minutes) from the total observation interval of 2.8 hours. The spectral line is observed to split, but does not disappear, upon increasing the number of filter weights. Such line splitting is usually due to insufficiencies in the data [2.181], although the signal-to-noise ratio in this case was relatively high. There can be no doubt, however, about the authenticity of the 70-minute oscillation. Other data sets show strong spectral peaks with gravest modes at periods from 1 to 2 hours. In still other cases, however, they are too weak to be detected. When present, the spectral power peaks are quite stable even upon increasing the number of filter prediction error coefficients m (where $m = 50$ seems to be a reasonable maximum bound for this specific application). In the example given, the improved spectral resolution reveals two additional peaks at 0.49 and 0.67 mHz (35 and 25 minute periods, respectively). The typical range of periods for the higher harmonics in the *Helios* time-delay data runs from 10 minutes on up to 40 minutes, with great variations in the intensities of these fluctuation "resonances". DRVID power spectra derived using conventional Fourier techniques are unable to resolve preferred periodicities and indicate only an increase in power at the lowest frequencies [2.33, 34]. There is strong evidence that the distribution of coronal electron density fluctuations is a continuous power-law spectrum at higher frequencies $f > 10^{-3}$ Hz [2.39, 268]. This topic is treated in more detail in Sect. 2.5.

Nonlinear Alfvén waves are excited in the strongly inhomogeneous photosphere and chromosphere by stochastic motions of the solar plasma at the upper boundary of the convective zone [2.91]. These waves are coupled to generate modified Alfvén waves of orthogonal polarization and fast waves, respectively. Fast hydromagnetic waves propagate isotropically and produce oscillating perturbations δN_e of the electron density. Such density oscillations in the coronal medium traversed by the *Helios* ray path are probably responsible for the periodicities observed in the electron content measurements. At a heliocentric distance of $10 R_\odot$, corresponding to the data of Fig. 2.10 (a), fast waves would be expected to travel at approximately the Alfvén velocity $v_A \simeq 600$ km/s (estimated from the electron density model of Table 2.2 and the magnetic field magnitude of Sect. 2.4). Assuming a wave frequency of 0.24 mHz, the wavelength of this hypothetical fast wave would be roughly $3.5 R_\odot$. The electron density and magnetic field fluctuations associated with this fast MHD wave can be shown to be of the order of $\delta N_e/N_e = 4\%$, and $\delta B/B = 10\%$, respectively [2.57]. The intermediate Alfvén wave, it should be mentioned, does not generate density fluctuations in the linearized approximation and would thus be undetectable with time-delay measurements. Such Alfvén waves, however, may have been detected by their associated magnetic field oscillations using a combination of *Helios* time-delay and Faraday rotation data [2.92].

2.4 Coronal Magnetic Structure

The physics of the solar corona is critically dependent on the magnetic field, which contains the dominant energy per unit volume in the solar wind all the way from the coronal base out to the super-Alfvénic transition at 20–30$R_\odot$ [2.180]. It is generally accepted that the magnetic field of the corona is generated primarily within the solar convection zone by differentially rotating plasma (e.g. the review by Gibson [2.73]). Meridional (poloidal) magnetic fields there are stretched and twisted into zonal (toroidal) fields, which occasionally well up through the solar surface to form a bipolar sunspot pair. Solar "activity", a term generally taken to mean production of intense and usually sporadic emission from the radio to the X-ray range, is concentrated in the regions containing sunspots. It is convenient, in fact, to quantify the degree of solar activity by a sunspot number R_z – a weighted sum of the observed spots on the solar disk. The most striking feature of the activity index R_z, which has been computed back in time to the middle of the 18th century, is its eleven-year quasi-periodic variation – the solar sunspot cycle. The first spots of a cycle appear near 40° latitude, even before the spots of the previous cycle have vanished. They steadily increase in number and slowly drift toward the equator as the cycle progresses. The most recent minimum in the sunspot curve, the start of the 22nd cycle since the count began, occurred in September 1986.

The emergence of additional magnetic flux in the form of loops can drastically rearrange the magnetic structure of the corona. Since the coronal material is an almost completely ionized plasma, constrained in its motion perpendicular to the magnetic field, it is forced to corotate and coevolve with the fields generated by the solar dynamo. The adherence of coronal material to magnetic lines of force is one of the more suggestive impressions one obtains from eclipse or coronagraph images in visible light (see Fig. 2.1). Two distinctly different magnetic field configurations are revealed:

(a) high-brightness (high-density) loops with both magnetic field footpoints on the solar surface, and
(b) low-brightness regions, where the magnetic field has one footpoint on the sun and the other dangling out into interplanetary space.

Magnetic regions of type (a) are said to be "closed" and those of type (b) are "open".

The coronal plasma, being free to travel along the magnetic field lines, will flow in a direction opposite to the pressure gradient. If the field line happens to be "open", the gas expands radially outward away from the high thermal and magnetic pressure. Solar gravity is incapable of retaining the gas, which continues outward as a solar wind until its ram pressure is counterbalanced by the external pressure of the local interstellar medium at the boundary of the heliosphere. If the field line is closed, the best the plasma can do is avoid the high (thermal and magnetic) pressure areas, and it therefore often collects in regions of low magnetic field strength, e.g. quasi-neutral sheets. Closed-field-line configurations rarely extend beyond $2R_\odot$ from the sun.

We are primarily interested here in the magnetic field in the source region of the solar wind, i.e. along the open field lines that extend radially outward and eventually form the interplanetary magnetic field (IMF). The IMF has been measured in the inner heliosphere as close to the sun as $62R_\odot$ on the *Helios* spacecraft. These results and other topics concerning the IMF are covered extensively in Chap. 4 [2.144]. The emphasis here will be placed on models and remote sensing observations of the coronal magnetic field, but restricting the discussion to the open-field topologies above ca. $2R_\odot$. Recent surveys of coronal magnetic fields and their extension into the heliosphere include those by Svalgaard and Wilcox [2.235], Dulk and McLean [2.52], MacQueen [2.138], and Hoeksema [2.86].

2.4.1 Coronal Magnetic Field Models

It was recognized in the early sixties [2.163] that the polarity of the IMF was highly correlated with the photospheric field measured at the earth's heliolongitude (central meridian) 4.5 days earlier. This implied a solar origin of the IMF with magnetic field footpoints approximately 60° west of the central meridian in good agreement with Parker's [2.169] Archimedian spiral model. The details of the qualitative connection between the IMF and the often extremely complex photospheric fields, however, were unclear.

Two different approaches were introduced for determining the solar magnetic field in the region between photosphere and the interplanetary spiral. Using a simple geometry with a solar dipole in an expanding plasma, Pneuman and Kopp [2.178] solved the applicable MHD equations to find a self-consistent numerical model of the fields and currents in and around a symmetric coronal "helmet" streamer. The resultant configuration featured closed magnetic regions that pinched down into a current sheet separating open coronal magnetic fields of opposite polarity. The other approach has come to be known as the "potential field" (PF) approximation, because the coronal magnetic field is assumed to be derivable from a scalar potential [2.2, 201, 202]. In this case, the magnetic field B is given by

$$B = -\nabla \Phi ; \quad \nabla \times B = \frac{4\pi}{c} J = 0 \qquad (2.17)$$

where Φ is the magnetic potential, and J is the current density. It is thus required that the region of interest be devoid of electric currents – a stark constrast to the massive current sheet inherent in the model of Pneuman and Kopp [2.178]. Nevertheless, if (2.17) is valid, the magnetic potential satisfies the Laplace equation $\nabla^2 \Phi = 0$ in the region above the photosphere (lower boundary condition) and below a so-called "source surface" of constant potential (upper boundary condition). The component of the photospheric magnetic field along the line-of-sight, measured for example by the Zeeman splitting of a neutral iron line, must be known over the entire solar surface. Although it would be possible to compute the potential field from these line-of-sight measurements [2.22], the conventional technique assumes the field to be purely radial at the photosphere. The exact shape and radius of the source surface are free parameters, but it is usually taken to be a sphere at about $2-3R_{\odot}$, upon which the field is assumed to be purely radial. Nonspherical source surfaces offer improvements over the spherical models [2.126, 204]. The resultant magnetic field is represented by a sum of spherical harmonics, providing the bonus that the individual contributions from all multipole components in the orthonormal expansion are known as well.

Although the PF and MHD calculations agree qualitatively in the region below the source surface, there are obvious differences in the region extending into interplanetary space. Among these is the reversal of the magnetic field across the coronal current sheet, which is much too gradual for the PF models. Recent efforts have moved away from the PF approach and have considered thermal and magnetic pressure distributions under magnetostatic equilibrium [2.100, 168, 263]. The dominant solar magnetic field was taken to be a dipole that becomes distended by the fields arising from an equatorial current sheet. The strength of the current in the sheet was found to be proportional to the thermal pressure difference between equator and pole [2.100, 168]. The stronger the current becomes, the more magnetic flux is shifted to open-field lines with correspondingly higher field strengths. Volume currents were shown by Wolfson [2.263] to be of only secondary importance, in agreement with previous results [2.125]. Allowing for a dynamic outflow was shown to have only minor effects on the quasi-static solution [2.167].

In spite of the nonphysical disregard of coronal currents, recent innovations to the PF model introduced by Hoeksema et al. [2.89, 90] have been quite successful in predicting the polarity of the IMF at 1 AU. These authors have computed a source-surface magnetic field at each 10° of solar rotation and then constructed contour plots using a weighted average of those computed fields within 30° of central meridian. These contour plots in heliolatitude and Carrington longitude always feature a wavy neutral line separating regions of toward and away polarity. This neutral line extends into interplanetary space as a warped heliospheric current sheet [2.254]. Adjusting for a 4.5-day solar wind transit time, it was found that the magnetic field polarity at the source surface agreed with the polarity of the IMF at earth about 80% of the time. Best correlations were achieved for source surfaces located between 2–$3 R_\odot$.

A similar success ratio could be achieved using synoptic K-coronagraph data displaying the limb brightness as a function of time (Carrington longitude) and heliolatitude. Associating the maximum brightness contour (MBC) with the heliospheric current sheet, Burlaga et al. [2.31] found general agreement between the coronal and IMF polarities. The MBC technique was determined to be marginally better than the PF predictor at solar minimum [2.29], but it cannot always be applied at solar maximum since the MBC is difficult to locate [2.255]. The PF model of Hoeksema et al. [2.89, 90] incorporates a moderately strong solar-cycle-dependent polar field [2.233], which reaches a maximum of slightly over 10 G at sunspot minimum and is thus particularly effective in reducing the latitudinal extent of the heliospheric current sheet. Polar fields of this magnitude are absolutely realistic, as demonstrated by magnetic flux transport calculations that account for both strong diffusion and poleward flow [2.247]. The constriction of magnetic flux in highly divergent polar coronal holes, such as that modeled by Munro and Jackson [2.159], has been shown to require surface fields of ca. 20 G [2.231]. An atlas of synoptic measurements of photospheric magnetic fields and the associated source surface calculations from 1976–1985 has been published by Hoeksema and Scherrer [2.87].

CORONAL BRIGHTNESS

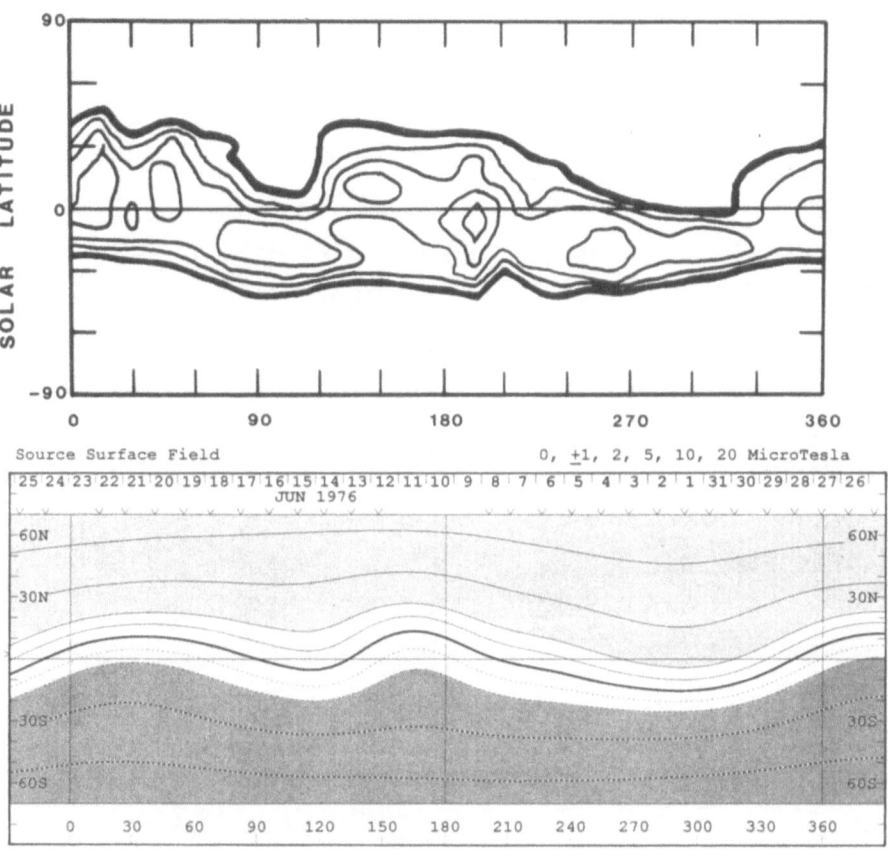

Fig. 2.12. Comparison of coronal brightness with source-surface magnetic field strengths obtained from potential field calculations during Carrington rotation 1642 (June 1976). *Upper panel:* Coronal polarization brightness (pB) contours recorded at $1.5R_\odot$ on west limb by the Mauna Loa Coronameter (courtesy R.T. Hansen, then at High Altitude Observatory). *Lower panel:* Magnetic field strength at source surface ($2.5R_\odot$) derived from photospheric field measurements taken at the Wilcox Solar Observatory (courtesy J.T. Hoeksema, Stanford University)

The remarkable similarity between the positions of the current sheet on the source surface and the band of maximum K-corona brightness is illustrated in Fig. 2.12. The data were recorded at extreme solar minimum in May–June 1976 (Carrington rotation 1642). The upper panel shows contours of the west limb polarization brightness pB (intensity component polarized perpendicular to the radius vector) at $1.5R_\odot$ recorded by the Mauna Loa Coronameter of the High Altitude Observatory (courtesy of R.T. Hansen). The lower panel displays the dis-

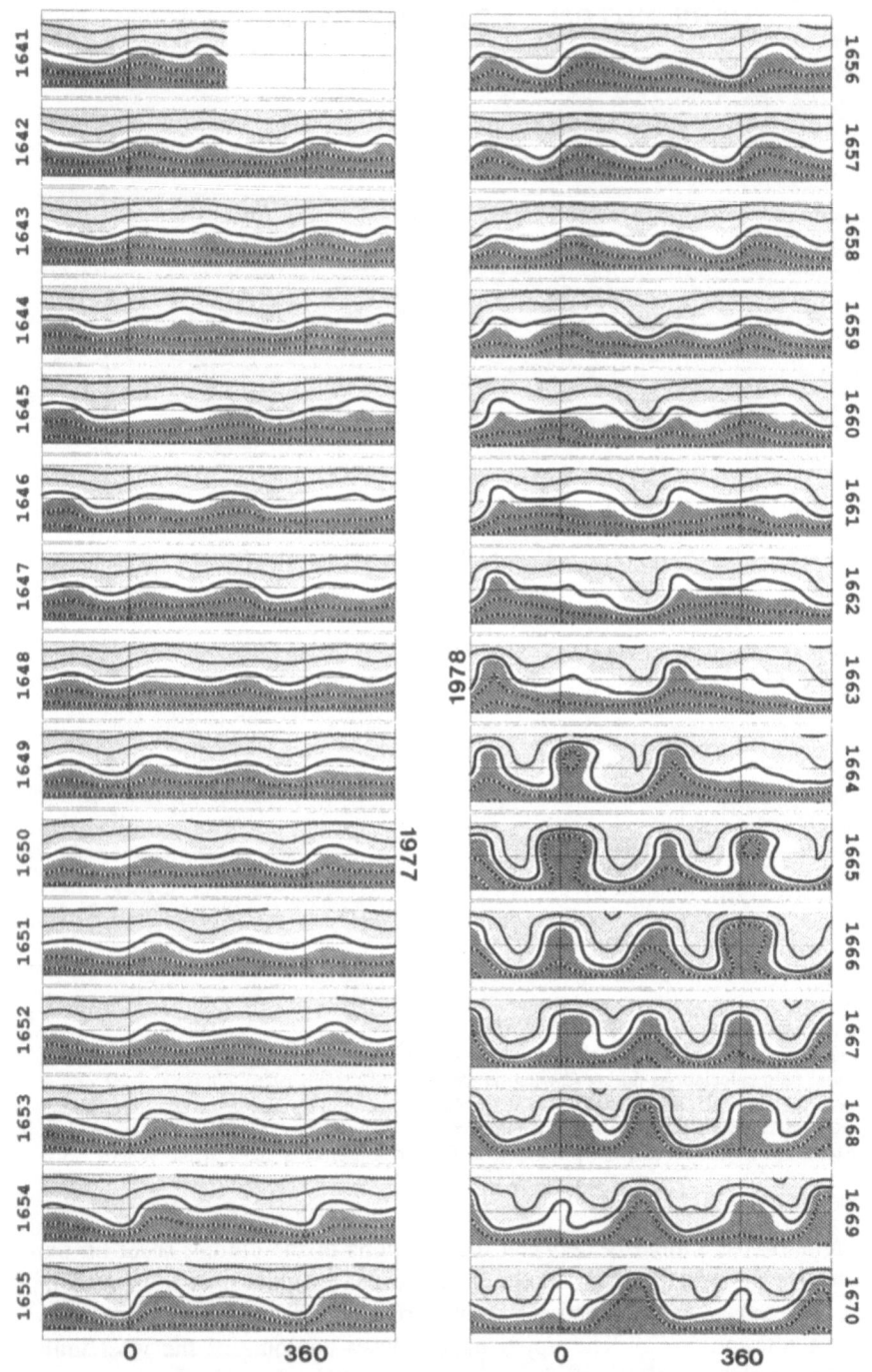

Fig. 2.13. Source-surface magnetic field strength over Carrington rotations 1641–1700 (May 1976 to October 1980). The heliospheric current sheet separates regions of away polarity (*light shading*) from regions of toward polarity (*dark shading*). The roughly equatorial location of this current sheet

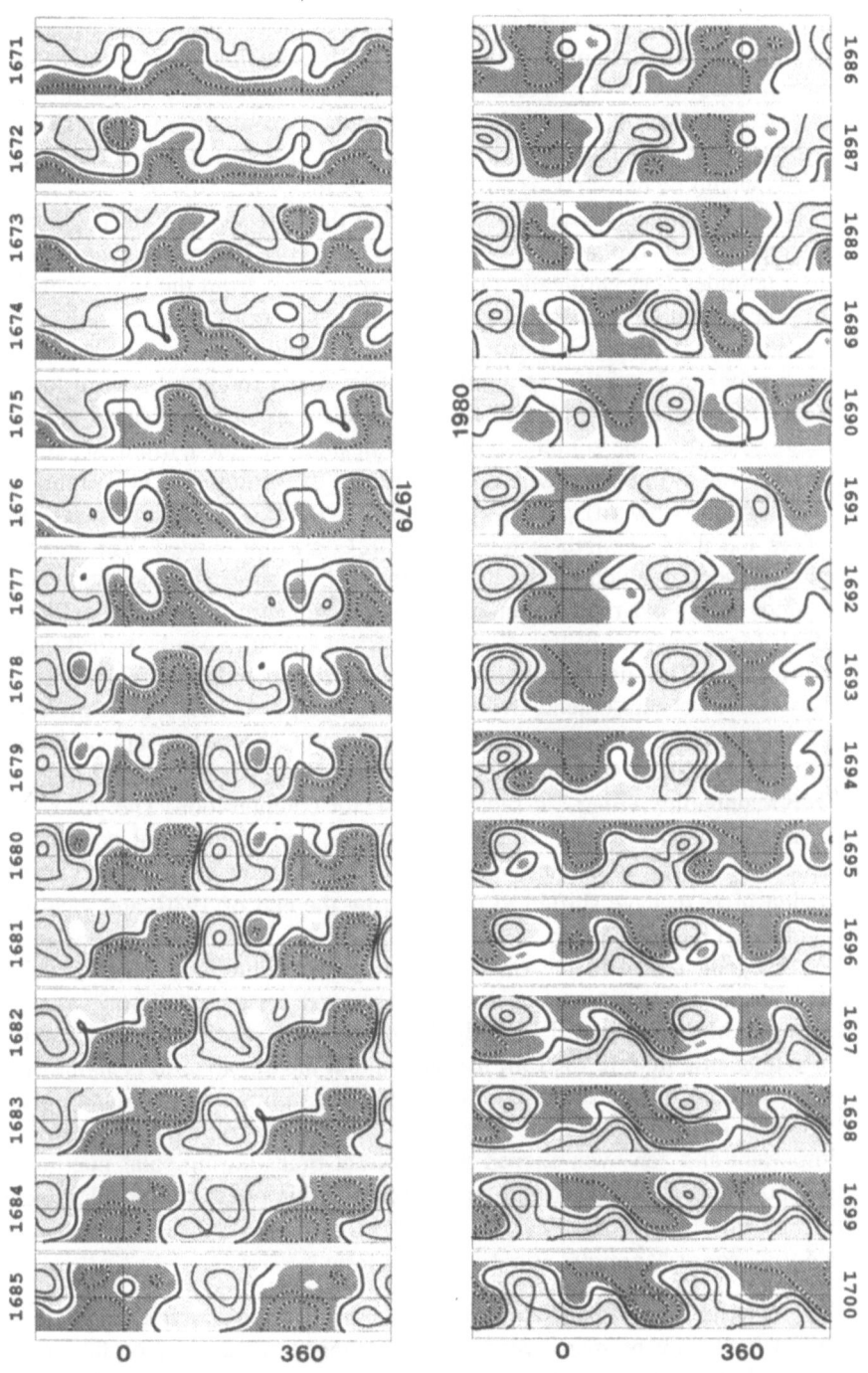

at solar minimum (*left panels*) contrasts with the chaotic structure at solar maximum (*right panels*). The reversal of the solar dipole component from left to right is readily apparent (courtesy of J.T. Hoeksema, Wilcox Solar Observatory, Stanford University)

tribution of radial magnetic field strength on a source surface at $2.5 R_\odot$ (courtesy of J.T. Hoeksema, Stanford University). The longitudinal form of the heliospheric current sheet features two extensions or "warps" to higher latitudes that produce the characteristic 4-sector IMF configuration. Depending upon the heliographic latitude of an interplanetary magnetic monitor, it is possible that only one of the warps would be detected. The resulting 2-sector structure is observed fairly often at the earth, which travels annually within the range $\pm 7.25°$ in heliolatitude. A monopolar IMF (no warps) is rare at the earth, but is probably commonplace at high latitudes (the verification is expected to come from the *Ulysses* mission).

The evolution of coronal magnetic structure from solar minimum to solar maximum is demonstrated in Fig. 2.13 (also kindly provided by J.T. Hoeksema). Each panel of Fig. 2.13 represents one Carrington rotation between the vertical lines at 0° and 360° with half-rotation extensions on either side for continuity. Field polarity at the source surface is indicated over the latitudinal range $\pm 70°$. Positive (away) polarity with a field strength exceeding 0.02 G is lightly shaded and negative (toward) polarity (with $B < -0.02\,\mathrm{G}$) is darkly shaded. Regions where the field strength is less than 0.02 G are not shaded. The quasi-neutral current sheet, denoted by the heavy contour line, stays close to the heliographic equator during solar minimum, high-latitude excursions being first evident in late 1977. Many features of this 4-sector recurrent structure can be traced throughout this interval. The situation at solar maximum (right-hand panels) is more complex. The current sheets extend over the entire computed range of latitude and are complemented by occasional secondary current sheets that appear as isolated "blobs". The sheet structure is still often recurrent over many solar rotations, the IMF maintaining either a 4-sector or 2-sector configuration. The polar fields are seen to flip sign over the course of the solar cycle, the actual polarity reversal occurring shortly after sunspot maximum. The away fields of the north polar region in 1976 have shifted to the south by the end of 1980, and vice versa for the toward polarity fields.

A cyclic return to the same solar magnetic configuration (a "Hale" cycle) thus requires two sunspot cycles (ca. 22 years) for completion. A study of the strengths of the lower-order multipole components contributing to the source surface magnetic field [2.87, 138] has shown that the dipole term is about five-times stronger than its nearest competitor at sunspot minimum. The quadrupole term, which is responsible for the "warp" in the heliospheric current sheet, becomes important and can even exceed the dipole at times during solar maximum. The octopole term should also not be neglected at this phase of the solar cycle.

2.4.2 Coronal Magnetic Field Observations

There are no *in situ* observations of the heliospheric magnetic field inside 0.3 AU and it well may be many years before a magnetometer can be flown into this region. The MHD and PF calculations of the previous subsection do provide one estimate of the coronal magnetic field configuration from the knowledge of the magnetic fields on the solar surface measured, for example, by Zeeman heliospec-

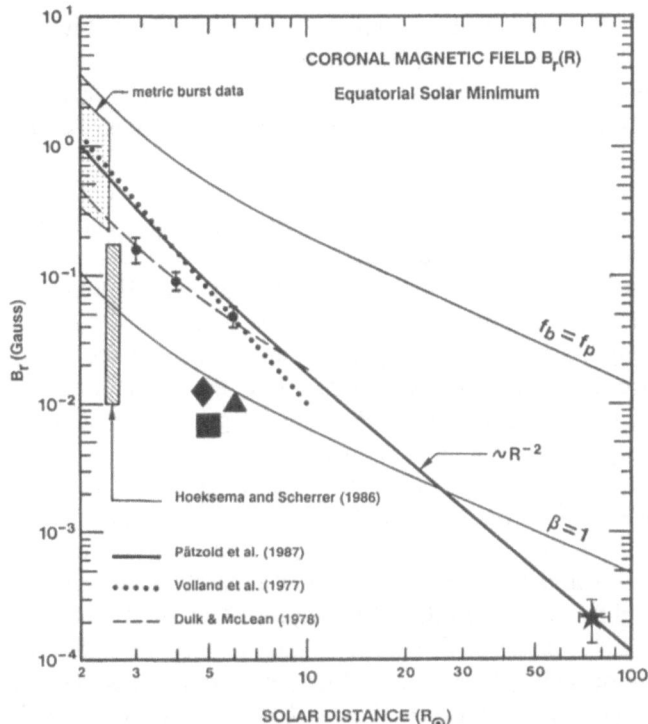

Fig. 2.14. Coronal magnetic field strength as a function of solar distance. Empirical models plotted are: *solid curve* – Pätzold et al. [2.173]; *dashed curve* – Dulk and McLean [2.52]; *dotted curve* – Volland et al. [2.244]. Also included are theoretical curves for $\beta = 1$ and $f_P = f_B$, the electron densities taken from an equatorial coronal model at solar minimum. The range of *in situ* values from *Helios (star)* appears in the diagram at the lower right [2.144]. The range of field strengths from PF calculations is shown by the vertical bar at $2.5 R_\odot$ [2.87]. Other magnetic field estimates in the diagram were made using observations of coronal gyrosynchrotron radiation (*solid circles* – Dulk et al. [2.53]), and Faraday rotation of natural radio sources (*triangle* – Sofue et al. [2.224]; *square* – Bird et al. [2.16], and *diamond* – Soboleva & Timofeeva [2.223]

troscopy. A complete MHD treatment using anything but a greatly simplified lower boundary condition is still beyond the reach of present-day computational capabilities. Since the Zeeman effect yields only the line-of-sight component of the continuously evolving photospheric magnetic field, we cannot even claim to know the exact state of the lower boundary at a given epoch. Nevertheless, some inroads toward inferring coronal magnetic field strengths from remote sensing observations have been established and are compiled here as a composite diagram of all available data (Fig. 2.14), covering the range in the solar equatorial plane from $2 R_\odot$ out to the innermost *in situ* measurements of *Helios* at $62 R_\odot$. The various features of Fig. 2.14 are discussed in turn below.

Optical and microwave polarization studies have been employed to deduce magnetic fields in the lower corona above active regions, i.e. in closed-field regions [2.52]. The possibility of exploiting the Hanle effect, which alters the polarization characteristics of resonantly scattered radiation in accordance with the strength and orientation of the local magnetic field, was investigated in detail [2.23, 199] for the UV emission lines of H I (Lyman-alpha = λ1215) and O VI (λ1032). Since the magnetic field strengths required for a detectable effect are of the order of 10 G, however, it is improbable that this technique could be extended to coronal heights beyond $2R_\odot$.

Under certain circumstances and appropriate assumptions, it is possible to infer the strength of the coronal field at greater heights from metric radio-burst emission. Two standard techniques include making either an estimate of the Alfvén velocity (primarily using the frequency drift rate of Type II bursts), or a measurement of the Faraday rotation of polarized Type III bursts [2.52]. Such derivations of field strength are invariably dependent upon one's choice of electron density mode. Since these emissions occur typically at the plasma frequency and its lower-order harmonics, bursts observable from the ground are cut off by the ionosphere when they reach a coronal height of about $2R_\odot$.

One particularly fruitful case study was described by Dulk et al. [2.53], who derived upper and lower bounds on the strength of the ambient magnetic field from the combined white-light and metric radio observations of a coronal mass ejection (CME). The estimate is based on the assumption that the CME produced a shock traveling at slightly faster than the Alfvén speed. The shock velocity at lower heights (1.5 to $3.0R_\odot$) could be determined from the onset of radio bursts at successively lower frequencies. This shock velocity was found to be the same as that of the leading edge of a compression region of the CME observed by the *Skylab* coronagraph higher up in the corona (4–$6R_\odot$). Using a reasonable value for the Alfvén Mach number $M_A = 1.6$, and an appropriate electron density model, the radial profile of ambient magnetic field could be determined out to $6R_\odot$. The magnetic field strengths of Dulk et al. [2.53] are plotted in Fig. 2.14 as solid dots with error bars covering the range from lower to upper bound. These estimates merge continuously with field estimates below $2R_\odot$ from other metric burst observations. Dulk and McLean [2.52] offer the following empirical formula for the radial dependence of magnetic field strength over the range $1.02R_\odot < R < 10R_\odot$:

$$B = 0.5(R/R_\odot - 1)^{-3/2} \quad \text{[gauss]} . \tag{2.18}$$

The radial profile (2.18) is shown in Fig. 2.14 as a dashed line.

Since many of the estimates in Fig. 2.14 involve an assumption concerning the electron density and are thereby solar cycle dependent, the diagram should be considered as valid only for solar minimum. Other data appearing in Fig. 2.14 include the following:

(a) *Helios in situ* measurements of B_r (solid star, see also Chap. 4 [2.144]).
(b) Magnetic field profile required to satisfy $f_B = f_P$, where f_B is the electron gyrofrequency, and f_P is the plasma frequency. For B given in gauss and N_e given in cm^{-3}, these two quantities are:

$$f_B = 2.8B \quad \text{[MHz]} \tag{2.19}$$

$$f_P = 9 \times 10^{-3} \sqrt{N_e} \quad \text{[MHz]} . \tag{2.20}$$

Since N_e goes approximately as r^{-2}, the magnetic field resulting from the condition $f_B = f_P$ falls off as r^{-1}. The Alfvén speed given by

$$v_A = \frac{B}{\sqrt{4\pi M N_e}} , \tag{2.21}$$

where $M = 1.9 \times 10^{-24}$ g is the mass ascribed to each electron in the plasma (includes 10% He), can be expressed in terms of (2.19) and (2.20) as

$$v_A = 6600 \frac{f_B}{f_P} \quad \text{[km/s]} . \tag{2.22}$$

Fundamental gyrosynchrotron radiation, which is rarely observed, cannot escape from the corona for magnetic field strengths below the curve of constant Alfvén velocity given by $f_B = f_P$. This curve thus represents an upper bound on the ambient magnetic field intensity, at least above $2R_\odot$.

(c) Magnetic field profile resulting from the condition $\beta = 1$. The plasma β is defined as the ratio of the thermal pressure $P = N k_B T$, where $N = N_i + N_e = 1.9 N_e$ and k_B is Boltzmann's constant, to the magnetic pressure $P_B = B^2/8\pi$. Assuming an isothermal corona at a temperature $T = 1.5 \times 10^6$ K, it can be shown that:

$$\beta = \frac{4 \times 10^{-4}}{v_A^2} = 10^{-3} \left(\frac{f_P}{f_B} \right)^2 . \tag{2.23}$$

The radial profile of the magnetic field for $\beta = 1$ therefore has the same form as that for $f_B = f_P$, but lowered by a factor $\sqrt{1000} = 31.6$. Magnetic fields above this curve are strong enough to control coronal dynamics. Again, this would seem to be the case in the sun's corona, at least out to the Alfvén point, where the kinetic component of the energy density assumes dominance.

The remaining data in Fig. 2.14 were obtained from Faraday rotation measurements using linearly polarized radio sources during solar occultation. Estimates of magnetic fields from Faraday rotation measurements also require an independent determination of the electron density, whether from observations or models. In contrast to the methods mentioned above, however, this magnetic field diagnostic is applicable only outside of ca. $2R_\odot$. This minimum elongation of the radio source from the sun is dictated by solar noise which overwhelms the occulted probe signal at smaller solar offsets. The outer bound of sensitivity for coronal Faraday rotation observations is typically $10R_\odot$, although definite coronal contributions have been detected beyond this distance. Coronal Faraday rotation has been measured at microwave frequencies using spacecraft [2.173, 229, 244], pulsars [2.14, 16], and the Crab Nebula [2.223, 224].

The applicable formula for the Faraday rotation FR of a linearly polarized signal in a magnetically biased medium with a magnetic field B and electron density N_e is given by the quasi-longitudinal approximation:

$$FR = \left(\frac{K}{f^2}\right) \int N_e \boldsymbol{B} \cdot d\boldsymbol{s} \quad \text{[radians]} , \tag{2.24}$$

where $K = 2.36 \times 10^4$ in either MKS or cgs units, f = frequency of radio probe signal, and the integral (2.24) is taken along the signal ray path with elemental increment $d\boldsymbol{s}$. The Faraday rotation can be either positive or negative, depending on the sign (polarity) of the magnetic field. FR is conventionally taken to be positive (electric vector rotates counterclockwise when viewed by an observer facing the source) if the magnetic field in the propagation medium points toward the observer. Only the component of the magnetic field along the ray path contributes to FR, which can also be expressed as

$$FR = \left(\frac{K}{f^2}\right) \langle B_L \rangle I_t \tag{2.25}$$

where $I_t = \int N_e ds$ is the total electron content, and $\langle B_L \rangle$ is the weighted mean longitudinal magnetic field.

The coronal magnetic field can be assumed to be primarily radial in the region of interest. The contribution to the integral (2.24) is thus zero at the solar proximate point along the ray path because $\boldsymbol{B}$ is perpendicular to $d\boldsymbol{s}$. If the magnetic field and electron density are approximated by simple power laws

$$B_r = B_0 \left[\frac{R_\odot}{r}\right]^{\alpha_1} ; \quad N_e = N_0 \left[\frac{R_\odot}{r}\right]^{\alpha_2} , \tag{2.26}$$

then the Faraday rotation (2.24) will vanish due to the symmetry of the integral. This happens if the entire region probed by the signal is occupied by a single monopolar magnetic sector. At the other extreme, if the coronal magnetic field were of one polarity in the region behind the solar limb and of opposite polarity in the region between sun and earth, then the integral (2.24) would attain a maximum value because all elemental contributions fo FR along the ray path would be of the same sign. This occurs when the signal penetrates an isolated magnetic sector boundary right at the solar limb. In this case, the Faraday rotation becomes

$$FR_{\max} = \left(\frac{KR_\odot}{f^2}\right) \frac{2N_0 B_0}{\alpha_1 + \alpha_2 - 1} \left[\frac{R_\odot}{R}\right]^{\alpha_1 + \alpha_2 - 1} . \tag{2.27}$$

A statistical study [2.173], using 460 hours of data recorded inside of $10R_\odot$ during the *Helios* solar conjunctions in 1975–76 (solar minimum), found that the null FR case was quite rare and that a large set of measurements tended to accumulate up under a maximal envelope that was most likely associated with the ideal case $FR_{\max}$. The radial decrease and level of this maximal envelope were used to determine the. parameters B_0 and α_1, for a given model of the coronal electron density parameters N_0 and α_2, appropriate for sunspot minimum.

An empirical formula that fits both the Faraday rotation data in the interval $2R_\odot < R < 10R_\odot$ and the *in situ* measurements starting at $62R_\odot$ [2.144] was derived:

$$B_r(r) = 6 \left[\frac{R_\odot}{r}\right]^3 + 1.18 \left[\frac{R_\odot}{r}\right]^2 \quad \text{[gauss]} \;. \tag{2.28}$$

The relation (2.28) is drawn in Fig. 2.14 as a solid curve.

Volland et al. [2.244] were able to derive the rough sector structure of the product $N_e B_r$ during two specific solar conjunctions of *Helios* in 1975. The power-law approximations (2.26) were used for the magnetic field and electron density, but the magnetic field polarities were required to match those observed in the IMF. Good agreement with the observations was found for the parametric values $\alpha_1 + \alpha_2 - 1 \simeq 4.5$ and $N_0 B_0 \simeq 10^7\,\text{G}\,\text{cm}^{-3}$. Using an electron density model for sunspot minimum with $N_0 = 1.4 \times 10^6\,\text{el/cm}^3$ and $\alpha_2 = 2.5$ (essentially a mean of the models considered by Pätzold et al. [2.173]), the results of Volland et al. [2.244] imply a magnetic field of the form

$$B_r(r) = 10 \left[\frac{R_\odot}{r}\right]^{3.0} \quad \text{[gauss]} \;. \tag{2.29}$$

This radial profile is plotted in Fig. 2.14 as a dotted line.

Selected values of the mean weighted longitudinal magnetic field $\langle B_L \rangle$ from (2.25), as derived from isolated coronal Faraday rotation measurements using natural sources, are also plotted in Fig. 2.14. Even if the coronal magnetic field assumes the configuration for FR_{max}, these values of $\langle B_L \rangle$ will only be a fraction of the actual magnetic field strength at the given solar offset R. With the observed sunspot-minimum parameters $\alpha_1 = 2.7$, and $\alpha_2 = 2.5$ [2.173], for example, it can be shown that $\langle B_L \rangle \simeq B_r(R)/5$. It is thus not surprising that the longitudinal magnetic fields derived from individual Faraday rotation measurements are much lower than the actual coronal field strengths.

Included in Fig. 2.14 is also a hatched bar showing the typical range of magnetic field strengths at the source surface found in the atlas compiled by Hoeksema and Scherrer [2.87]. The bar has been adjusted upward by a factor of 1.8 to account for the underestimate of the photospheric fields due to saturation of the Fe ($\lambda5250$) line [2.87]. It can be seen that the top of the bar lies about a factor of five below the estimates derived from metric burst and Faraday rotation observations. In fact, much of the range of field strengths is below the curve $\beta = 1$, indicating that these fields are too weak to fulfill the condition of magnetic energy density dominance. The same discrepancy was found in one specific case study [2.186], which concluded that the coronal Faraday rotation measured during the first occultation of *Helios 1* [2.244] was a factor of five larger than that expected from models of coronal electron density and PF calculations of coronal magnetic fields. If the PF calculations are in error, there are two possible remedies available:

(a) more photospheric magnetic flux must be channeled into open field lines, and/or

(b) the photospheric fields appropriate for the lower boundary must be stronger than indicated by the Zeeman measurements.

One of the tasks remaining for the coming years is to resolve this discrepancy between the computed and observed magnetic field strengths in the corona.

2.5 Coronal Velocities and Turbulence

It is appropriate that the velocity and turbulence of the coronal plasma be considered together because the solar wind acceleration and heating are merely two different manifestations of energy deposition from a common energy source [2.258, 259]. The exact nature and dissipation mechanisms of this energy source are still topics of unresolved debate, but they are undoubtedly associated with the manifold of MHD waves that are generated in the sun's convection zone [2.91, 98]. It is estimated that only about 10% of this wave energy is eventually converted to outward-directed convection and conduction. Most of the energy dissipated in the corona, in fact, is quickly conducted back down to lower heights due to the steep thermal gradient in the transition region [2.180].

In principle, a radial profile of the solar wind velocity could be constructed only from measurements taken by a string of plasma detectors appropriately spaced along a solar radial – an idealized configuration that probably will not be realized for many a year. The best examples of statistical studies of the bulk velocity of the solar wind in the ecliptic plane were performed with data from the two *Helios* spacecraft (see Chap. 3 [2.208]). It was found that the solar wind's acceleration to a terminal velocity was already more or less complete at the solar distance 0.3 AU, particularly for the high-speed streams. This has made it apparent that the velocity profile out to this inner boundary of *in situ* observations is required in order to distinguish between the various solar wind acceleration mechanisms [2.122].

A particularly elusive quest has been waged over the past few years to pinpoint that particular characteristic of coronal holes responsible for driving the solar wind to the high velocities observed in the *in situ* region of the inner heliosphere. It has been known for some time that such high speeds cannot be attained if the only available force arises from the coronal thermal pressure gradient [2.121, 122]. One way to achieve the high solar wind velocities would be simply to increase the nonradiative energy input (viz. MHD waves) into the coronal holes from the underlying source regions [2.63, 121]. If this were the case, however, it is difficult to explain why coronal holes are nearly indistinguishable from the neighboring "quiet" regions at chromospheric heights. Another deleterious side effect of increasing the wave input energy in model calculations is that the coronal temperature and density are increased, in stark contrast to the cooler, less dense plasma found in coronal holes [2.258, 259].

A much more promising hypothesis is that the MHD-wave energy flux is essentially isotropic over the sun's surface, but that waves are dissipated at different coronal heights [2.258, 259]. In particular, it is recognized that the highly divergent magnetic flux tube geometry of coronal holes leads naturally to a lower plasma density. If the characteristic damping length of the mechanical wave energy input is then proportional to the density, the dissipation of the MHD waves and their energy transfer to the kinetic component of the solar wind will be spread out to larger solar distances in coronal holes. Indeed, much of the dissipation here would be expected to occur beyond the sonic point. On the other hand, the same input flux into less divergent field regions with higher densities would dissipate primarily inside the sonic point. It is exactly this distribution of energy deposition which is capable of producing the high-velocity streams in coronal holes as well as the higher mass fluxes observed in the slow-speed solar wind [2.121, 208, 258, 259].

Remote sensing methods that measure coronal bulk velocity on the solar disk are possible only at transition layer heights. EUV observations at the solar limb do potentially have the advantage of measuring velocities at many coronal heights along a solar radial, thereby yielding an actual velocity profile rather than a statistical average, but have been limited in practice by instrument sensitivity. The solar wind velocity in the corona can also be obtained from radio occultation experiments. These observations, while yielding line-of-sight averages over an extended segment of the solar limb, have nevertheless provided a rough indication of the radial increase in the solar wind velocity from an almost stationary inner corona out to its asymptotic value in interplanetary space.

The turbulence of the solar corona is quantitatively represented by its spectrum of spatial (electron) density inhomogeneities. The form of this spectrum, in fact, was once the subject of some controversy [2.80, 93, 106]. Many radio scintillation data appeared to be consistent with a Gaussian distribution [2.183]. It has since been accepted that the radio scintillation data are best explained by a power-law spectrum of density fluctuations, whereby the power-law exponent is a function of solar distance [2.5, 39, 268]. The evolution of the spectrum from the coronal base out into interplanetary space is intimately connected with the dissipation of MHD waves and can be used to constrain solar wind models [2.187]. Since the dissipated energy is eventually converted into the kinetic component of the solar wind, the radial distance where maximum dissipation occurs is likely to be where the solar wind undergoes its greatest acceleration.

The story of coronal turbulence, however, is not finished with a description of the electron density fluctuations. The coronal plasma can be vigorously stirred around by the same Alfvénic fluctuations that are regularly observed and spectrally analyzed beyond 0.3 AU (see Chap. 4 [2.144]). A study of Faraday rotation and electron content fluctuations in the corona has provided evidence that the relative variance of the magnetic field $(\delta B/B)^2$ is considerably greater than the corresponding quantity for the electron density $(\delta N_e/N_e)^2$ [2.92].

Keeping within the previously mentioned scope of this chapter, this section surveys the measurements of the coronal plasma velocity and turbulence obtained from remote sensing (EUV and radio-sounding) techniques.

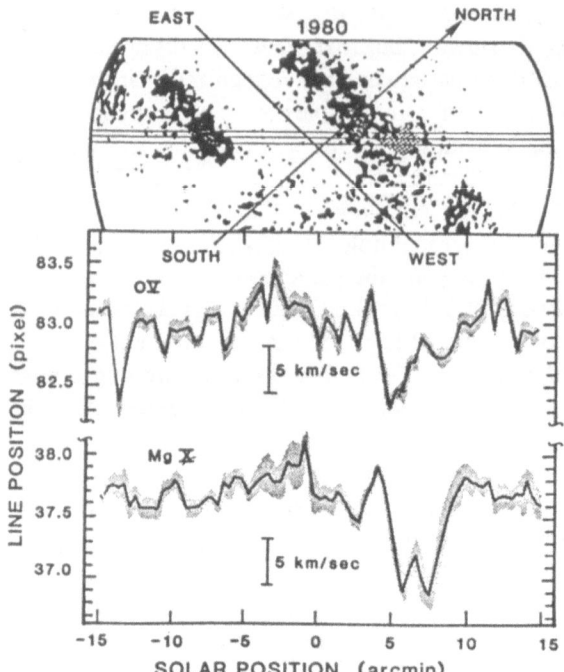

Fig. 2.15. Relative Doppler shifts of EUV emission lines across solar disk. The lines of O V (λ629) and Mg (λ625) are seen to be significantly blue-shifted at the position of a known coronal hole. These are interpreted as outward-flowing solar wind at two different transition region heights [2.190]

2.5.1 EUV Observations

Spectroscopic observations capable of determining plasma flow velocities in the acceleration region of the solar wind have been performed at EUV and XUV wavelengths during rocket flights over the past 15 years. Two of the more successful methods, to be described in the following, exploit the Doppler effect for these velocity measurements. The two techniques, however, are fundamentally different in their actual observables and regions of applicability.

The first method uses the relative Doppler shifts of coronal and transition region emission lines on the solar disk [2.190], an example of which is shown in Fig. 2.15. Spectra of two EUV emission lines, O V (λ629) and Mg X (λ625) were recorded at ca. 85 points spaced across the solar disk along a specified chord. The chord selected for this rocket flight (15 July 1980) is shown as three parallel lines in the upper panel of Fig. 2.15. The spatial resolution is given by the spacing of the outer two parallel lines. This scan line was chosen so that it passed through a compact low-latitude coronal hole (cross-hatched area), as determined from synoptic He I (λ10830) spectroheliograms. The coronal hole was nestled in among active regions bright in the K-line of Ca II (dark area on disk). The lower panel of Fig. 2.15 shows the spectral position of the mean emission line (in pixels; one pixel = 14 km/s) at each location across the solar disk. The shaded band indicates the statistical error (rms) from 9 scans along the chord.

The most striking feature of Fig. 2.15 is the pronounced relative blue shift of both emission lines associated with the coronal hole. The mean Doppler shifts of the two lines were 7 km/s for the O V and 12 km/s for the Mg X. The most straightforward interpretation of this blue shift is that the O V and Mg X ions throughout this region are flowing outward toward the observer, wafted by a solar wind that has already formed at transition region heights! The formation temperatures of these two ions are quite different. O V exists only in a layer of the transition region near $T = 2.5 \times 10^5$ K, while Mg X is formed higher in the corona at $T = 1.4 \times 10^6$ K. The larger blue shift for Mg X would indicate that the outflow has accelerated over the radial difference in heights between the main emission zones of these ions. Depending on the preferred model, accelerations of some 5–9 km/s^2 would be necessary to explain the observations of Fig. 2.15. This acceleration is comparable to that estimated for the solar wind in the outer corona. This blue shift effect was first observed in the coronal XUV line of Si XI ($\lambda 303$) [2.46], and has now been verified in five separate sounding rocket experiments with the O V and Mg X emission lines [2.191]. Large-scale outflow has also been observed in polar coronal holes [2.166]. The somewhat smaller blue shifts observed in the polar regions may be due to projection effects.

In contrast to these direct Doppler shift measurements of EUV emission lines on the solar disk, the second method for determining solar wind flow velocities infers the Doppler shift indirectly by its effect on the brightness of resonantly scattered emission above the solar limb. The ultraviolet lines of the E-corona, which can be excited either by collisions or by resonance scattering, are generally very faint and must be observed from coronagraphs above the earth's atmosphere. Resonance scattering becomes the more important excitation process for most coronal ions beyond about $2R_{\odot}$ [2.113], and is dominant at virtually all coronal heights for the Lyman-alpha line of neutral hydrogen [2.261]. H I is the only neutral species of importance, its abundance with respect to protons being about 10^{-7} at the temperatures and densities typical of the solar corona. Nevertheless, the high intensity of the chromospheric Lyman-alpha emission line gives rise to a strong resonantly scattered component in the overlying corona that can be measured with a UV coronagraph. Since the resonant absorption and re-emission can occur only if the wavelength of the incoming radiation is tuned to that of the applicable atomic transition, the brightness of a given resonance line will be dependent upon the relative velocity (Doppler shift) of the scattering gas parcel with respect to the sun. The solar EUV/XUV spectrum contains numerous peaks and valleys, so that even a small apparent Doppler shift of the incident solar irradiation could mean a large variation in the amount of flux at a particular resonance line. Scattering atoms with large radial velocities are Doppler shifted away from the center of the usually rather narrow solar lines so that the incident solar flux at the resonance wavelength is diminished and less resonant scattered emission is produced. This "Doppler dimming" effect becomes more severe for increasing radial outflow velocities v_0. To a first approximation, the decrease in resonantly scattered brightness is contained in a Doppler dimming factor $D(v_0)$ given by:

$$D(v_0) = \exp[-(v_0/v_{\mathrm{rms}})^2] , \qquad \text{where} \qquad (2.30)$$

$$v_{\text{rms}}^2 = v_{\text{e}}^2 + v_{\text{s}}^2 \,,$$

v_{e}^2 = velocity dispersion of solar emitting species,

v_{s}^2 = velocity dispersion of coronal scattering species.

The velocity dispersion of the coronal scattering species is usually higher than that of the solar emitting species so that one can often use the approximation $v_{\text{s}}^2 \gg v_{\text{e}}^2$. This is true not only because of the higher coronal temperature, but also because the coronal line can be subject to nonthermal broadening from waves and turbulence at spatial scales greater than the mean free path but less than the region being observed. For a coronal species of mass M and nonthermal velocity dispersion component v_{n}^2, the total coronal velocity dispersion can be written:

$$v_{\text{s}}^2 = \frac{2k_{\text{B}}T}{M} + v_{\text{n}}^2 \,. \tag{2.31}$$

The mass dependence in (2.31) means that each coronal species will have a different radial profile of Doppler dimming in (2.30), some examples of which are shown in Fig. 2.16 for an isothermal corona with $T = 1.5 \times 10^6$ K and no nonthermal broadening. The Lyman-alpha profile (curve 1) indicates that a significant decrease in resonant scattered intensity will occur at outflow velocities $v_0 > 100$ km/s. Doppler dimming of other coronal ions occurs at smaller velocities. The Si XII ($\lambda 499$) line (curve 3) differs somewhat from the others because the width of its solar line is comparable to that of the scattering ions ($v_{\text{e}}^2 \simeq v_{\text{s}}^2$). The dashed line (curve 6) results from a peculiarity in the circumstances associated with the O VI ($\lambda 1037$) resonance. For velocities $v_0 < 120$ km/s, both the O VI ($\lambda 1037.6$) and O VI ($\lambda 1031.8$) lines follow the same Doppler dimming profile (curve 4). At higher velocities, coronal O VI is red shifted into a neighboring chromospheric line of C II ($\lambda 1037.0$) that significantly enhances the intensity

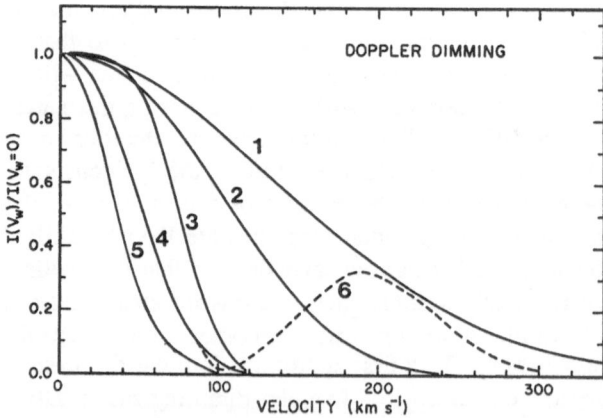

Fig. 2.16. Doppler dimming factor as a function of solar wind velocity for various ionic species. The individual curves correspond to (after [2.113]): (1) H I ($\lambda 1216$) Lyman-alpha; (2) He II ($\lambda 304$); (3) Si XII ($\lambda 499$); (4) O VI ($\lambda 1032$, $\lambda 1038$); (5) Fe XII ($\lambda 195$); (6) O VI ($\lambda 1038$); pumped by C II ($\lambda 1037$)

of the O VI (λ1037.6) resonance, but leaves the O VI (λ1031.8) line unaffected [2.113].

In principle, the calibration curves of Fig. 2.16 could be used to derive a high-resolution solar wind velocity profile $v_0(R)$ over the range from 30 to 300 km/s from measurements of $D(R)$ for a number of coronal emission lines. The Doppler dimming factor one actually derives from the remote sensing measurements, $\langle D(R) \rangle$, is a weighted mean over the line-of-sight through the optically thin corona. It is one of many factors affecting the brightness B_{ij} of a particular resonance line of the species j in charge state i as given by [2.262]

$$B_{ij} = \text{const.} A_j R_{ij} \langle D(R) \rangle I_t , \tag{2.32}$$

where

A_j = abundance fraction of atomic species j with respect to hydrogen,

R_{ij} = fraction of charge state i with respect to all charge states of atomic species j, and

I_t = total electron content.

One may determine $\langle D(R) \rangle$ from (2.32) by simultaneously measuring the intensity of coronal features that are not dependent on the radial outflow velocity. One example of this would be the brightness of the white-light K-corona B_K, which is only dependent on the total electron content I_t (see also Sect. 2.2.2). Each of the solar resonance lines has a small K-corona component that is typically much weaker than the resonantly scattered brightness. In the case of the Lyman-alpha line with no Doppler dimming, the narrow E-corona line ($\Delta\lambda \simeq 1$ Å) is about a thousand times brighter than the broad electron scattered (K-corona) component ($\Delta\lambda \simeq 50$ Å). Nevertheless, measurements of the Lyman-alpha K-corona may be feasible [2.261] and could thus be used instead of the optical K-corona intensities to obtain a single-instrument estimate of the Doppler dimming factor, i.e. independent of the absolute calibration. Another possibility would be to measure the intensity ratio of a resonance line pair such as the O VI (λ1031) and (λ1037), which is not only sensitive to outflow velocity but is also independent of possible height variations in A_j and R_{ij} [2.113].

Actual brightness observations of the resonantly scattered coronal Lyman-alpha line are shown in Fig. 2.17. Measurements were taken along three radial scans during two separate rocket flights [2.260–262]. The intensity at all heights for the 1979 flight was found to be greater in a "quiet" region of the corona than in a polar coronal hole. The brightness in a different polar coronal hole observed in 1980 was found to lie between the two radial scans of 1979. A theoretical brightness curve (solid line) has been fitted to each radial scan, corresponding to a stationary solar atmosphere with no Doppler dimming ($v_0 = 0$). For the lower curve only, additional curves have been drawn that demonstrate the expected drop in brightness due to Doppler dimming as based on solar wind models. The

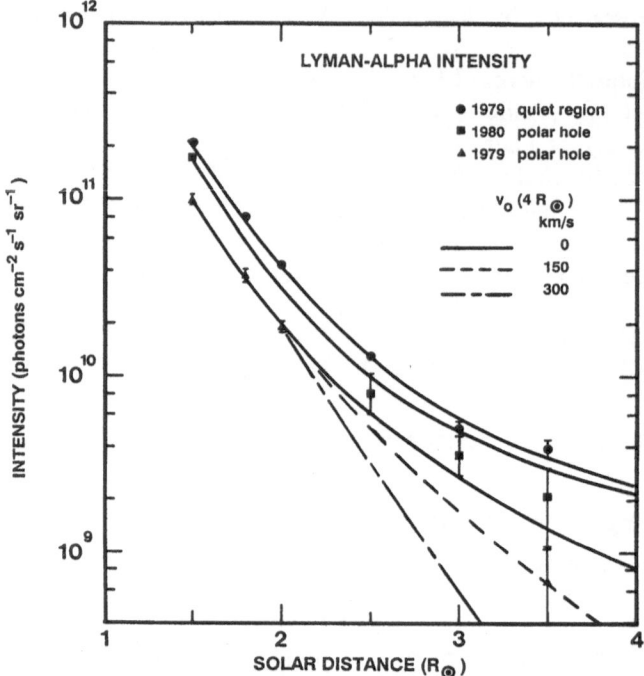

Fig. 2.17. Lyman-alpha intensities in the corona as a function of solar distance. Three radial scans are presented for a quiet region in 1979 (*circles*), a polar hole in 1980 (*squares*), and a polar hole in 1979 (*triangles*). The theoretical effects of Doppler dimming for three solar wind models, parameterized by their velocities at $4R_\odot$, are indicated for this last data scan. The measurements are consistent with a solar wind velocity $v_0 < 100\,\text{km/s}$ at all heights (after [2.260–262])

two additional cases plotted are for solar wind velocities of 150 and 300 km/s, respectively, at the solar distance of $4R_\odot$. Although there is a tendency for the measured points to lie below the $v_0 = 0$ curves, the simplest interpretation that can be deduced from these Lyman-alpha observations is that the solar wind was most likely still subsonic, with speeds of less than 100 km/s, over the radial range of the observations [2.260].

An important assumption inherent to these observations is that the velocity of the neutral hydrogen producing the Lyman-alpha resonantly scattered emission is equal to that of the solar wind. This will be true if the lifetime of the H I atoms τ_H, which is dictated primarily by the mean time between charge exchange interactions τ_{ex}, is small compared with the solar wind expansion time τ_{sw} defined by

$$\tau_{sw} = \frac{N_e}{v_0 dN_e/dr} . \tag{2.33}$$

The value of τ_{sw} is roughly constant over the coronal heights from 2 to $10R_\odot$ for most solar wind models, but is considerably shorter in low-density, high-velocity coronal holes than in "quiet" coronal regions. Since the charge exchange time τ_{ex} is inversely proportional to the density, it increases with height and will eventually exceed τ_{sw}, thereby decoupling the outflow motions of the H I and proton (solar wind) populations. The height at which this decoupling occurs has been estimated at ca. $8R_\odot$ in the "quiet" corona, but at around $3R_\odot$ in coronal holes [2.113, 261]. Since Doppler dimming for H I is most sensitive at relatively high velocities ($v_0 = 100$–$300\,$km/s), the Lyman-alpha line may not be the optimum diagnostic for many outflow geometries. Emission lines of the highly ionized species in Fig. 2.16, which are sensitive to smaller outflow velocities, would be more appropriate in these cases.

2.5.2 Radio-Sounding Observations

Observations of interplanetary scintillations (IPS) have been used effectively to determine the latitudinal variation of the solar wind velocity at various phases of the solar cycle [2.42, 218]. Higher speeds ($v_0 > 600\,$km/s) are found to be associated with polar coronal holes at and around solar minimum [2.185]. The minimum speed regions are generally distributed along a belt that closely follows the warped magnetic neutral sheet from potential field calculations [2.114]. The most commonly employed IPS velocity measurement technique consists of recording the fluctuating intensity of a compact radio source at two or more ground stations. The lag time τ derived from a cross correlation of these data is then used to deduce the "pattern velocity" of refractive index fluctuations across the known distance between the interplanetary (coronal) ray paths. This pattern velocity v_p is usually taken to be equal to the bulk velocity v_0 of the propagation medium, i.e. the electron density inhomogeneities responsible for radio scattering are assumed to be entrained in the solar wind (the "frozen-in" hypothesis). At least 3 stations are required to obtain a 2D representation of the velocity, i.e. its projection onto the plane of the sky. This velocity is usually directed radially outward from the sun to within a few degrees [2.7]. IPS-determined velocities were found to compare quite favorably with *in situ* measurements from interplanetary spacecraft [2.38, 41], even reproducing many of the observed daily fluctuations in solar wind velocity. This rigorous validation of the IPS data instilled confidence that they would provide a good estimate of the solar wind velocity at distances inside 0.3 AU.

Conventional IPS observations, which can be conducted on a synoptic basis, employ nonsteerable antenna arrays operating at rather low radio frequencies, e.g. 70–80 MHz, and are typically only feasible at large solar elongation angles. Radio-sounding measurements in the acceleration regime of the solar wind must be made with large, fully articulating parabolic dishes at higher radio frequencies (small beamwidths), because otherwise the sun's own radio noise overwhelms the occulted source signals. Unfortunately, the flux density of the sun increases and that of most natural radio sources decreases upon moving to higher radio frequencies. Many other radio sources are excluded from the near-sun region because they are located too far from the ecliptic plane. The additional requirements on the instrumentation and the decreasing number of suitable radio sources have

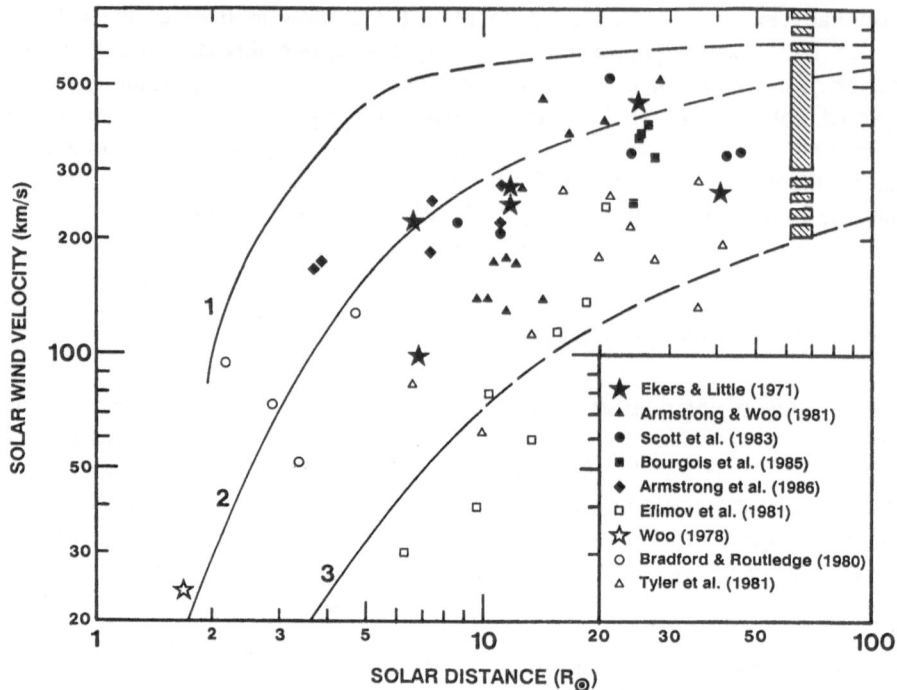

Fig. 2.18. Solar wind velocities in the solar corona: an overview of remote sensing measurements. Refer to text for explanation of various techniques. The range of actually measured values on *Helios*, 75% of which lie between 300 and 600 km/s, are indicated by the bars at $62 R_\odot$. Theoretical coronal velocity profiles with their probable extension to the *in situ* region of the inner heliosphere are included for comparison: (1) polar coronal hole from 2 to $5 R_\odot$ [2.159]; (2) solar wind with MHD-wave pressure [2.63]; (3) solar wind without MHD-wave pressure [2.63]

tended to restrict the experimental opportunities for IPS velocity measurements near the sun.

An overview diagram of velocity measurements inside $60 R_\odot$, representing four different radio-sounding techniques, is shown in Fig. 2.18. These four data types, discussed in more detail below, are:

1) Multistation intensity correlation measurements shown by the solid symbols as follows: stars [2.61], triangles [2.8], circles [2.210], squares [2.25], and diamonds [2.7]

2) Phase-correlation measurements of a spacecraft signal across two widely spaced ground stations shown by the open squares [2.60]

3) Combination of spectral and angular broadening of spacecraft signals shown by the open star at $1.7 R_\odot$ [2.266], and the open circles [2.26]

4) Determinations of the break frequency (proportional to the pattern velocity) caused by the "Fresnel filter" in the power spectrum of intensity fluctuations shown by the open triangles [2.240].

Three solid curves with dashed extensions have been drawn in Fig. 2.18. Curve 1 is a solar wind velocity profile in a rapidly diverging polar coronal hole of cross-sectional area $A(r)$, obtained from *Skylab* white-light observations [2.159]. The characteristically strong acceleration of this velocity profile, which is valid from 2 to $5R_\odot$, is derived by invoking the conservation of mass flux, $N_e(r)v_o(r)A(r) =$ const. The radial profile above $5R_\odot$ was taken from a solar wind model for a high-speed stream [2.122]. Curves 2 and 3 are models for the solar wind bulk velocity with and without additional momentum input from MHD waves, respectively [2.63]. The vertical bars at the inner boundary of the *in situ* measurements of *Helios* (62–$70R_\odot$) indicate the range of observed solar wind velocities, ca. 75% of which fall within the solid bar from 300–600 km/s (see Chap. 3 [2.208]). The velocity measurements of Fig. 2.18 all fall distinctly short of the values expected in a large coronal hole (curve 1). This may be because many of the points in Fig. 2.18 were taken very close to the ecliptic plane, and are thus associated with regions very close to the belt of slow solar wind [2.114]. Exceptions to this rule are the solid squares and diamonds and the open circles, for which the occulted sources passed over the north pole. It is satisfying that the vast majority of the velocities lie between the model curves that represent solar winds with and without a boost from MHD-wave pressure. It is not expected that an "average" velocity profile could be extracted from Fig. 2.18, because the measurements were taken at many different heliolatitudes under widely varying coronal conditions. Indeed, since considerable variations in acceleration and asymptotic velocity are expected theoretically (coronal holes, streamers, etc.), it is more appropriate to analyze separately each of the uniquely determined velocity profiles given in the original references.

Multipath radio scintillation intensity correlations (data type 1 above) are considered to be the most reliable radio-sounding velocity estimators [2.6–8]. Based on this confidence, these measurements are all plotted with "solid" symbols in Fig. 2.18. A few caveats should be heeded, however, concerning the association of the measured pattern velocity to the solar wind velocity. One obvious over-simplification is to assign the measured velocity to that solar distance R defined by the proximate point along the signal ray path from the radio source to earth. Since the scintillation strength is directly proportional to the mean electron density fluctuation (a rapidly decreasing function of solar distance r), the proximate point does generally provide the greatest contribution, but other coronal regions along the ray path at larger solar distances may often be quite important.

In the case of a purely radial outflow of a perfectly spherical distribution of coronal inhomogeneities at constant speed, the line-of-sight effect results in an underestimate of the true velocity at the proximate point by 15–20% [2.6, 106]. The error results because the correlation time lag for regions behind and in front of the solar limb is affected only by the component of the medium's radial velocity projected onto the plane of the sky.

An error working in the other direction would be an acceleration of the solar wind. The total projection effect would be nullified, for example, if the radial bulk velocity increases linearly with r:

$$v(r) = v(R)\frac{r}{R} \, , \tag{2.34}$$

where R is the solar proximate distance of the radio ray path. The velocity profile (2.34) may indeed be a fair approximation in the acceleration region of the solar wind [2.146].

The assumption of spherical symmetry of the mean density fluctuations is most certainly invalid in view of the considerable longitudinal structuring of the solar wind (see Chap. 3 [2.208]). If the scintillations are strongest in the regions of greatest electron density, where the radial flow is known to be generally slow, then the radio sounding of many solar wind streams along the solar limb would result in an underestimate of the velocity at the proximate distance [2.6].

The final, and perhaps most important consideration, is that the pattern velocity can be affected by a rearrangement of the scattering inhomogeneities arising from waves [2.61]. In this case, the effective pattern velocity v_p is a vectorial sum of the plasma bulk velocity v_o, and the group velocity of the waves v_g, which can be different for the various inhomogeneity scale sizes. In order to account for a superposed wave spectrum, one must allow for an entire distribution of pattern velocities $P_v(v_p)$. If the pattern is then considered to be "frozen" for each wave packet, the space–time correlation function can be written as [2.8]

$$C(\varrho, \tau) = \int C(\varrho - v_p\tau, 0)P_v(v_p)dv_p \, , \tag{2.35}$$

where $\varrho = r_1 - r_2$ is the separation vector between the observation positions, and $\tau = t_1 - t_2$ is the time lag between the two time series. The autocorrelation results when $\rho = 0$. The distribution $P_v(v_p)$, which has been taken variously as a Gaussian [2.61] or a "box function" [2.8], is characterized by a "random velocity" Δv that describes the spread in pattern velocities from the various scale sizes contributing to the scattering. This random velocity is of only minor importance in the super-Alfvénic solar wind, but can reach values $\Delta v \simeq v_o$ at solar distances $R < 30R_\odot$ [2.8, 209, 210]. The pattern velocity systematically overestimates the solar wind velocity in this case, because the cross correlation function (2.35) will be distorted to form a peak at a smaller value of τ [2.40, 130].

Although most of the previous velocity determinations have been made using intensity correlations, it is possible to use any available signal parameter such as phase, frequency, Faraday rotation, etc., as long as the occulted radio source emission is coherent. Spacecraft signals fulfill this prerequisite, but most natural radio sources are not compact enough to maintain the necessary coherency. Intensity correlations are best obtained from ground stations spaced at distances comparable to the size of the first Fresnel zone of the diffraction pattern d_F given by:

$$d_F = 773\sqrt{\frac{\lambda}{L}} \quad [\text{km}] \, , \tag{2.36}$$

where $\lambda = c/f =$ is the radio wavelength in meters, and

$$L = \frac{\Delta}{\Delta - 1} \, , \tag{2.37}$$

with Δ = distance from earth to radio source (AU). Intensity scintillations become uncorrelated at stations separated by distances ϱ greater than d_F. Scintillations of phase or frequency, on the other hand, remain correlated at all distances up to the largest scale in the propagation medium [2.265].

Only one attempt to derive coronal velocities from phase correlations (data type 2 in Fig. 2.18: open squares) has been reported in the literature [2.60]. The lag times of the correlation maxima are of the order of 10–100 s for the widely spaced ground stations used in the study, much longer than the typical lags (0.1–1.0 s) obtained from correlated intensity scintillations [2.8]. The velocities derived from the phase correlations thus correspond to much larger coronal features with scale sizes of the order 10^3–10^4 km. These measurements were taken with only 2 ground stations under the assumption of radially outward propagation. The correlations were found to be quite weak at times, perhaps indicating the absence of large-scale inhomogeneities. The velocities deduced from phase correlations are among the slowest in Fig. 2.18. It is unclear whether this can be attributed to a slower bulk velocity for the larger scale sizes [2.60], or if the correlations are biased by the random velocity or other effects described earlier.

The above correlation techniques require two or more separate radio links through the medium under investigation. It is also conceivable that multilink correlations could be achieved using only one ground station (e.g. dual-frequency, dual-spacecraft, up- and downlink [2.15, 56]). In the case of *Helios*, the uplink and downlink were typically separated by a distance $\simeq 10^4$ km. Plasma inhomogeneities, whether they are propagating outward through the corona on their own or are merely entrained in a solar wind moving at a comparable speed, therefore intercept the two links at different times. The time lag is derived from an autocorrelation analysis of the time-delay data. Examples of this technique using *Helios 2* integrated Doppler measurements have shown sharp peaks in the autocorrelation function that imply pattern speeds quite comparable to the assumed solar wind speeds, occasionally even exceeding 10^3 km/s [2.56].

The data type 3 in Fig. 2.18 (open star and open circles) were obtained at a single ground station from simultaneous measurements of spectral and angular broadening of spacecraft signals [2.26, 266]. Both the signal linewidth Δf and the apparent angular diameter $\Delta\theta$ exhibit the same dependence on the mean electron density fluctuation at the solar proximate point $\sigma_N(R)$. The bandwidth Δf, however, depends additionally on the drift velocity v_d of the medium with respect to the radio ray path. In the case when the radio source is close to the ecliptic, this velocity is given by the remarkably simple formula (adapted from [2.266])

$$v_d = v_o \pm v_{rp} = 6.3 \frac{\Delta f}{\Delta\theta} \quad [\text{km/s}] \, , \tag{2.38}$$

where L is given in (2.37), Δf is in Hz and $\Delta\theta$ is in arcmin ($\Delta\theta$ = major axis, assumed to be aligned along solar radial). It should be noted that the velocity v_d may also contain a contribution v_{rp} from the motion of the ray path with respect to the sun which is generally of the order of a few km/s. This velocity will add to the solar wind bulk velocity v_o during the ingress phase of a solar occultation, at which time the plus sign in (2.38) is appropriate. The minus sign holds during occultation egress. Close to the sun, where v_o and v_{rp} can be of comparable magnitude, v_o can be derived from the difference in the linewidth measurements during egress and ingress provided the mean electron density fluctuation $\sigma_N(R)$ is only a function of solar distance (axial symmetry). Rough radial profiles of solar wind velocity have been obtained in this way from data taken during the 1976 solar occultation of *Venera 10* [2.115, 273].

The remaining measurements of v_o in Fig. 2.18 (data type 4: open triangles) were obtained from single station intensity power spectra [2.240]. The power spectrum of intensity scintillations $W_I(\nu)$ can be shown to be flat at frequencies below a break frequency ν_b given by

$$\nu_b = \frac{v_d}{d_F} \tag{2.39}$$

where d_F is the Fresnel zone size defined by (2.36). At higher frequencies the power spectrum falls off as a power law. Since d_F depends only on the geometry, the break frequency in $W_I(\nu)$ is a direct measure of the plasma drift velocity v_d. All measurements of this type in Fig. 2.18 were taken from spectra with distinct break frequencies [2.240]. The situation is different for the phase spectra $W_S(\nu)$, which are sensitive to coronal irregularities of all scale sizes and do not display the characteristic knee at ν_b. Model spectra demonstrating the effects of varying v_o ($v_o \gg v_{rp}$) on W_I and W_S are displayed in Fig. 2.19 [2.146]. The break frequency in $W_I(\nu)$ is seen to move linearly to higher values upon increasing v_o. The phase spectra retain a single power-law decrease with ν over the entire range of interest, the overall power increasing linearly with v_o.

Under conditions of weak, single scattering (random phase changes less than one radian), the standard IPS scattering theory (e.g. Jokipii [2.106], and references therein) yields the following expressions for the scintillation intensity and phase power spectra W_I and W_S (generally following the notation of Woo [2.264] and Martin [2.146], again assuming $v_o \gg v_{rp}$):

$$W_I(\nu, v_p) = 8\pi\lambda^2 r_e^2 \int \sin^2[q^2\lambda/4\pi L]\Phi_N(q)\, dq , \tag{2.40}$$

$$W_S(\nu, v_p) = 2\pi\lambda^2 r_e^2 \int \cos^2[q^2\lambda/4\pi L]\Phi_N(q)\, dq \tag{2.41}$$

where L is again defined in (2.37) and r_e is the classical electron radius. The spatial spectrum of electron density fluctuations as a function of wavenumber q and distance from the sun r is usually approximated by a power law of the form [2.5]

$$\Phi_N(q, r) = C^2(r)\frac{\exp(-q^2/q_i^2)}{[q_0^2 + q^2]^{p/2}} , \tag{2.42}$$

where the wavenumber q is associated with a "scale size" of the density fluctuation L_q by the relation

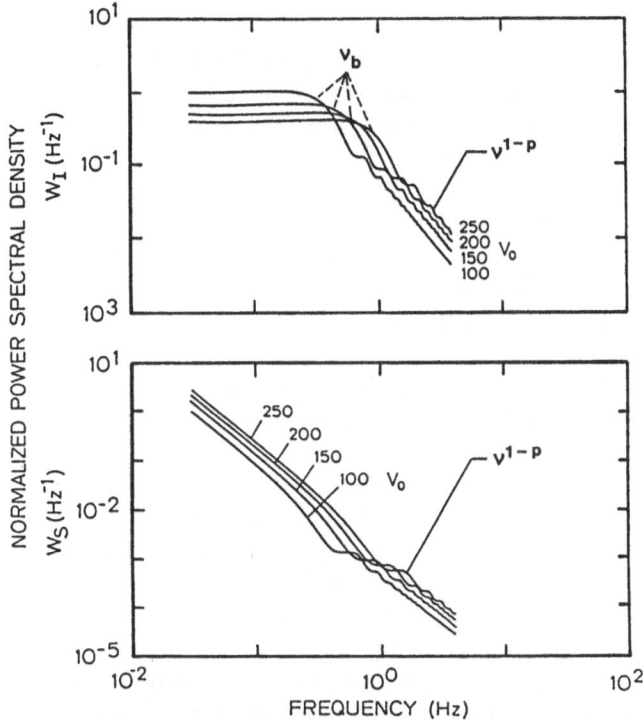

Fig. 2.19. Theoretical scintillation spectra for various values of the solar wind velocity v_o. The spectra are "red", with greater power at lower frequencies (greater scale sizes). The power-law falloff goes as ν^{1-p}, where p is the spectral index of the coronal electron density fluctuation spectrum. The intensity spectrum (*upper panel*) features a "knee" at a frequency ν_b that is proportional to the solar wind velocity. The power in the phase spectrum (*lower panel*) continues to rise toward lower frequencies (after [2.146])

$$L_q = \frac{2\pi}{q} . \qquad (2.43)$$

The spectrum of spatial density fluctuations (2.42) features a high wavenumber cutoff at an "inner scale" $L_i = 2\pi/q_i$ and a low wavenumber turnover or "outer scale" $L_o = 2\pi/q_o$, which is conceivably as large as, but surely not too much larger than, a solar radius. For a density fluctuation spectrum of the form (2.42) with a power-law falloff (spectral index) denoted by p, it can be shown that the spectral index of the intensity or phase spectra (2.40) and (2.41) is simply given by $1 - p$. The level of the turbulence at a given value of q is controlled to a large extent by the factor $C^2(r) = \sigma_N(r)h(p, q_o)$, where the function h depends only on p and q_o, and $\sigma_N(r)$ is the rms electron density fluctuation.

If the scattering occurs in the "thin screen" at the solar limb plane containing the solar proximate point along the ray path, one can simplify the spectrum $\Phi_N(q, r)$ by letting $r = R$ and using the approximation

$$q^2 = q_x^2 + (q_y/Q)^2 ; \quad q_x = 2\pi\nu/v_p ; \quad q_z = 0 ; \qquad (2.44)$$

65

where the density fluctuations move radially outward in the x direction at the pattern velocity v_p. The anisotropy of the medium, quantified by the axial ratio Q in (2.44), describes the shape of the inhomogeneities in the scattering medium. The usual approach is to consider the diffraction pattern to be elliptical in shape over the entire range of scale sizes. An isotropic medium ($Q = 1$) would presumably consist of absolutely random orientations and shapes. Larger axial ratios correspond to inhomogeneities elongated in the radial direction, i.e. along the direction of the coronal magnetic field. Previously thought to be rather weak [2.210], the anisotropy now appears to be quite strong near the sun, reaching values as high as $Q = 4$–8 for $R < 4R_\odot$ [2.7]. As Q increases, the intensity scintillation spectrum tends to flatten, thereby rounding the characteristic knee in Fig. 2.19. Although the break frequency ν_b remains unchanged by a variation in Q, the velocity estimate from (2.39) becomes increasingly more difficult. The same knee-rounding tendency is produced by the presence of an appreciable random velocity Δv. In this case, the break frequency also shifts to higher values [2.146], thus causing the same overestimate of v_0 that results from the cross correlation analysis (2.35). Since both the anisotropy and the random velocity increase near the sun, the use of intensity spectra for determining the solar wind bulk velocity v_0 becomes considerably less reliable at small solar offsets.

If an ensemble of velocities are present in the coronal plasma, then (2.40) and (2.41) are modified by summing the contributions over the velocity distribution $P_v(v_p)$ in the same manner as done for the correlation function (2.35), i.e.

$$W(\nu) = \int W(\nu, v_p)P_v(v_p)\, dv_p . \qquad (2.45)$$

The spectra W_I (2.40) and W_S (2.41), aside from the factor 4 in magnitude, differ only by the factors $\sin^2$ and $\cos^2$ in the integrand, which function as filters that block out certain ranges of the spatial density spectrum $\Phi_N(q, R)$. The intensity scintillations are therefore not affected by inhomogeneities of scale size $L_q > d_F$, the Fresnel zone size defined in (2.36). The phase scintillations, on the other hand, are modulated by a low-pass filter and are thus most sensitive to the largest coronal scale sizes [2.146].

In addition to the bulk velocity of the propagation medium, the scintillation spectra are useful for deriving information about the magnitude and radial dependence of the parameters describing the spatial spectrum of electron density fluctuations (2.42). For example, the spectral index p, as demonstrated by Armand et al. [2.5], is a crucial parameter for defining the total power contained in the coronal turbulence. Woo and Armstrong [2.268] concluded that phase scintillation spectra could be modeled quite adequately with a single power exponent p over the frequency range from 10^{-3} to $10\,\mathrm{Hz}$ (corresponding to spatial scales of 10–$10^5\,\mathrm{km}$). The mean value of the spectral index was found to be $\langle p \rangle = 3.65$ for $R > 20R_\odot$, i.e. close to the Kolmogorov value for fully developed isotropic turbulence. At smaller solar distances (2–$7R_\odot$), the value fell to $\langle p \rangle = 3.07$, indicating the possible existence of a superimposed spectrum of smaller-scale turbulence such as might be expected from MHD waves [2.268].

This increase in p with solar distance would also imply that the dissipation of higher-frequency MHD waves occurs at lower coronal heights while waves of lower frequency propagate further out before providing their energy to the solar wind. This concept would be a natural extension of the trend observed *in situ* on *Helios* [2.47]. The spectral index for phase fluctuations at the lowest observable frequencies (10^{-4} to 10^{-2} Hz) was found to be approximately constant ($p = 3.6$) over the range of solar distances $R = 5$–$30R_\odot$ [2.13, 115]. The flattening of the spectrum close to the sun may well be restricted to the higher frequency range, corresponding to scale sizes $L_q < 1000$ km [2.39].

The inner scale L_i affects the spectral slope at the higher frequencies and cuts off the spectrum at frequencies $\nu > v_0/L_i$. This cutoff frequency is not always well defined in the intensity spectra, but there is mounting evidence that L_i increases more or less linearly with R [2.5, 39, 210]. The inner scale evidently represents the lower bound of scale sizes not subject to dissipation. It is interesting that the inner scale is of the same size and exhibits the same radial dependence as the proton Larmor radius, thus suggesting cyclotron damping as an effective dissipation mechanism [2.39].

Theoretical electron density fluctuation spectra at three different solar distances, reflecting the results of many radio propagation studies, are displayed in Fig. 2.20 [2.39]. The relative spectral density over a wide range of fluctuation scale sizes is presented in terms of wavenumber q (lower scale) and scale size L_q

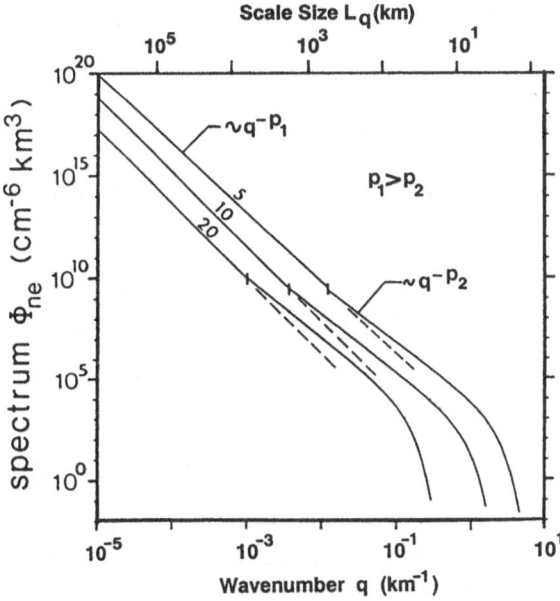

Fig. 2.20. Theoretical electron density spatial spectra over six decades of fluctuation scale sizes at solar distances of 5, 10, and $20R_\odot$. The spectra are characterized by: (a) steep increase toward the largest scale sizes with spectral index close to Kolmogorov value ($p_1 \simeq 11/3$); (b) flattening of spectra at intermediate coronal scales ($p_2 < p_1$); (c) cutoff at smaller scale sizes (inner scale). The inner scale increases approximately linearly with distance from the sun (after [2.39])

(upper scale) according to (2.43). At the lowest wavenumbers the spectra feature a spectral index close to the Kolmogorov value ($p_1 = 11/3$), but become considerably flatter (spectral index $p_2 < p_1$) at a scale size $L_q \simeq 1000\,\text{km}$, the actual break point depending on solar distance. This flattening, which produces an increase in the turbulence level by an order of magnitude, is made more evident in Fig. 2.20 by the dashed-line continuations of the spectra at their low-wavenumber falloff rate. At higher wavenumbers, the electron density fluctuation spectra are cut off at an inner scale L_i, which increases monotonically with distance from the sun.

Even without detailed analysis of the intensity and phase scintillation spectra (2.40) and (2.41), the radial variation of the turbulence level in the corona can be quantified by calculating the "scintillation index" m, defined by

$$m^2 = \frac{\langle \delta I^2 \rangle}{\langle I \rangle^2} \tag{2.46}$$

where $\langle I \rangle$ and $\langle \delta I^2 \rangle = \langle [I(t) - \langle I \rangle]^2 \rangle$ are the mean and variance of the intensity of the occulted radio source, respectively. A corresponding quantity for the mean frequency fluctuation of the signal is the "Doppler noise" σ_D defined by [2.266]

$$\sigma_D = \langle [f(t) - \langle f \rangle]^2 \rangle . \tag{2.47}$$

The quantity m (and analogously for σ_D) may be interpreted as a measure of intensity (phase) scintillation strength integrated over a specific range of fluctuation frequencies, i.e.

$$m^2 = \int_{\nu_0}^{\nu_N} W_I(\nu) \, d\nu \tag{2.48}$$

where the lower frequency $\nu_0 = 1/2T_0$ corresponds to the averaging time T_0 as indicated by the angular brackets in (2.46) and (2.47). The upper frequency $\nu_N = 1/2T_N$ is the Nyquist frequency defined by the sampling time T_N. Under the conditions of isotropic turbulence ($Q = 1$) and "frozen-in" convection ($\Delta v = 0$), it can be shown [2.5, 266] that these two measures of scintillation strength can be written as:

$$m = h_m(p)\lambda^{(2+p)/4}\sigma_N \sqrt{L_{\text{eff}}} \, L^{(2-p)/4} , \tag{2.49}$$

$$\sigma_D = h_\sigma(p)\lambda\sigma_N \sqrt{L_{\text{eff}}} \, v^{(p-2)/2} , \tag{2.50}$$

where h_m and h_σ are functions of p only, L is defined by (2.37), and L_{eff} is the effective thickness of the scattering layer, usually considered to be linearly proportional to the solar distance R. The expression (2.49) is valid only if the scintillation index m is less than unity (weak scintillations). In order to investigate the behavior of σ_N at smaller solar distances with intensity scintillations, one must observe at increasingly higher frequencies. It can be shown [2.5] that the weak scattering condition $m < 1$ is fulfilled when

$$R > R_{cr} \; ; \quad R_{cr} = 32\lambda^{0.6} \quad \text{[solar radii]}, \tag{2.51}$$

where the wavelength λ is given in meters.

According to (2.49), the scintillation index would increase as $\lambda^{1.4}$ if the spectral index were to maintain the Kolmogorov value $p = 11/3$. Using scintillation measurements at many different frequencies under the criterion (2.51), the actual behavior is known to be approximately linear $m \sim \lambda$ [2.5, 183]. This apparent discrepancy can be explained by allowing the spectral index p to be a function of solar distance R [2.5].

The radial variations of the scintillation index and Doppler noise measured on *Helios* at a single wavelength (S-band: $\lambda = 0.13\,\text{m}$) are shown in Fig. 2.21. At S-band, the criterion (2.51) for weak intensity scintillations is met (upper panel of Fig. 2.21) for solar distances $R > 10 R_{\odot}$. The radial decrease in m is described by a power law of index 1.55 [2.8]. A similar radial dependence with $m \sim R^{-1.6}$ was found to represent adequately scintillation data using spacecraft [2.5] and water maser sources [2.129]. This implies that the mean electron density fluctuation in (2.49) must decrease with distance as $\sigma_N(R) \sim R^{-2.1}$, as one would expect if it were simply proportional to the electron density N_e. A systematic enhancement in $m(R)$ has been detected in the radial range from 10–25 $R_{\odot}$ and attributed to

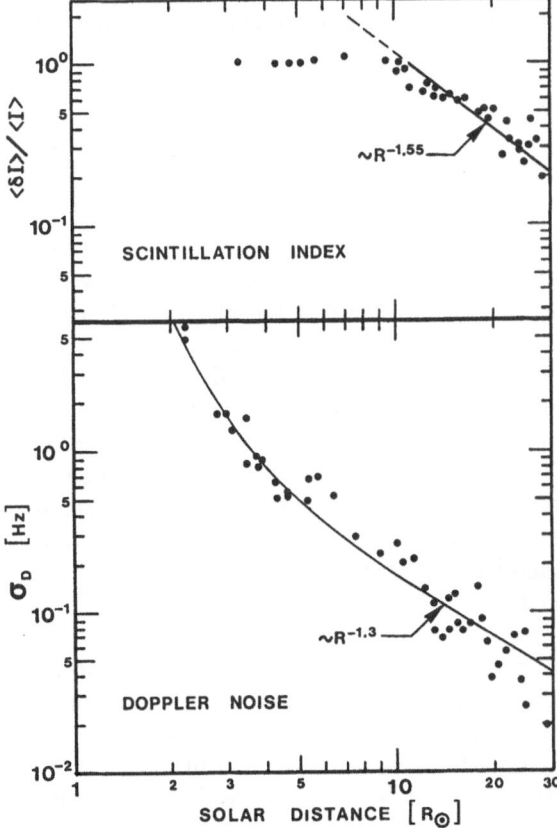

Fig. 2.21. Scintillation strength of *Helios* S-band (2.3 GHz) signal as a function of solar distance. The intensity scintillation index m (*upper panel*) saturates at solar distances inside of R_{cr} and falls off as a power law $\sim R^{-1.55}$ at greater distances [2.8]. The Doppler noise (*lower panel*) exhibits a slightly different power law falloff at large R, but it continues its increase at an even steeper rate toward smaller solar distances (adapted from [2.12, 266])

an excess of MHD-wave dissipation that also provides the primary acceleration of the solar wind [2.5, 129]. There is evidence that the radial location and width of this "zone of enhanced turbulence" is variable, being thinner and closer to the sun for solar wind high-speed streams [2.129]. The same increases above the expected $R^{-1.6}$ radial falloff have been observed in the apparent broadening of quasars at metric wavelengths [2.131]. The radial range of this anomalously high level of scintillation has been associated with the transonic region of the solar wind, over which both subsonic and supersonic flow regimes are traversed by the radio ray path [2.128].

The Doppler noise (2.50) in the lower panel of Fig. 2.21 is not subject to a restriction of the form (2.51). These data [2.12, 266] are adequately fit by a power law with index 1.3 for distances $R > 10 R_\odot$. Referring to (2.50) and assuming that $\sigma_N \sim R^{-2.1}$, this would imply that the bulk velocity has the radial dependence $v_o \sim R^{0.3}$ [2.266]. The Doppler noise exhibits a steeper increase for $R < 10 R_\odot$, primarily a result of more rapidly increasing rms electron density fluctuations, although the effect is somewhat mitigated by the simultaneously decreasing solar wind velocity. It should be cautioned that the effects of anisotropy and velocity fluctuations, both of which will add to the observed Doppler noise, also cannot be neglected in this radial range. The smooth curve is a two-term empirical fit to the Doppler noise data, based on the assumption that the radial variation of σ_D is approximately the same as that of the total electron content I_t [2.12].

2.6 Coronal Mass Ejections

The coronal mass ejection, certainly one of the most scrutinized solar phenomena since its somewhat unexpected revelation to the scientific community less than 20 years ago, exemplifies the archetype of coronal dynamics. As the name implies, the coronal mass ejection is an event where relatively dense, but discretely bounded coronal material (bright as observed in white light) is propelled irretrievably into interplanetary space. Since nonejective coronal motion such as streamer displacements can also be transient in nature, the designation "coronal transient" will be suppressed in favor of "coronal mass ejection", often abbreviated in the following with the acronym "CME". In addition to the visual feature observed in the outer corona, it has since been appreciated that many other "transient" phenomena near the solar surface (e.g. flares, eruptive prominences) and in interplanetary space (shocks, magnetic clouds) can all be associated with the CME event.

The fact that CMEs were only recently discovered is most likely due to the spatial and temporal scales of the phenomenon. After having now seen many examples, it is apparent that the size of a CME is so large (typically a few solar radii in extent), that it can usually only be recognized in its "full glory" in the region above $2 R_\odot$, i.e. the approximate outer limit of observation for ground-based coronagraphs. The best view of CMEs is thus obtained above the earth's atmosphere from an orbiting coronagraph. Those instuments used previously for studies of CMEs were already introduced in Sect. 2.2.1 (Table 2.1). The time scales applicable to the discernible evolution of a CME are also not favorable for a serendipitous sighting. CMEs are now known to occur at a rate of about one event per day, so there must have been at least occasional opportunities to catch one in progress during a solar eclipse. Their apparent outward motion over the few minutes of totality is simply insufficient, however, to have been registered by naked-eye or low-power telescopic observations. Now that we are aware of their existence, CMEs have been reported from eclipse observations of 1980 [2.193] and even 1860 [2.55].

It would be futile to attempt a complete overview of CMEs in this section. The interested reader is referred here to the many recent reviews of the subject [2.48, 83, 99, 107, 108, 136, 207, 245]. This section is thus restricted to a brief phenomenological description of CMEs, including their classifications, white-light morphology, three-dimensional structure, statistical distributions, associations with other solar events, acceleration mechanisms, and origins. A somewhat more detailed treatment is presented for investigations of mass ejections in the corona and interplanetary space using radio-sounding techniques.

2.6.1 CME Morphology, Phenomenology and Statistics

CMEs come in a variety of shapes and sizes and various *classifications* have been introduced to provide a morphological overview of the phenomenon. The *Skylab* coronal mass ejections, a total of 77 events, have been grouped into 7 classes according to their white-light appearance [2.160]. The most common CME variety of the *Skylab* era (20 examples) was the "loop". Couting also the "filled bottle" type (8 events), coronal mass ejections with rounded leading edges comprise about 40% of the total. This percentage agrees with that of the combined "curved front" and "loop" structural classes as selected from 998 Solwind CMEs observed in 1979–81 [2.96]. Another important group of CMEs are those with pointed leading edges such as the "ray" and "injected streamer" classes of *Skylab* and the "spikes" (single, double, multiple) of Solwind. They were the most common variety seen by Solwind (over 50%), and were generally less massive and slower than the CMEs with rounded leading edges. Their statistically different dynamical properties indicate that the "spikes" are truly a separate class and not merely looplike CMEs observed in an "edge-on" perspective [2.151]. An unusual class of CMEs was recognized by both surveys and referred to as a "streamer separation" (*Skylab*) or "streamer blowout" (Solwind). This type of CME surprisingly became the most common single class during the years of reduced solar activity 1984–85 [2.97].

One picture is certainly worth more than 10^3 words in describing the *white-light morphology* of a typical coronal mass ejection. In view of the wide variety of CMEs in the structural surveys, even one picture would be insufficient. Much

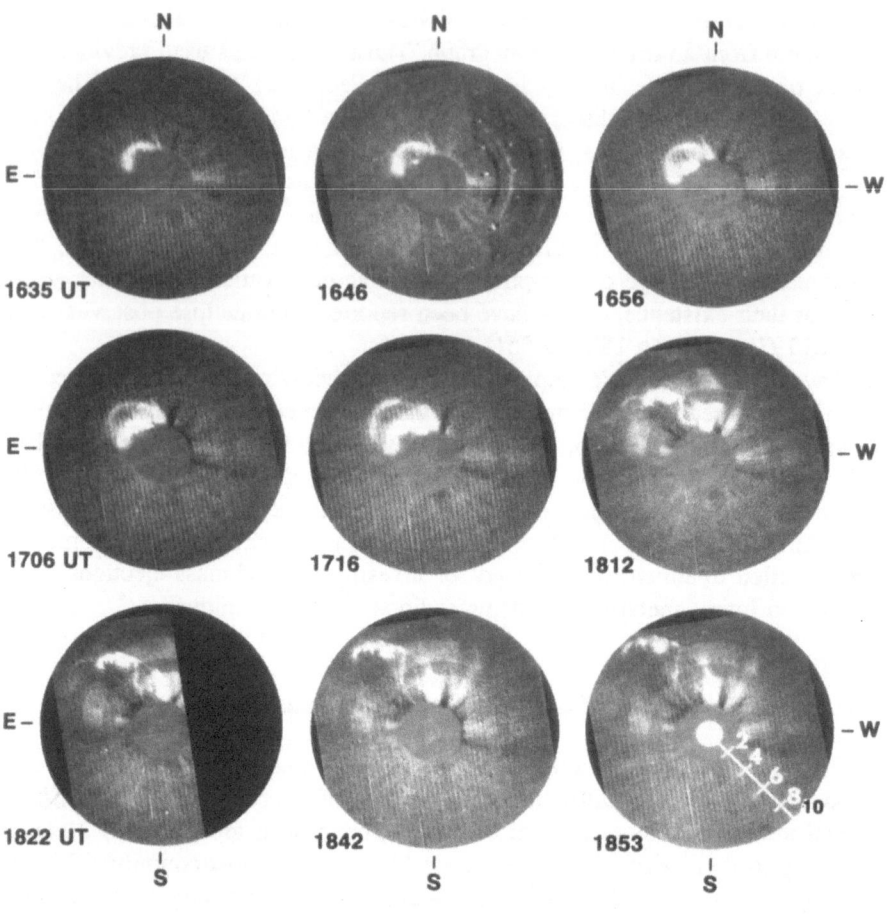

Coronal Mass Ejection   24 May 1979   SOLWIND (P78-1)

Fig. 2.22. The coronal mass ejection of 24 May 1979. This time sequence of images taken by the "Solwind" coronagraph on the *P78-1* satellite shows the outward progression of a particularly massive and energetic CME (projected velocity estimate of the curved white-light front: ca. 900 km/s at $10 R_\odot$). An eruptive prominence, bright in the H-alpha line of neutral hydrogen, follows the white-light front at a lower velocity. A distance scale and the size of the photosphere (centered white spot) are indicated in the last image at the lower right (from [2.215])

better would be a time-lapse video (unfortunately not yet available) to illustrate the dynamical interaction of the many classes of CME with the ambient corona. Lacking this technical breakthrough, however, it is possible to display a series of chronologically ordered coronal images that record the evolution of the CME event. Figure 2.22 shows such a time sequence of Solwind coronagraph images for the CME of 24 May 1979 [2.215]. The large mass and high speed, as well as the unusual appearance of prominence material at large solar distances, would not justify labeling this a "typical" CME, but it is an aesthetically pleasing example for illustrative purposes. It does contain the classical three-part structure generally attributed to CMEs: bright coronal loop, underlying dark prominence cavity, and bright prominence core [2.107, 108]. The scale for the images is given in the last image at the lower right. The centered white spot indicates the actual size of the photosphere, which is blocked out by an occulting disk of radius $2.6R_\odot$. All of these images are "difference images" produced by displaying only the difference in coronal brightness between the given time and an earlier "base" time.

The leading edge of the CME in Fig. 2.22 has the appearance of a curved loop. The brightness of the loop is diminished considerably as it traverses the annular polarizer ring (best seen at 1716 UT), indicating that the emission is highly polarized and must therefore be produced by Thomson scattering from electrons. This material is almost completely ionized and was originally located in the lower corona [2.84, 99]. Underneath the curved loop is a dark void, perhaps the single commonmost feature of all CMEs [2.99], which is most likely a region of relatively strong magnetic field. Below the dark cavity is the brightest white-light feature of the CME (just emerging from behind the occulting disk at 1646 UT). This is the remnant of an erupted prominence containing much cooler material with strong (unpolarized) emission in the H-alpha line of neutral hydrogen. The apparent radial velocity of the prominence projected against the plane of the sky increases up to 900 km/s by the time it leaves the field of view at $10R_\odot$ [2.104]. Even though the size of the prominence grows by an order of magnitude, its contorted shape is roughly preserved throughout the observation interval [2.151]. The prominence is presumably wrapped in its own strong magnetic field, which shields the neutral gas from the hot surrounding corona.

Coronagraph images provide a view from only one aspect and are thus quite limited in their ability to distinguish differences in location, extent, or motion along the line-of-sight. In particular, it has been difficult to determine the *three-dimensional structure* of a CME. Is it "loop" or "bubble"? Similar to the example of Fig. 2.22, the most impressive CMEs of the *Skylab* era exhibited a looplike appearance that implied a more or less planar structure, possibly bent into that shape by a magnetic flux tube. In recent years, however, the controversy loop vs. bubble has swung heavily in favor of the bubble [2.107, 108, 207, 245]. The most convincing arguments along this line are based on the following evidence:

(a) Polarization Studies. As shown in Sect. 2.2.2, Thomson scattered light from coronal electrons is emitted preferentially perpendicular to the incident photon moving out along a solar radial. Since the electric vector of the scattered photon is orthogonal to its propagation direction and also orthogonal to the propagation direction of the incident photon at the scattering site, the K-corona was shown to be rather highly polarized in a direction tangential to the solar limb. The degree of polarization attains a broad maximum in the plane of the sky and decreases symmetrically for solar longitudes either in front of or behind this plane. It is not possible to determine whether the scattering region is behind or in front of the solar limb from polarization measurements. Nevertheless, the K-coronal polarization distribution of a CME will be different for a two-dimensional loop than for a three-dimensional bubble. Using loop/bubble models of the electron

density distribution for the well-documented *Skylab* CME of 10 August 1973, Crifo et al. [2.44] concluded that the observed polarization was more consistent with a bubble.

(b) CMEs Propagating Toward Earth. The necessity of an occulting disk and the above-mentioned angular scattering dependence of the K-corona greatly limit the sensitivity of white-light coronagraphs in regions away from the solar limb plane. About 2% of the CMEs detected by Solwind, however, were unusual in that the brightness distribution was virtually uniform over all position angles around the solar disk. A detailed case study of one such "halo" event [2.95], interpreted the observations as a fairly massive, axially symmetric CME that propagated directly toward earth. The three-dimensional structure inferred for the CME contrasted with the essentially two-dimensional structure of an associated disappearing filament near the center of the solar disk. Part of the brightness excess from these halo events may not necessarily arise from material directed toward earth, but rather from a deflection toward the coronal limb, where Thomson scattering is more efficient [2.225].

(c) Stereoscopic Views from Helios. Transient white-light brightness enhancements followed occasionally out to 0.5 AU by the zodiacal light photometers on board the two *Helios* spacecraft were found in many cases to be the same CMEs observed in the corona by earth-orbiting coronagraphs [2.104, 105]. *Helios* was often positioned well away from the sun–earth line and thus provided a different viewing perspective from which one could better assess the three-dimensional structure of CMEs. Low-resolution maps of the CME brightness distribution from *Helios* clearly showed angular extents comparable to or only slightly less than those determined from earth, thus providing further support for the bubble model.

Important clues about the physical nature of coronal mass ejections can be provided by considering their *statistical distributions* with respect to such parameters as occurrence frequency, speed, heliolatitude/longitude, angular width, etc. Table 2.3, presenting data from the continuously operating Solwind coronagraph [2.96, 97], shows that the average properties of CMEs are dependent upon the level of solar activity. Solwind saw many more CMEs per day during the solar maximum years 1979–81 than in the years of lower activity 1984–85. Furthermore, a very low percentage of CMEs were observed at high heliolatitudes during these latter years. The speeds of CMEs at solar maximum averaged about 470 km/s over the Solwind field of view from 2.6 to $10R_\odot$, dropping to less than half this value in the low activity years. Since the angular widths (proportional to total mass) decreased by a factor of two over this same time span, the energies of typical CMEs at solar maximum were a good order of magnitude higher than those in 1984–85. The differences between low vs. high solar activity are reflected in the estimated CME mass flux, which approaches 10% of the total solar wind mass flux at solar maximum.

The *Skylab* observations had instilled the concept that CMEs occur more frequently over regions of higher sunspot activity [2.85]. A fourfold increase in the number of CMEs per day was predicted for the subsequent solar maximum, not inconsistent with the rates later inferred from the Solwind data (Table

Table 2.3. Average properties of CMEs from Solwind observations [2.96,97]. (a) No. of CMEs per day observed by continuous solar limb monitoring (100% duty cycle) over all position angles. (b) Radial velocity projected onto plane of the sky. (c) Values at the heliographic equator. Mean solar wind flux is assumed to vary with solar cycle [2.207].

Time Period		1979–1981 high solar activity	1984–1985 low solar activity
Sunspot number		150	25
No. events recorded		998	59
Occurrence frequency (CMEs per day)	(a)	1.8	0.3
Fraction of CMEs above 45 deg heliolatitude (%)		31	7
Speed (km/s)	(b)	472	208
Angular span (degrees)		45	24
Mass (10^{15} g)		4.1	2.1
Kinetic energy (10^{30} erg)		3.5	0.3
Equatorial CME mass flux (10^6 proton cm^{-2} s^{-1})		22	8.4
CME fraction of solar wind flux (%)	(c)	8	3

2.3). This trend could not be substantiated by the *SMM* C/P observations in 1980 [2.101], during which the incidence rate was found to be 0.9 CMEs per day, only 20% higher than during the *Skylab* epoch. Determination of CME occurrence frequency requires careful consideration of the duty-cycle, sensitivity threshold, and field of view of the coronagraph, as well as the image processing and scanning techniques. All of these factors were different for the *SMM* C/P and Solwind coronagraphs. A comparison of CME lists during the common observation times in 1980 [2.83] indicated that Solwind was seeing considerably more "minor" events than *SMM* C/P, possibly because of using the more sensitive difference images for CME detection [2.96], rather than the direct images [2.101]. In spite of the more frequent detection of CMEs at solar maximum, the Solwind data could not confirm the *Skylab* correlation [2.85] of CME occurrence frequency with the short-term mean sunspot number.

The agreement between the two coronagraphs was much better with respect to the observed latitudinal distribution. Whereas the CMEs of the *Skylab* epoch were never observed farther than 50° from the equator, the CMEs at solar maximum were distributed more uniformly in solar latitude [2.96, 101]. This finding is consistent with the notion that CMEs occur more often above active regions on the sun. On the other hand, the sunspot number is evidently not the best indicator of CME onset probability since these two parameters display significantly different latitudinal distributions [2.251]. In view of this, it would probably be correct to say that CMEs are not likely to emerge from large monopolar open-field-line regions (coronal holes). Even here, however, one should be aware of a "minority opinion" that suspects coronal holes to be among the most prevalent CME sources [2.81].

The projected speeds of the *Skylab* CMEs ranged from 100 to 1200 km/s and averaged 470 km/s [2.77]. The speeds usually remained constant in the field of view out to $6R_{\odot}$, but occasional examples of acceleration were also noted. The same basic distribution with a remarkably similar average speed was found for the Solwind observations [2.96] at solar maximum (Table 2.3). The distribution of CME speeds was found to be correlated with the associated surface activity (see below). The *Skylab* CMEs accompanied by solar flares were found to travel considerably faster than those without associated flares. The rather clear discrimination in the acceleration and average speed of the observed mass ejections suggested that fundamentally different physical processes may be acting to generate separate categories of CMEs [2.206]. Studies in the inner solar corona [2.141] substantiated this apparent division, pointing out the different behavior of CMEs associated with flares and CMEs associated with eruptive prominences. The flare events were observed to maintain high constant speed over the K-coronameter's field of view, presumably having been accelerated at lower coronal heights. CMEs associated with eruptive prominences, however, exhibited large accelerations of up to $50 \, \mathrm{m/s^2}$ in the region from 1.3 to $2.0R_{\odot}$. Acceleration is also implied in the speeds determined for 18 CMEs observed jointly by the *SMM* C/P and by Solwind. The Solwind speeds tended to be about twice those determined by the *SMM* C/P, most likely because Solwind observed the (postaccelerated) CMEs higher in the corona [2.83].

The most often named CME *associations with other solar events* are H-alpha solar flares (HF), eruptive prominences (EP), and X-ray bursts (XB). It is still not entirely clear, however, if the CMEs are the cause or the effect of these accompanying "surface" phenomena [2.107, 245]. This cause/effect ambiguity does not exist for the various interplanetary associations such as fast shocks [2.206, 212] and magnetic bubbles [2.32], for which a one-to-one causal relationship has been established [2.207].

Type II radio bursts, which are assumed to be generated at the local plasma frequency and its second harmonic by fast MHD shocks that start in the corona and propagate radially outward [2.50], have been interpreted both ways. The validity of their association is uncontested [2.109, 217] and they are certainly effects of CMEs when they are emitted at kilometer wavelengths in interplanetary space [2.36]. The situation is less obvious in the corona where examples of Type II metric bursts have been found to originate from heights well below the white-light front of a CME [2.69]. It should be cautioned, however, that such observations are subject to very difficult corrections for projection effects [2.227] as well as radio-path ducting and refraction in the corona. The considerably rarer moving radio bursts of Type IV have also been mentioned in association with CMEs. Only 5% of the CMEs have accompanying moving Type IV bursts [2.71], compared with the some 40% of CMEs having associated Type II emission [2.217]. The inverse correlations are stronger: 70% (40%) of Type II (moving Type IV) bursts were observed to have associated CMEs [2.71, 217]. A CME minimum velocity threshold of some 400–500 km/s has been cited as a possible necessary criterion for the association with Type II emission in the corona [2.77] or in interplanetary space [2.36]. This will depend critically on the radial profile of the local Alfvén velocity. Slower CMEs will reach the Alfvén speed too far from the sun so that their Type II emission is (a) at frequencies below the ionospheric cutoff, and (b) quite weak due to the small shock Mach number.

A search for surface events occurring near the projected onset times of the 77 definite *Skylab* CMEs [2.158] found that about twice as many were associated with eruptive prominences (EP) than with H-alpha flares (HF). A follow-up study using the 1980 *SMM* data base of recorded CMEs during solar maximum [2.251] confirmed the overwhelmingly close association (70–90%) between CMEs and EPs. Although about 40% of the CMEs were found to have HF associations, it was perhaps significant that no instance of a CME association with only an H-alpha flare could be found in either the *Skylab* or *SMM* data! Simultaneous X-ray bursts (XB) have also been observed in connection with CMEs. The XB association becomes increasingly stronger for X-ray events of longer duration [2.211, 251].

The existence of a class of "spontaneous" CMEs without any associations whatsoever was proposed by Wagner [2.245]. This possibility, which is still controversial [2.107], would have profound implications for models of the CME onset. Some CMEs may have no association because they were strongly accelerated before emerging above the coronagraph occulting disk [2.137]. Still others are surely connected with unobserved events on the solar "backside". Since about half the observed CME trajectories project back to points behind the solar limb, it is remarkable that 70–80% of all CMEs with calculable onset times were still found to have probable surface associations [2.251].

Although theorists have offered a number of attractive models, the *acceleration mechanisms and origins* of CMEs are still not fully understood [2.83, 99, 107, 108, 207]. There have been three distinct approaches toward a theoretical description of the initiation and motion of coronal mass ejections [2.48, 189, 245], roughly divided into (a) numerical analyses, (b) semiempirical force balance models, and (c) quasi-steady evolution into a nonequilibrium state.

The first line of attack on the problem has been a time-dependent numerical integration of the coupled MHD moment equations to simulate the coronal and interplanetary reaction to an energetic pressure pulse. The pressure increase could conceivably be of many varieties, a purely thermal pulse near the solar surface (flare), a sudden jump in the mass flux (eruptive prominence), or the rapid emergence of new magnetic flux [2.245]. In practice, most of the simulations have concentrated on the evolution in a solar meridional plane (i.e. 2D) and have been triggered with a thermal pulse in various ambient field line geometries

[2.49, 228, 272]. The resulting coronal disturbance, a fast shock followed by a dense kidney-shaped plasmoid, is a reasonable facsimile of a CME with credible values for its extent, density enhancement factor, and speed. Three shortcomings of the earlier models, which were not in good agreement with the white-light observations [2.222], were (a) not enough material at the flanks of the CME, (b) no density depletion underneath the CME, and (c) too much lateral motion late in the event. These discrepancies have been diminished in more recent simulations of CMEs initiated in existing streamer configurations [2.226].

A second group of theorists has tried to explain the physics of CMEs by examining their acceleration and deformation, i.e. their driving forces, during transit through the corona. A general consensus has now been reached that the forces are most likely magnetic in nature [2.83]. An early model for looplike CMEs featuring twisted magnetic flux ropes [2.154] was difficult to accept because the applicable forces were internal, i.e. derived from the gradient of magnetic pressure of a self-induced magnetic field [2.48]. Another MHD model [2.274] was also aimed at explaining a looplike transient, but concluded that the outward propulsion of a CME arises from the (external) buoyancy force exerted by the ambient medium. Only deformations of the loop (broadening and stretching) were attributed to self-induced magnetic forces and pressure imbalance between internal and external gas pressures. Crudely visualized, the buoyancy force arises simply from the difference in pressure at the bottom and at the top of the CME. The magnetic component of this difference in pressures is the essential element in theories invoking the magnetic reconnection process to explain the initiation of CMEs [2.4, 176]. If the coronal mass ejection is a diamagnetic plasmoid so that the fields of the ambient corona are unable to penetrate the structure, the plasmoid will be constricted at its lower end by the external field and squirted radially upward in a process called the "melon seed" mechanism [2.177]. The actual reconnection process, which implies the conversion of magnetic energy into kinetic energy, is assumed to be active for at least the initial phase of the event. This scenario accounts for the CME as being caused by upward moving prominence material entrained in a magnetic looplike cloud. At the same time, material sunward of the magnetic neutral line is accelerated down both legs of the remnant arch into the chromosphere where it is quickly thermalized and is witnessed as a two-ribbon flare. An example of an inverted coronal loop possibly disconnected by the reconnection process was found in the *SMM* C/P data [2.102].

The third approach to the problem of CME origins has been the concept of a slow coronal evolution to nonequilibrium, e.g. by constantly stretching and differentially rotating magnetic field footpoints. This process is continued until a CME is triggered, when the magnetic tension and gravity are no longer capable of restraining the plasma in a given bound coronal structure [2.133, 134]. In accordance with the strong association with eruptive prominences, the actual CME acceleration mechanism in this approach can again be found in the coordination of magnetic reconnection and buoyancy, similar to that proposed in the second approach described above.

After many years of search for the CME "driver", the present consensus now appears to be moving away from either the solar flare or the eruptive prominence. Cases have been cited where the CME is observed in the corona "before" the associated prominence erupts [2.107, 108, 207]. Other evidence comes from well-documented examples of weak "precursor" X-ray bursts that were found to occur about ten minutes prior to the projected launch time of CMEs [2.79]. Stronger X-ray bursts were initiated close to this launch time, accompanied by an associated HF event that occurred in one footpoint of the same coronal arch that appeared to be the source of all observed X-rays. This chain of events would imply that flares are not a cause, but rather only a side effect of the same process that drives CMEs. Consistent with all observations to date, the one fundamental process capable of expelling CMEs of all shapes and sizes, with or without surface associations, appears to be a large-scale (global) disruption of the solar magnetic field [2.99]. Our understanding of the exact nature of this process, however, may have to wait for the "quantum jump" in observational capability expected from the *SOHO* mission in the mid nineties.

2.6.2 Radio-Sounding Investigations of Coronal and Interplanetary Transients

In addition to providing information on the physical state of the quiet corona as described in previous sections of this chapter, coronal and interplanetary transients have been investigated by observing their effects on radio signals from spacecraft and natural sources. Spacecraft signal parameters sensitive to changes in the propagation medium include frequency ("Doppler"), group delay time ("ranging"), bandwidth, amplitude/phase scintillations, and Faraday rotation [2.15]. Interplanetary disturbances can be monitored by IPS observations of their velocity or amplitude scintillation [2.250].

Radio-sounding investigations of this type have been carried out for many years now. The size and density of solar wind plasma clouds near 1 AU were estimated by their increased group delay time difference between two radio signals from earth to the *Pioneer* spacecraft as early as the 1960s [2.112, 119]. Interplanetary disturbances caused by the 1972 August flares were also observed by *Pioneer* [2.45] and their extent in and out of the ecliptic plane was estimated from IPS observations of enhancements in solar wind velocity [2.110, 184] and scintillation strength [2.249]. Anomalous spectral broadening [2.74] and "Faraday rotation transients" [2.127] were observed in the solar corona during the 1968 occultation of *Pioneer 9* at solar maximum. Additional Faraday rotation events were recorded during solar minimum using the *Helios* spacecraft [2.20]. Subsequent simultaneous observations of Faraday rotation transients on *Helios* and CMEs on Solwind are evidence that there may well be a one-to-one correspondence between the two events [2.17]. In retrospect, it is somewhat of a novelty that CMEs were evidently witnessed as a radio signal disturbance in the corona before they were recognized as such in visible light.

Three specific advantages of radio-sounding observations of CMEs will be singled out for further elaboration in the following. The first is the possibility of deriving the magnetic field in and around a coronal mass ejection using combined white-light and Faraday rotation observations [2.18]. Another advantage is the ability to determine CME velocities both in the corona and in the intermediate region between corona and *in situ* spacecraft [2.270, 271]. Finally, the full extent of interplanetary disturbances resulting from CMEs can be mapped from synoptic IPS observations [2.68, 243].

From both observational and theoretical considerations, the coronal magnetic field has emerged as the most probable active agent in the initiation and acceleration of CMEs [2.83]. Estimates of the magnitudes of magnetic fields near CMEs, based on the assumption that associated radio bursts were produced by gyrosynchrotron radiation, were of the order of 1 gauss at a nominal height of $2.5R_\odot$ [2.53, 72, 230]. These field strengths are about an order of magnitude greater than the mean ambient coronal field at that height (see also Fig. 2.14, Sect. 2.4.2). Directional information on the magnetic field cannot be obtained from such measurements, which are also only applicable to the radio-burst site in each event.

It had been suspected that the Faraday rotation transients seen during the solar occultations of the *Pioneer* and *Helios* spacecraft were the signatures of CMEs passing through the signal ray path. As such, it was proposed that the "magnetic profile" of a CME could be deduced from simultaneous Faraday rotation and electron content (e.g. white-light coronagraph) measurements [2.19]. The most extensive joint measurements of this type were taken during the solar conjunctions of *Helios 2* (October 1979) and *Helios 1* (November 1979). Five well-defined coronal mass ejections were observed to pass through the line-of-sight from *Helios* to earth, where they produced definite disturbances in the observed Faraday rotation [2.17]. Unfortunate gaps in both data sets limited the resolution and coverage of the events, but values for the mean longitudinal component of the transient magnetic field, as well as some qualitative information on the field orientation, could be deduced [2.18].

An example from this analysis, observed on 24 October 1979, is shown in Fig. 2.23. Three Solwind difference images of the event are displayed at the top; the size of the photosphere is again indicated by the large centered white circle in the first image. The radio-sounding data, including both Faraday rotation (*FR*) and spectral broadening (*SB*) measurements, are plotted in the bottom panel. The west limb CME was seen to emerge from behind the occulting disk at 0400 UT and move out toward the *Helios 2* ray path, indicated by the smaller white dot. Two features in the CME, the leading edge (LE) and a following bright core (BC) were tracked in their radial expansion to determine the expected penetration time at the *Helios*/earth line-of-sight. These instants are indicated in the lower panel by the vertical dashed lines. The widths of the boxes LE and BC give the estimated errors in these arrival times. The projected velocities of these two features were 160 ± 50 km/s (LE) and 120 ± 40 km/s (BC). The increase in signal bandwidth, marked "SB" in the lower panel, was seen to occur before the

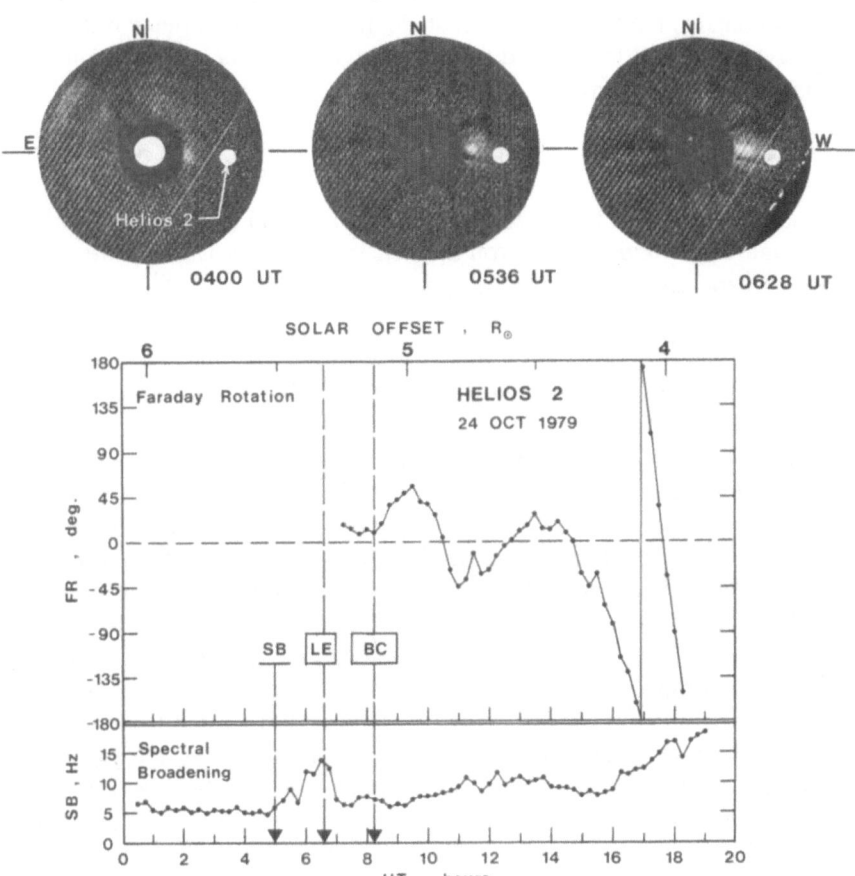

Fig. 2.23. Faraday rotation and spectral broadening profiles of the coronal mass ejection of 24 October 1979. This is the best example of radio-sounding measurements made during the observed passage of a white-light CME through the signal ray path of a spacecraft in solar occultation. The three Solwind difference images at the top indicate the geometrical relationship between the CME and the apparent position of *Helios* off the solar west limb. Large amplitude excursions of Faraday rotation and spectral broadening (*lower diagrams*) were recorded, possibly caused by magnetic field reversals and enhanced turbulence, respectively (adapted from [2.17])

arrival of the white-light front at the signal ray path. The *FR* data, which are unfortunately only available after 0700 UT, display wild positive and negative deviations of ±50 deg about their mean value, indicating magnetic field reversals in and underneath the bright core region of the CME. The scale of the magnetic structures causing these polarization reversals is $2R_\odot$ or even larger.

The Faraday rotation *FR* (2.25) is proportional to

$$FR \sim \langle B_\mathrm{L} \rangle I_\mathrm{t} \tag{2.52}$$

where $\langle B_\mathrm{L} \rangle$ is the weighted mean magnetic field component along the radio ray path, and I_t is the total coronal electron content. Assuming that the regions along the ray path outside the CME are quasi-static, the change in Faraday rotation due to an intervening CME is obtained from an integral over the region occupied by the transient coronal material:

$$\Delta FR = (K/f^2) \int [N_{et} B_t - N_{e0} B_0] \cdot ds , \qquad (2.53)$$

where the indices "t" and "0" refer to the transient and pretransient (ambient) values of magnetic field and electron density, respectively. Under most circumstances, the expression (2.53) can be approximated by

$$\Delta FR = (K/f^2)\langle B_L \rangle \Delta I_t , \qquad (2.54)$$

where $\langle B_L \rangle$ is now the weighted mean magnetic field component along the signal ray path *within the CME*, and ΔI_t is the excess electron content of the CME at the *Helios* ray path, a quantity determined from Solwind difference images. The change in FR and the excess electron content ΔI_t can therefore be used in (2.54) to compute values for $\langle B_L \rangle$. The magnitudes of $\langle B_L \rangle$ were found to lie in the range 0.01 to 0.1 G at the reference height $2.5 R_\odot$, i.e. considerably lower than the total magnetic field strength estimates from radio-burst observations. The values of $\langle B_L \rangle$, which refer to only one component of the total CME magnetic field, were nevertheless estimated to be greater than the actual total field strength in the ambient corona ahead of the CME.

The signal parameters FR and SB are not expected to vary in tandem because they are sensitive to quite different physical parameters along the propagation path. The SB was relatively placid, for example, during the interval of field reversals seen in the FR data. In contrast to (2.52), the spectral broadening SB is governed by

$$SB \sim \sigma_N v_d , \qquad (2.55)$$

where σ_N is the rms electron density fluctuation and v_d is the plasma velocity perpendicular to the ray path. Since σ_N is presumably proportional to N_e, changes in FR without a simultaneous change in SB are most easily attributed to a fluctuating magnetic field with a more or less constant electron density. It may be noted from (2.55) that SB is sensitive to more than just the changes in the electron density along the signal ray path. A number of examples were observed where the SB was definitely increased prior to the passage of the white-light CME through the *Helios* line-of-sight. These are best explained by a faster-moving plasma of only moderately higher density, such as a shock wave running out ahead of the region of greatest white-light brightness [2.271].

The properties of coronal shocks can be derived from combining various radio-sounding observations using a single radio link. The technique is best exemplified by the large coronal shock of 18 August 1979, which passed through the dual-frequency downlinks from *Voyager 1* at a solar proximate distance of $13.1 R_\odot$ [2.269]. In this case, the speed of the pre- and postshock solar wind flows could be derived using two different methods: by determining the "spectral knee" of the amplitude scintillations, and by combining SB data with the amplitude scintillation data (see also Sect. 2.5.2). Using the simultaneously recorded differential phase measurements to obtain the change in electron density across the shock front, Woo and Armstrong [2.269] then derived a shock velocity of ca. 3500 km/s. This estimate was in good agreement with a shock velocity profile derived from an associated interplanetary Type II burst [2.35] and is consistent with acceleration of the shock out to $20R_\odot$. The extended SB and scintillation strength time profiles of this event displayed a second enhancement following the shock that was identified as the turbulent "driver gas" of the shock wave.

Radio metric parameters such as Faraday rotation, spectral broadening and amplitude scintillations lose sensitivity at large solar elongations. Since interplanetary spacecraft generally do not spend too much time at superior conjunction, there are only a few examples comparable to the above *Voyager 1* event or the October/November CMEs observed with *Helios*. One radio metric parameter not subject to this restricted range in elongation is the dispersion in residual frequency, referred to simply as "Doppler noise" [2.267, 270].

Doppler noise depends on the physical parameters of the propagation medium in a manner very similar to the SB as given by (2.55). A recent synoptic survey of some 40 000 hours of S-band Doppler noise resulted in the detection of 148 separate transients, assumed to be interplanetary shocks, 26 of which were observed to pass through more than one radio link [2.267]. Speeds based on the radial transit time between radio links could be derived for these multispacecraft observations. The Doppler noise enhancement, which varied by factors of 2 to 100, was found to be highly correlated with the computed shock propagation speeds, which ranged up to ca. 2000 km/s. The enhancement factors and the shock speeds both displayed very similar distributions with heliocentric distance. This substantiated the results of an earlier analysis [2.270] which indicated a much larger dispersion of shock velocities close to the sun than near 1 AU. The implied deceleration of the shocks is largest for the fastest shocks. It was also noted that the shock velocities deduced from transit times, which are typically an average speed over the distance out to 0.3 AU, were considerably higher than those of their associated CMEs from white-light observations in the corona. This finding would appear to require acceleration of the shocks prior to the deceleration out to the *in situ* detections near 1 AU. An alternative interpretation is that the CME speeds substantially underestimate the speed of the shock. A comparison of CME speeds with shock velocities derived from Type II bursts [2.70] determined that the ratio of the CME speed to the shock speed was about 0.6, a value consistent with the results of Woo et al. [2.270].

With only a limited number of spacecraft for *in situ* and radio-sounding coverage, the best method for determining the evolving large-scale morphology of interplanetary disturbances is to map their effects on the scintillation of natural sources. An efficient instrument for this task is the 81.5 MHz array at Cambridge [2.68], where synoptic measurements of the scintillation level of some 900 sources are recorded at their daily meridian transit. A mean scintillation index is determined for each source as a function of solar elongation. Maps of the daily variations in the scintillation relative to the mean level along the line-of-sight to each source have been shown to be accurate representations of the interplanetary weather. Scintillation enhancements from solar events can be clearly distinguished from corotating solar wind streams and tracked over heliocentric distances from 0.6 to 1.5 AU [2.236]. Three-station IPS observations can be used to detect interplanetary disturbances by their increases in flow velocity as well as their enhanced scintillation strength [2.250]. A study of 17 such events during the last solar maximum indicated that the disturbances often had oblate configurations that were flattened in their latitudinal extent. On the average, the events expanded more or less symmetrically about a radial line extended upward from the solar event. These systems are quite efficient in forecasting the arrival times of interplanetary disturbances at earth [2.82], providing typically 12–24 hours lead time to prepare for the ensuing geomagnetic storm.

2.7 Appendix: Radio-Sounding Measurement Techniques

Two basic modes of operation, generally referred to as "one-way" and "two-way", have been utilized for radio sounding observations of the solar corona [2.148]. For the case of one-way measurements, the signal generator is a monochromatic ultrastable crystal oscillator (USO) on board the spacecraft, and only the propagation medium along the ray path from spacecraft to earth (downlink) can affect the received signal. Two-way measurements, on the other hand, are conducted with a signal generated at the ground tracking station by a high-stability atomic oscillator (e.g. spectral line of rubidium vapor). The radio signal from earth to spacecraft (uplink, usually at S-band) is turned around by the spacecraft's transponder, shifted in frequency, and retransmitted back to Earth.

The more recent interplanetary spacecraft have been equipped with S/X-band dual-frequency downlinks. If these downlinks are derived from the same source, whether it be an on-board USO or a ground-generated uplink, they are said to be "phase coherent". This condition is particularly useful for isolating frequency-dependent (i.e. dispersive) effects imposed by the propagation medium. Using two phase-coherent downlinks, the nondispersive changes in phase (e.g. inherent oscillator drift, orbital perturbations, terrestrial troposphere, gravitational Doppler shift, etc.) can be separated out by applying an appropriate differencing technique.

A significant enhancement of radio science experiments can be achieved when the spacecraft can be tracked by more than one ground station, thereby enabling correlation studies along multiple downlink propagation paths. Such simultaneous observations have been performed both with closely spaced stations at the same deep space tracking complex and with very long baselines of the order of the earth's radius. Solar wind velocity measurements based on such studies are presented in Sect. 2.5.

The characteristic parameters of the radio signals, amplitude, phase (or frequency, which is the time derivative of the phase), linewidth, and polarization, can all be utilized for remote sensing purposes. These measurements yield, respectively, the attenuation, time delay, spectral broadening, and Faraday rotation imposed by the propagation medium. Since the solar corona is almost completely ionized, all of the above signal parameters are subject to dispersion, the effects generally being more pronounced at lower radio frequencies.

Time-delay measurements can be used most easily to determine the coronal electron density N_e from the columnar electron content $I_t(t)$:

$$I_t(t) = \int_{s_1}^{s_2} N_e(r, \theta, \phi, t) \, ds \quad [\text{el/m}^2] \tag{2.56}$$

where $s_1(t)$ and $s_2(t)$ are the positions of the radio source and receiver, respectively [2.58]. The actual observable for measurements of time delay can be either the group time delay (ranging), phase time delay (Doppler), or a combination of these two (DRVID - Differenced Range Versus Integrated Doppler). In all cases the following dispersion relations hold for the phase (index ph) and group (index g) velocities:

$$v_{ph} = \frac{c}{\sqrt{1 - (f_P/f)^2}} \simeq c\left(1 + \frac{40.3 N_e}{f^2}\right) , \tag{2.57}$$

$$v_g = c\sqrt{1 - (f_P/f)^2} \simeq c\left(1 - \frac{40.3 N_e}{f^2}\right) , \tag{2.58}$$

where c is the velocity of light, f is the radio frequency in Hz, f_P is the plasma frequency and N_e is given in el/m^3. The approximations in (2.57) and (2.58) are valid for $f \gg f_P$, which is the case for deep-space radio signal propagation in the corona.

Doppler measurements are made with the carrier signal (typical sample rate: one point per second), the observable quantity being the phase delay time τ_{ph} given by

$$\tau_{ph}(t) = \int_{s_1}^{s_2} \frac{ds}{v_{ph}} \simeq \frac{s_2 - s_1}{c} - \frac{40.3}{cf^2} I_t(t) . \tag{2.59}$$

The first term of (2.59) is the light time in vacuum between the source and the receiver, a purely geometrical quantity that can be estimated to a high degree of precision from orbit determination programs. If dual-frequency Doppler measurements at $f_1 < f_2$ are available along the link, it is possible to extract $I_t(t)$ from the differential phase delay time $\Delta\tau_{ph}$ according to

$$\Delta\tau_{ph}(t) = \frac{40.3}{c}\left(\frac{1}{f_1^2} - \frac{1}{f_2^2}\right) I_t(t) , \tag{2.60}$$

thereby eliminating the uncertainties from the orbit determination. It should be noted that the quantities $\tau_{\mathrm{ph}}(t)$ and $\Delta\tau_{\mathrm{ph}}(t)$ in (2.59) and (2.60) cannot be measured in an absolute sense, but rather only relative to the phase of a local oscillator at the start of the measurements. As a result, $I_{\mathrm{t}}(t)$ represents here only the changes in electron content since the start of the tracking pass, to which an unknown constant electron content $I_{\mathrm{t}0}$ must be added.

Ranging measurements require a modulated carrier signal (typical sample rate: one point every few minutes). Generally a phase modulation (square wave) is applied, using a binary coded system of sequentially arranged pulse chains [2.147]. The group time delay is determined by measuring the elapsed time between the transmission of an encoded signal and its reception at the tracking station after having been relayed back to earth by the spacecraft's transponder. Such measurements can only be performed in two-way mode. In analogy with (2.59), the obsevable is the group delay time τ_{g} given by

$$\tau_{\mathrm{g}}(t) = \int_{s_1}^{s_2} \frac{ds}{v_{\mathrm{g}}} \simeq \frac{s_2 - s_1}{c} + \frac{40.3}{cf^2} I_{\mathrm{t}}(t) . \tag{2.61}$$

The first term in (2.61) must again be estimated from orbit determination. For dual-frequency operation with $f_1 < f_2$, one can compute a differential group time delay $\Delta\tau_{\mathrm{g}}$ given by

$$\Delta\tau_{\mathrm{g}}(t) = \frac{40.3}{c} \left(\frac{1}{f_1^2} - \frac{1}{f_2^2} \right) I_{\mathrm{t}}(t) . \tag{2.62}$$

In contrast to (2.60), the differential range from (2.62) can provide an absolute value of $I_{\mathrm{t}}(t)$ to within the ambiguity of the lowest frequency ranging component. For the commonly used S/X-band radio system, a columnar electron content of $I_{\mathrm{t}} = 4 \times 10^{16}$ el/m^2 produces a group time delay of $\Delta\tau_{\mathrm{g}} = 1$ ns.

DRVID *measurements* basically represent the integrated rate of change of the electron content expressed by the quantity

$$\tau_{\mathrm{D}}(t) = \tau_{\mathrm{g}}(t) - \tau_{\mathrm{ph}}(t) = \frac{80.6}{cf^2} I_{\mathrm{t}}(t) . \tag{2.63}$$

The orbital uncertainties and other nondispersive contributions are eliminated from expression (2.63) using essentially the same trick available for dual-frequency radio links. Typical plasma delay times of the order of 30 μs are recorded at solar offsets of $R = 5R_\odot$. The outer boundary of sensitivity for detecting a significant increase in coronal electron content was about $30R_\odot$ for the *Helios* S-band experiment.

Faraday rotation measurements can be used to determine either the coronal plasma density or the coronal magnetic field [2.244]. In the first case, the coronal magnetic field must be modeled by taking a spatial average along the signal ray path (see Sect. 2.4.2). In the second case, the coronal plasma density must either be modeled (e.g. spherically symmetric, radial power law falloff), or the electron content along the same propagation path must be obtained by using some independent measurement technique (e.g. time-delay measurements). If simultaneous Faraday rotation and time-delay measurements are taken, one may separate the contributions to the Faraday rotation due to electron density and magnetic field. Using such data from *Helios 2* in 1976, for example, Hollweg et al. [2.92] were able to provide convincing evidence for the existence of coronal Alfvén waves.

Faraday rotation is most easily measured if the downlink is linearly polarized. This was the case for the two *Helios* spacecraft, for which the orientation of the incident electric vector was determined using a special S-band mechanical polarimeter [2.165]. Since the rotation angle of the incident electric vector can only be measured MOD 180°, it is impossible to distinguish how many half-rotations the downlink signal may have undergone along the propagation path. Coronal Faraday rotation at S-band only begins to be measureable for solar offsets $R < 10$–$20R_\odot$. Nevertheless, measurement ambiguities of $\pm n\pi$ can arise closer to the sun if large data gaps in the recorded Faraday rotation occur between tracking passes.

References

2.1   Allen, C.W., Interpretation of electron densities from corona brightness, Mon. Not. R. Astron. Soc., **107**, 426–432, 1947.
2.2   Altschuler, M.D., G. Newkirk, Jr., Magnetic fields and the structure of the solar corona, Solar Phys., **9**, 131–149, 1969.
2.3   Anderson, J.D., T.P. Krisher, S.E. Borutzki, M.J. Connally, P.M. Eshe, H.B. Hotz, S. Kinslow, E.R. Kursinski, L.B. Light, S.E. Matousek, K.I. Moyd, D.C. Roth, D.N. Sweetnam, A.H. Taylor, G.L. Tyler, D.L. Gresh, P.A. Rosen, Radio range measurements of coronal electron densities at 13 and 3.6 centimeter wavelengths during the 1985 solar occultation of Voyager 2, Ap. J., **323**, L141–L143, 1987.
2.4   Anzer, U., G.W. Pneuman, Magnetic reconnection and coronal transients, Solar Phys., **79**, 129–147, 1982.
2.5   Armand, N.A., A.I. Efimov, O.I. Yakovlev, A model of the solar wind turbulence from radio occultation experiments, Astron. Astrophys., **183**, 135–141, 1987.
2.6   Armstrong, J.W., W.A. Coles, Analysis of three station interplanetary scintillations, J. Geophys. Res., **77**, 4602–4609, 1972.
2.7   Armstrong, J.W., W.A. Coles, M. Kojima, B.J. Rickett, Solar wind observations near the Sun, in *The Sun and the Heliosphere in Three Dimensions*, ed. by R.G. Marsden, 59–64, Reidel Publ. Co., Dordrecht, 1986.
2.8   Armstrong, J.W., R. Woo, Solar wind motion within $30R_\odot$: spacecraft radio scintillation observations, Astron. Astrophys., **103**, 415–421, 1981.
2.9   Axford, W.I., The three-dimensional structure of the interplanetary medium, in *Study of Travelling Interplanetary Phenomena 1977*, ed. by M.A. Shea et al., 145–164, D. Reidel Publ. Co., Dordrecht, 1977.
2.10 Beard, D.B., The solar corona, Ap. J., **234**, 696–706, 1979.
2.11 Beard, D.B., Infrared coronal polarization and the size of interplanetary dust particles, Astron. Astrophys., **132**, 317–320, 1984.
2.12 Berman, A.L., DSN telecommunications interfaces, solar corona and solar wind effects, in *Deep Space Network/Flight Project Interface Design Handbook*, JPL Doc. 810-5, Rev. D, Vol. I, TCI-50, 1–19, 1979.
2.13 Berman, A.L., A.D. Conteas, Voyager 1979: update to the radial and solar cycle variations in the solar wind phase fluctuation spectral index, DSN Prog. Rep **42–54**, 71–81, 1979.

2.14   Bird, M.K., Coronal sounding with pulsars, in *Solar Wind 4*, ed. by H. Rosenbauer, MPAE-W-100-81-31, 78–83, 1981.

2.15   Bird, M.K., Coronal sounding with occulted spacecraft signals, Space Sci. Rev., **33**, 99–126, 1982.

2.16   Bird, M.K., E. Schrüfer, H. Volland, W. Sieber, Coronal Faraday rotation during solar occultation of PSR 0525+21, Nature, **283**, 459–460, 1980.

2.17   Bird, M.K., H. Volland, R.A. Howard, M.J. Koomen, D.J. Michels, N.R. Sheeley, Jr., J.W. Armstrong, B.L. Seidel, C.T. Stelzried, R. Woo, Coronal transients observed during solar occultation of the Helios spacecraft in STIP Interval VIII, in *STIP Symposium on Solar/Interplanetary Intervals*, ed. by M.A. Shea et al., 101–112, Book Crafters, Chelsea, Michigan, 1984.

2.18   Bird, M.K., H. Volland, R.A. Howard, M.J. Koomen, D.J. Michels, N.R. Sheeley, Jr., J.W. Armstrong, B.L. Seidel, C.T. Stelzried, R. Woo, White-light and radio sounding observations of coronal transients, Solar Phys., **98**, 341–368, 1985.

2.19   Bird, M.K., H. Volland, B.L. Seidel, C.T. Stelzried, The cross sectional magnetic profile of a coronal transient, in *Solar and Interplanetary Dynamics*, ed. by M. Dryer, E. Tandberg-Hanssen, 475–481, D. Reidel Publ. Co., Dordrecht, 1980.

2.20   Bird, M.K., H. Volland, C.T. Stelzried, G.S. Levy, B.L. Seidel, Faraday rotation transients observed during solar occultation of the Helios spacecraft, in *Contributed Papers to the Study of Travelling Interplanetary Phenomena*, ed. by. M.A. Shea et al., AFGL-TR-77-0309, 63–75, 1977.

2.21   Blums, D.F., N.A. Lotova, On the recurrence of solar wind as a function of heliographic latitude, Sov. Astron., **23**(4), 491–492, 1979.

2.22   Bogdan, T.J., The determination of coronal potential magnetic fields using line-of-sight boundary conditions, Solar Phys., **103**, 311–315, 1986.

2.23   Bommier, V., S. Sahal-Bréchot, The Hanle effect of the coronal Lα line of hydrogen: theoretical investigation, Solar Phys., **78**, 157–178, 1982.

2.24   Bougeret, J.-L., J.H. King, R. Schwenn, Solar radio burst and in situ determination of interplanetary electron density, Solar Phys., **90**, 401–412, 1984.

2.25   Bourgois, G., W.A. Coles, G. Daigne, J. Silen, T. Turunen, P.J. Williams, Measurements of the solar wind velocity with EISCAT, Astron. Astrophys., **144**, 452–462, 1985.

2.26   Bradford, H.M., D. Routledge, Coronal occultation of Voyager 2, 1979 August, Mon. Not. R. Astr. Soc., **190**, 73P–77P, 1980.

2.27   Brecher, K., A. Brecher, P. Morrison, I. Wasserman, Is there a ring around the sun?, Nature, **282**, 50–52, 1979.

2.28   Brueckner, G.E., J.-D.F. Bartoe, Observations of high-energy jets in the corona above the quiet sun, the heating of the corona, and the acceleration of the solar wind, Ap. J., **272**, 329–348, 1983.

2.29   Bruno, R., L.F. Burlaga, A.J. Hundhausen, K-coronameter observations and potential field comparison in 1976 and 1977, J. Geophys. Res., **89**, 5381–5385, 1984.

2.30   Bumba, V., R. Howard, Solar activity and recurrences in magnetic field distribution, Solar Phys., **7**, 28–38, 1969.

2.31   Burlaga, L.F., A.J. Hundhausen, X. Zhao, The coronal and interplanetary current sheet in early 1976, J. Geophys. Res., **86**, 8893–8898, 1981.

2.32   Burlaga, L.F., L. Klein, N.R. Sheeley, Jr., D.J. Michels, R.A. Howard, M.J. Koomen, R. Schwenn, H. Rosenbauer, A magnetic cloud and a coronal mass ejection, Geophys. Res. Lett., **9**, 1317–1320, 1982.

2.33   Callahan, P.S., Interpretation of columnar content measurements of the solar wind turbulence, Ap. J., **187**, 185–190, 1974.

2.34   Callahan, P.S., Columnar content measurements of the solar wind turbulence near the sun, Ap. J., **199**, 227–236, 1975.

2.35   Cane, H.V., R.G. Stone, R. Woo, Velocity of the shock generated by a large east limb flare on August 18, 1979, Geophys. Res. Lett., **9**, 897–900, 1982.

2.36   Cane, H.V., N.R. Sheeley, Jr., R.A. Howard, Energetic interplanetary shocks, radio emission, and coronal mass ejections, J Geophys. Res., **92**, 9869–9874, 1987.

2.37   Chipman, E.G., The Solar Maximum Mission, Ap. J., **244**, L113–L115, 1981.

2.38   Coles, W.A., Interplanetary scintillation, Space Sci. Rev., **21**, 411–425, 1978.

2.39   Coles, W.A., J.K. Harmon, Propagation observations of the solar wind near the sun, Ap. J., **337**, 1023–1034, 1989.

2.40   Coles, W.A., J.J. Kaufman, Solar wind velocity estimation from multi-station IPS, Radio Sci., **13**, 591–597, 1978.

2.41   Coles, W.A., B.J. Rickett, V.H. Rumsey, Interplanetary scintillations, in *Solar Wind Three*, ed. by. C.T. Russell, 351–367, Inst. Geophys. Planet. Phys., UCLA, Los Angeles, 1974.

2.42   Coles, W.A., B.J. Rickett, V.H. Rumsey, J.J. Kaufman, D.G. Turley, S. Ananthakrishnan, J.W. Armstrong, J.K. Harmon, S.L. Scott, D.G. Sime, Solar cycle changes in the polar solar wind, Nature, **286**, 239–241, 1980.

2.43   Counselman III, C.C., J.M. Rankin, Density of the solar corona from occultations of NP 0532, Ap. J., **175**, 843–856, 1972.

2.44   Crifo, F., J.P. Picat, M. Cailloux, Coronal transients: loop or bubble?, Solar Phys., **83**, 143–152, 1983.

2.45   Croft, T.A., Traveling regions of high solar wind density observed in early August 1972, J. Geophys. Res., **78**, 3159–3166, 1973.

2.46   Cushman, G.W., W.A. Rense, Evidence of outward flow of plasma in a coronal hole, Ap. J., **207**, L61–L62, 1976 [Erratum, Ap. J., **211**, L57, 1977].

2.47   Denskat, K.U., H.J. Beinroth, F.M. Neubauer, Interplanetary magnetic field power spectra with frequencies from 2.4×10^{-5} Hz to 470 Hz from HELIOS-observations during solar minimum conditions, J. Geophys., **54**, 60–67, 1983.

2.48   Dryer, M., Coronal transient phenomena, Space Sci. Rev., **33**, 233–275, 1982.

2.49   Dryer, M., S.T. Wu, R.S. Steinolfson, R.M. Wilson, Magnetohydrodynamic models of coronal transients in the meridional plane. II. Simulation of the coronal transient of 1973 August 21, Ap. J., **227**, 1059–1071, 1979.

2.50   Dulk, G.W., Radio and white-light observations of coronal transients, in *Radio Physics of the Sun*, ed. by M.R. Kundu, T.E. Gergely, 419–432, Reidel Publ. Co., Dordrecht, 1980.

2.51   Dulk, G.A., Radio emission from the sun and stars, Ann. Rev. Astron. Astrophys., **23**, 169–224, 1985.

2.52   Dulk, G.A., D.J. McLean, Coronal magnetic fields, Solar Phys., **57**, 279–295, 1978.

2.53   Dulk, G.A., S.F. Smerd, R.M. MacQueen, J.T. Gosling, A. Magun, R.T. Stewart, K.V. Sheridan, R.D. Robinson, S. Jacques, White light and radio studies of the coronal transient of 14–15 September 1973, Solar Phys., **49**, 369–394, 1976.

2.54   Dürst, J., Two colour photometry and polarimetry of the solar corona of 16 February 1980, Astron. Astrophys., **112**, 241–250, 1982.

2.55   Eddy, J.A., A nineteenth-century coronal transient, Astron. Astrophys., **34**, 235–240, 1974.

2.56   Edenhofer, P., Helios-Okkultationsexperiment: Plasma-Fernerkundung mit Laufzeitmessungen, BMFT-FB-W **81-051**, 1981 (in German).

2.57   Edenhofer, P., M.K. Bird, H. Volland, Comparison of time delay and Faraday rotation measurements from the HELIOS spacecraft, Kleinheubacher Ber., **21**, 305–312, 1978.

2.58   Edenhofer, P., P.B. Esposito, R.T. Hansen, S.F. Hansen, E. Lüneburg, W.L. Martin, A.I. Zygielbaum, Time delay occultation data of the Helios spacecrafts and preliminary analysis for probing the solar corona, J. Geophys. **42**, 673–698, 1977.

2.59   Edenhofer, P., P.B. Esposito, E. Lüneburg, Hydromagnetic wavelike phenomena from Helios time delay measurements by remote sensing, Kleinheubacher Ber., **23**, 267–273, 1980.

2.60   Efimov, A.I., O.I. Yakovlev, V.K. Shtrykov, V.I. Rogal'sky, V.F. Tikhonov, Observations from different points of fluctuations in frequency and phase of radio waves that are scattered by plasma around the sun, Sov. Radio Engg. Electron., **26**, 311–318, 1981.

2.61   Ekers, R.D., L.T. Little, The motion of the solar wind close to the sun, Astron. Astrophys., **10**, 310–316, 1971.

2.62   Esposito, P.B., P. Edenhofer, E. Lüneburg, Solar corona electron density distribution, J. Geophys. Res., **85**, 3414–3418, 1980.

2.63   Esser, R., E. Leer, S.R. Habbal, G.L. Withbroe, A two-fluid solar wind model with Alfvén waves: parameter study and application to observations", J. Geophys. Res., **91**, 2950–2960, 1986.

2.64   Fisher, R.R., L.B. Lacey, K.A. Rock, E.A. Yasukawa, N.R. Sheeley, Jr., D.J. Michels, R.A. Howard, M.J. Koomen, A. Bagrov, The solar corona on 31 July, 1981, Solar Phys., **83**, 233–242, 1983.

2.65   Fisher, R.R., R.H. Lee, R.M. MacQueen, A.I. Poland, New Mauna Loa coronagraph systems, Appl. Opt., **20**, 1094–1101, 1981.

2.66   Fisher, R., D.G. Sime, Solar activity cycle variation of the K corona, Ap. J., **285**, 354–358, 1984.

2.67   Fisher, R., D.G. Sime, Rotational characteristics of the white-light corona: 1965–1983, Ap. J., **287**, 959–968, 1984.

2.68   Gapper, G.R., A. Hewish, A. Purvis, P.J. Duffett-Smith, Observing interplanetary disturbances from the ground, Nature, **296**, 633–636, 1982.

2.69   Gary, D.E., G.A. Dulk, L. House, R. Illing, C. Sawyer, W.J. Wagner, D.J. McLean, E. Hildner, Type II bursts, shock waves, and coronal transients: the event of 1980 June 29, 0233 UT, Astron. Astrophys., **134**, 222–233, 1984.

2.70   Gergely, T.E., On the relative velocity of coronal transients and Type II bursts, in *STIP Symposium on Solar/Interplanetary Intervals*, ed. by. M.A. Shea et al. 347–357, Book Crafters, Chelsea, Michigan, 1984.

2.71   Gergely, T.E., Type IV bursts and coronal mass ejections, Solar Phys., **104**, 175–178, 1986.

2.72   Gergely, T.E., M.R. Kundu, R.H. Munro, A.I. Poland, Radio and white-light observations of the 1973 August 21 coronal transient, Ap. J., **230**, 575–580, 1979.

2.73   Gibson, E.G., *The quiet Sun*, NASA SP-303, 1973.

2.74   Goldstein, R.M., Superior conjunction of Pioneer 6, Science, **166**, 598–601, 1969.

2.75   Gosling, J.T., G. Borrini, J.R. Asbridge, S.J. Bame, W.C. Feldman, R.T. Hansen, Coronal streamers in the solar wind at 1 AU, J. Geophys. Res., **86**, 5438–5448, 1981.

2.76   Gosling, J.T., E. Hildner, R.M. MacQueen, R.H. Munro, A.I. Poland, C.L. Ross, Mass ejections from the sun: a view from Skylab, J. Geophys. Res., **79**, 4581–4587, 1974.

2.77   Gosling, J.T., E. Hildner, R.M. MacQueen, R.H. Munro, A.I. Poland, C.L. Ross, The speeds of coronal mass ejection events, Solar Phys., **48**, 389–397, 1976.

2.78   Hansen, R.T., C.J. Garcia, S.F. Hansen, H.G. Loomis, Brightness variations of the white light corona during the years 1964–67, Solar Phys., **7**, 417–433, 1969.

2.79   Harrison, R.A., Solar coronal mass ejections and flares, Astron. Astrophys., **162**, 283–291, 1986.

2.80   Hewish, A., Observations of the solar plasma using radio scattering and scintillation methods, in *Solar Wind*, ed. by C.P. Sonett et al., NASA SP-308, 477–493, 1972.

2.81   Hewish, A., S. Bravo, The sources of large-scale heliospheric disturbances, Solar Phys., **106**, 185–200, 1986.

2.82   Hewish, A., P.J. Duffett-Smith, A new method of forecasting geomagnetic activity and proton showers, Planet. Space Sci., **35**, 487–491, 1987.

2.83   Hildner, E., Do we understand coronal mass ejections yet?, Adv. Space Res., **6**(6), 297–306, 1986.

2.84   Hildner, E., J.T. Gosling, R.T. Hansen, J.D. Bohlin, The sources of material comprising a mass ejection coronal transient, Solar Phys., **45**, 363–376, 1975.

2.85   Hildner, E., J.T. Gosling, R.M. MacQueen, R.H. Munro, A.I. Poland, C.L. Ross, Frequency of coronal transients and solar activity, Solar Phys., **48**, 127–135, 1976.

2.86   Hoeksema, J.T., Extending the sun's magnetic field through the three-dimensional heliosphere, Adv. Space Res., **9**(4), 141–152, 1989.

2.87   Hoeksema, J.T., P.H. Scherrer, The solar magnetic field – 1976 through 1985, WDC-A Report **UAG-94**, 1986.

2.88   Hoeksema, J.T., P.H. Scherrer, Rotation of the coronal magnetic field, Ap. J., **318**, 428–436, 1987.

2.89   Hoeksema, J.T., J.M. Wilcox, P.H. Scherrer, Structure of the heliospheric current sheet in the early portion of sunspot cycle 21, J. Geophys. Res., **87**, 10331–10338, 1982.

2.90   Hoeksema, J.T., J.M. Wilcox, P.H. Scherrer, The structure of the heliospheric current sheet 1978–1982, J. Geophys. Res., **88**, 9910–9918, 1983.

2.91   Hollweg, J.V., Alfvén waves in the solar atmosphere, Solar Physics, **56**, 305–333, 1978.

2.92   Hollweg, J.V., M.K. Bird, H. Volland, P. Edenhofer, C.T. Stelzried, B.L. Seidel, Possible evidence for coronal Alfvén waves, J. Geophys. Res., **87**, 1–8, 1982.

2.93   Hollweg, J.V., J.R. Jokipii, Wavelength dependence of the interplanetary scintillation index, in *Solar Wind*, ed. by C.P. Sonett et al., NASA SP-308, 494–505, 1972.

2.94   House, L.L., W.J. Wagner, E. Hildner, C. Sawyer, H.U. Schmidt, Studies of the corona with the Solar Maximum Mission coronagraph/polarimeter, Ap. J., **244**, L117–L121, 1981.

2.95   Howard, R.A., D.J. Michels, N.R. Sheeley, Jr., M.J. Koomen, The observation of a coronal transient directed at Earth, Ap. J., **263**, L101–L104, 1982.

2.96   Howard, R.A., N.R. Sheeley, Jr., M.J. Koomen, D.J. Michels, Coronal mass ejections: 1979–1981, J. Geophys. Res., **90**, 8173–8191, 1985.

2.97   Howard, R.A., N.R. Sheeley, Jr., D.J. Michels, M.J. Koomen, The solar cycle dependence of coronal mass ejections, in *The Sun and Heliosphere in Three Dimensions*, ed. by R.G. Marsden, 107–111, D. Reidel Publ. Co., Dordrecht, 1986.

2.98   Hundhausen, A.J.: *Coronal Expansion and Solar Wind*, Springer-Verlag, Berlin, 1972.

2.99   Hundhausen, A.J., The origin and propagation of coronal mass ejections, in *Proc. 6th International Solar Wind Conf.*, ed. by V.J. Pizzo et al., 181–214, NCAR/TN-306, HAO-NCAR, Boulder, CO/USA, 1988.

2.100 Hundhausen, J.R., A.J. Hundhausen, E.G. Zweibel, Magnetostatic atmospheres in a spherical geometry and their application to the solar corona, J. Geophys. Res., 86, 11117–11126, 1981.

2.101 Hundhausen, A.J., C.B. Sawyer, L. House, R.M.E. Illing, W.J. Wagner, Coronal mass ejections observed during the Solar Maximum Mission: latitude distribution and rate of occurrence, J. Geophys. Res., 89, 2639–2646, 1984.

2.102 Illing, R.M.E., A.J. Hundhausen, Possible observation of a disconnected magnetic structure in a coronal transient, J. Geophys. Res., 88, 10210–10214, 1983.

2.103 Isobe, S., T. Hirayama, N. Baba, N. Miura, Optical polarization observations of circumsolar dust during the 1983 solar eclipse, Nature, 318, 644–646, 1985.

2.104 Jackson, B.V., R.A. Howard, N.R. Sheeley, Jr., D.J. Michels, M.J. Koomen, R.M.E. Illing, Helios spacecraft and Earth perspective observations of three looplike solar mass ejection transients, J. Geophys. Res., 90, 5075–5081, 1985.

2.105 Jackson, B.V., C. Leinert, Helios images of solar mass ejections, J. Geophys. Res., 90, 10759–10764, 1985.

2.106 Jokipii, J.R., Turbulence and scintillations in the interplanetary plasma, Ann. Rev. Astron. Astrophys., 11, 1–28, 1973.

2.107 Kahler, S., Coronal mass ejections, Rev. Geophys., 25, 663–675, 1987.

2.108 Kahler, S., Observations of coronal mass ejections near the sun, in Proc. 6th Internat. Solar Wind Conf., ed. by V.J. Pizzo et al., 215–231, NCAR/TN-306, HAO-NCAR, Boulder, CO/USA, 1988.

2.109 Kahler, S., N.R. Sheeley, Jr., R.A. Howard, M.J. Koomen, D.J. Michels, Characteristics of flares producing metric type II bursts and coronal mass ejections, Solar Phys., 93, 133–141, 1984.

2.110 Kakinuma, T., T. Watanabe, Interplanetary scintillation of radio sources during August 1972, Space Sci. Rev., 19, 611–627, 1976.

2.111 Kawashima, N., Analysis of fluctuations in the interplanetary magnetic field obtained by IMP 2, J. Geophys. Res., 74, 225–230, 1969.

2.112 Koehler, R.L., Radio propagation measurements of pulsed plasma streams from the sun using Pioneer spacecraft, J. Geophys. Res., 73, 4883–4894, 1968.

2.113 Kohl, J.L., G.L. Withbroe, C.A. Zapata, G. Noci, Spectroscopic measurements of solar wind generation, in Solar Wind 5, ed. by M. Neugebauer, NASA CP-2280, 47–60, 1983.

2.114 Kojima, M., T. Kakinuma, Solar cycle evolution of solar wind speed structure between 1973 and 1985 observed with the interplanetary scintillation method, J. Geophys. Res., 92, 7269–7279, 1987.

2.115 Kolosov, M.A., O.I. Yakovlev, A.I. Efimov, V.I. Rogal'sky, V.M. Razmanov, V.K. Shtrykov, Decimeter radio wave propagation in the turbulent plasma near the sun, using Venera 10 spacecraft, Radio Sci., 17, 664–674, 1982.

2.116 Koomen, M.J., C.R. Detweiler, G.E. Brueckner, H.W. Cooper, R. Tousey, White light coronagraph in OSO-7, Appl. Opt., 14, 743–751, 1975.

2.117 Koutchmy, S., P.L. Lamy, The F-corona and the circumsolar dust evidence and properties, in Properties and Interactions of Interplanetary Dust, ed by R.H. Giese, P. Lamy, 63–74, D. Reidel Publ. Co., Dordrecht, 1985

2.118 Krieger, A.S., Temporal behavior of coronal holes, in Coronal Holes and High Speed Wind Streams, ed. by J.B. Zirker, 71–102, Colorado Asso. Univ. Press, Boulder, 1977.

2.119 Landt, J.A., T.A. Croft, Shape of a solar wind disturbance on July 9, 1966, inferred from radio signal delay to Pioneer 6, J. Geophys. Res., 75, 4623–4630, 1970.

2.120 Lebecq, C., S. Koutchmy, G. Stellmacher, The 1981 total solar eclipse corona, II. Global absolute photometric analysis, Astron. Astrophys., 152, 157–164, 1985.

2.121 Leer, E, Wave acceleration mechanisms for the solar wind, in Proc. 6th International Solar Wind Conf., ed. by V.J. Pizzo et al., 89–106, NCAR/TN-306, HAO-NCAR, Boulder, CO/USA, 1988.

2.122 Leer, E., T.E. Holzer, T. Fla, Acceleration of the solar wind, Space Sci. Rev., 33, 161–200, 1982.

2.123 Leinert, C., E. Grün, Interplanetary dust, in Physics of the Inner Heliosphere (this volume).

2.124 Leinert, C., M. Hanner, H. Link, E. Pitz, Search for a dust free zone around the sun from the Helios 1 solar probe, Astron. Astrophys. 64, 119–122, 1978.

2.125 Levine, R.H., M.D. Altschuler, Representations of coronal magnetic fields including currents, Solar Phys., 36, 345–350, 1974.

2.126 Levine, R.H., M. Schulz, E.N. Frazier, Simulation of the magnetic structure of the inner heliosphere by means of a non-spherical source surface, Solar Phys., 77, 363–392, 1982.

2.127 Levy, G.S., T. Sato, B.L. Seidel, C.T. Stelzried, J.E. Ohlson, W.V.T. Rusch, Pioneer 6: measurement of transient Faraday rotation phenomena observed during solar occultation, Science, 166, 596–598, 1969.

2.128 Lotova, N.A., The solar wind transsonic region, Solar Phys., 117, 399–406, 1988.

2.129 Lotova, N.A., D.F. Blums, K.V. Vladimirskii, Interplanetary scintillation and the structure of the solar wind transonic region, Astron. Astrophys., 150, 266–272, 1985.

2.130 Lotova, N.A., I.V. Chashey, The fine structure of solar wind velocity, in Solar Wind Three, ed. by C.T. Russell, 375–381, Inst. Geophys. and Planet. Phys., UCLA, Los Angeles, 1974.

2.131 Lotova, N.A., Y.V. Nagelys, The investigation of the solar wind transsonic region at meter wavelengths, Solar Phys., 117, 407–414, 1988.

2.132 Loucif, M.L., S. Koutchmy, Solar cycle variations of coronal structures, Astron. Astrophys. Suppl. Ser., 77, 45–66, 1989.

2.133 Low, B.C., Eruptive solar magnetic fields, Ap. J., 251, 352–363, 1981.

2.134 Low, B.C., On the large-scale magnetostatic coronal structures and their stability, Ap. J., 286, 772–786, 1984.

2.135 MacQueen, R.M., Infrared observations of the outer solar corona, Ap. J., 154, 1059–1076, 1968.

2.136 MacQueen, R.M., Coronal transients: a summary, Phil. Trans. R. Soc. Lond., A297, 605–620, 1980.

2.137 MacQueen, R.M., Coronal mass ejections: acceleration and surface associations, Solar Phys., 95, 359–361, 1985.

2.138 MacQueen, R.M., Coronal magnetic fields – a mini survey, in The Sun and Heliosphere in Three Dimensions, ed. by R.G. Marsden, 5–18, D. Reidel Publ. Co., Dordrecht, 1986.

2.139 MacQueen, R.M., A. Csoeke-Poeckh, E. Hildner, L. House, R. Reynolds, A. Stanger, H. Tepoel, W. Wagner, The High Altitude Observatory Coronagraph/Polarimeter on the Solar Maximum Mission, Solar Phys., 65, 91–105, 1980.

2.140 MacQueen, R.M., J.A. Eddy, J.T. Gosling, E. Hildner, R.H. Munro, G.A. Newkirk, Jr., A.I. Poland, C.L. Ross, The outer solar corona as observed from Skylab: preliminary results, Ap. J., 187, L85–L88, 1974.

2.141 MacQueen, R.M., R.R. Fisher, The kinematics of solar inner coronal transients, Solar Phys., 89, 89–102, 1983.

2.142 Mampaso, A., C. Sanchez-Magro, M.J. Shelby, A.D. MacGregor, Infrared observations of the thermal emission from the corona, Rev. Mex. Astron. Astrof., 8, 3–6, 1983.

2.143 Maran, S.P., B.E. Woodgate, A second chance for Solar Max, Sky & Telescope, 67, 498–500, 1984.

2.144 Mariani, F., F.M. Neubauer, Interplanetary magnetic field, in Physics of the Inner Heliosphere (this volume).

2.145 Marshallm, E., Working solar monitor shot down by ASAT, Science, 230, 44–45, 1985.

2.146 Martin, J.M., Voyager microwave scintillation measurements of solar wind plasma parameters, Sci. Rept. D208-1985-1, Center for Radar Astronomy, Stanford University, 1985.

2.147 Martin, W., A binary-coded sequential acquisition ranging system, JPL Space Prog. Sum., 37–57/II, 72–81, 1969.

2.148 Martin, W., A.I. Zygielbaum, Mu-II ranging, JPL Tech. Memorandum, 33–768, 1977.

2.149 McLean, D.J., N.R. Labrum (eds.), Solar Radiophysics, Cambridge University Press, Cambridge, 1985.

2.150 Michels, D.J., R.A. Howard, M.J. Koomen, N.R. Sheeley, Jr., Satellite observations of the outer corona near sunspot maximum, in Radio Physics of the Sun, ed. by M.R. Kundu, T.E. Gergely, 439–442, Reidel Publ. Co., Dordrecht, 1980.

2.151 Michels, D.J., N.R. Sheeley, Jr., R.A. Howard, M.J. Koomen, R. Schwenn, K.H. Mühlhäuser, H. Rosenbauer, Synoptic observations of coronal transients and their interplanetary consequences, Adv. Space Res., 4(7), 311–321, 1984.

2.152 Misconi, N.Y., E.T. Rusk, Brightness contribution of zodiacal dust along the line of sight in and out of the ecliptic plane and in the F-corona, Planet. Space Sci., 35, 1571–1574, 1987.

2.153 Mizutani, K., T. Maihara, N. Hiromoto, H. Takami, Near-infrared observation of the circumsolar dust emission during the 1983 solar eclipse, Nature, 312, 134–136, 1984

2.154 Mouschovias, T.C., A.I. Poland, Expansion and broadening of coronal loop transients: a theoretical explanation, Ap. J., 220, 675–682, 1978.

2.155 Muhleman, D.O., J.D. Anderson, Solar wind electron densities from Viking dual-frequency radio measurements, Ap. J., 247, 1093–1101, 1981.

2.156 Muhleman, D.O., R.D. Ekers, E.B. Fomalont, Radio interferometric test of the general relativistic light bending near the sun, Phys. Rev. Lett., 24, 1377–1380, 1970.

2.157 Muhlemen, D.O., P.B. Esposito, J.D. Anderson, The electron density profile of the outer corona and the interplanetary medium from Mariner 6 and Mariner 7 time-delay measurements, Ap. J., **211**, 943–957, 1977.

2.158 Munro, R.H., J.T. Gosling, E. Hildner, R.M. MacQueen, A.I. Poland, C.L. Ross, The association of coronal mass ejection transients with other forms of solar activity, Solar Phys., **61**, 201–215, 1979.

2.159 Munro, R.H., B.V. Jackson, Physical properties of a polar coronal hole from 2 to $5R_\odot$, Ap. J., **213**, 874–886, 1977.

2.160 Munro, R.H., D.G. Sime, White-light coronal transients observed from Skylab May 1973 to February 1974: a classification by apparent morphology, Solar Phys., **97**, 191–201, 1985.

2.161 Nash, A.G., N.R. Sheeley, Jr., Y.-M. Wang, Mechanisms for the rigid rotation of coronal holes, Solar Phys., **117**, 359–389, 1988.

2.162 Neugebauer, M., R.W. Davies (eds.), *A Close-up of the Sun*, JPL Publ. **78-70**, 1978.

2.163 Ness, N.F., J.M. Wilcox, Solar origin of the interplanetary magnetic field, Phys. Rev. Lett., **13**, 461–464, 1964.

2.164 Newkirk, G., Jr., Structure of the solar corona, Ann. Rev. Astr. Astrophys., **5**, 213–266, 1967.

2.165 Ohlson, J.E., G.S. Levy, C.T. Stelzried, A tracking polarimeter for measuring solar and ionospheric Faraday rotation of signals from deep space probes, IEEE Trans. Instru. Meas., **23**, 167–177, 1974.

2.166 Orrall, F.Q., G.J. Rottman, J.A. Klimchuk, Outflow from the sun's polar corona, Ap. J., **266**, L65–L68, 1983.

2.167 Osherovich, V.A., E.B. Gliner, I. Tzur, Theoretical model of the solar corona during sunspot minimum. II. Dynamic approximation, Ap. J., **288**, 396–400, 1985.

2.168 Osherovich, V.A., I. Tzur, E.B. Gliner, Theoretical model of the solar corona during sunspot minimum. I. Quasi-static approximation, Ap. J., **284**, 412–421, 1984.

2.169 Parker, E.N, Dynamics of the interplanetary gas and magnetic fields, Ap. J., **128**, 664–676, 1958.

2.170 Parker, G.D., A comparison of coronal rotation and interplanetary recurrence during 1964–1976, J. Geophys. Res., **92**, 7235–7240, 1987.

2.171 Parker, G.D., Radial variation of differential rotation in the solar electron corona, Solar Phys., **108**, 77–87, 1987.

2.172 Parker, G.D., R.T. Hansen, S.F. Hansen, Coronal rotation during solar cycle 20, Solar Phys., **80**, 185–198, 1982.

2.173 Pätzold, M., M.K. Bird, H. Volland, G.S. Levy, B.L. Seidel, C.T. Stelzried, The mean coronal magnetic field determined from Helios Faraday rotation measurements, Solar Phys., **109**, 91–105, 1987.

2.174 Peterson, A.W., Thermal radiation from interplanetary dust, Ap. J., **138**, 1218–1230, 1963.

2.175 Phinney, R.A., D.L. Anderson, On the radio occultation method for studying planetary atmospheres, J. Geophys. Res., **73**, 1819–1827, 1968.

2.176 Pneuman, G.W., Ejection of magnetic fields from the sun: acceleration of a solar wind containing diamagnetic plasmoids, Ap. J., **265**, 468–482, 1983.

2.177 Pneuman, G.W., The 'melon seed' mechanism and coronal transients, Solar Phys., **94**, 387–411, 1984.

2.178 Pneuman, G.W., R.A. Kopp, Gas-magnetic field interactions in the solar corona, Solar Phys., **18**, 258–270, 1971.

2.179 Pneuman, G.W., F.Q. Orrall, Structure, dynamics and heating of the solar atmosphere, in *Physics of the Sun*, Vol. II, ed. by P.A. Sturrick et al., 71–134, D. Reidel Publ. Co., Dordrecht, 1986.

2.180 Priest, E.R., *Solar Magnetohydrodynamics*, D. Reidel Publ. Co., Dordrecht, 1982.

2.181 Radowski, H.R., P.F. Fougere, E.J. Zawalick, A comparison of power spectral estimates and applications of the maximum entropy method, J. Geophys. Res., **80**, 619–625, 1975.

2.182 Rao, U.R., T.K. Alex, V.S. Iyengar, K. Kasturirangan, T.M.K. Marar, R.S. Mathur, D.P. Sharma, IR observations of the solar corona – a ring around the sun?, Nature, **289**, 779–780, 1981.

2.183 Readhead, A.C.S., M.C. Kemp, A. Hewish, The spectrum of small-scale density fluctuations in the solar wind, Mon. Not. R. Astr. Soc., **185**, 207–225, 1978.

2.184 Rickett, B.J., Disturbances in the solar wind from IPS measurements in August 1972, Solar Phys., **43**, 237–247, 1975.

2.185 Rickett, B.J., W.A. Coles, Solar cycle evolution of the solar wind in three dimensions, in *Solar wind 5*, ed. by M. Neugebauer, NASA **CP-2280**, 315–321, 1983.

2.186   Riesebieter, W., F.M. Neubauer, A comparison of 3D solar wind predictions with observations, in *Pleins Feux sur la Physique Solaire*, Proc. Euro. Solar Phys., Toulouse, France 331–340, CNRS, Paris, 1978.

2.187   Roberts, D.A., Interplanetary observational constraints on Alfvén wave acceleration of the solar wind, J. Geophys. Res., **94**, 6899–6905, 1989.

2.188   Rosenbauer, H., R. Schwenn, E. Marsch, B. Meyer, H. Miggenrieder, M.D. Montgomery, K.H. Mühlhäuser, W. Pillip, W. Voges, S.M. Zink, A survey on initial results of the Helios plasma experiment, J. Geophys., **42**, 561–580, 1977.

2.189   Rosner, R., B.C. Low, T.E. Holzer, Physical processes in the solar corona, in *Physics of the Sun*, Vol. II, ed. by P.A. Sturrock et al., 135–180, D. Reidel Publ. Co., Dordrecht, 1986.

2.190   Rottman, G.J., F.Q. Orrall, Observational evidence for solar wind acceleration at the base of coronal holes, in *Solar Wind 5*, ed. by M. Neugebauer, NASA **CP-2280**, 199–210, 1983.

2.191   Rottman, G.J., F.Q. Orrall, EUV observations of solar coronal outflows, in *Cool Stars, Stellar Systems, and the Sun*, ed. by M. Zeilik, D.M. Gibson, 439–441, Springer-Verlag, Berlin, 1986.

2.192   Rubtsov, S.N., O.I. Yakovlev, A.I. Efimov, Plasma density, inhomogeneity, and kinetic energy of the solar wind: radio occultation data from Venera 15 and 16, Cosmic Res., **25**, 475–480, 1987.

2.193   Rušin, V., M. Rybanský, Structure of the solar corona during the solar eclipse of 1980 February 16, Bull. Astron. Inst. Czech., **34**, 257–264, 1983.

2.194   Rušin, V., M. Rybanský, Absolute photometry of the corona during the solar eclipse of 1980 February 16, Bull. Astro. Inst. Czech., **34**, 265–276, 1983.

2.195   Rušin, V., M. Rybanský, The white-light and emission corona on July 31, 1981, Bull. Astron. Inst. Czech., **35**, 347–353, 1984.

2.196   Rušin, V., M. Rybanský, Variations of the total brightness of the white-light corona with the phase of the solar cycle, Bull. Astron. Inst. Czech., **36**, 77–81, 1985.

2.197   Rušin, V., M. Rybanský, The structure of the white-light corona on June 11, 1983, Bull. Astron. Inst. Czech., **36**, 281–283, 1985.

2.198   Rybanský, M., V. Rušin, The total brightness of the corona during the solar eclipse of February 16, 1980, Bull. Astron. Inst. Czech., **36**, 73–77, 1985.

2.199   Sahal-Bréchot, S., M. Malinovsky, V. Bommier, The polarization of the O VI 1032 Å line as a probe for measuring the coronal vector magnetic field via the Hanle effect, Astron. Astrophys., **168**, 284–300, 1986.

2.200   Saito, K., A.I. Poland, R.H. Munro, A study of the background corona near solar minimum, Solar Phys., **55**, 121–134, 1977.

2.201   Schatten, K.H., Coronal magnetic field models, Rev. Geophys. Space Phys., **13**, 589–592, 1975.

2.202   Schatten, K.H., J.M. Wilcox, N.F. Ness, A model of interplanetary and coronal magnetic fields, Solar Phys., **6**, 442–455, 1969.

2.203   Schmidt, R., V. Domingo, S.D. Shawhan, D. Bohlin, Cluster and SOHO: a joint endeavor by ESA and NASA to address problems in solar, heliospheric and space plasma physics, EOS, **69**, 177–190, 1988.

2.204   Schulz, M., E.N. Frazier, D.J. Boucher, Jr., Coronal magnetic field model with non-spherical source surface, Solar Phys., **60**, 83–104, 1978.

2.205   Schwenn, R., The "average" solar wind in the inner heliosphere: structure and slow variations, in *Solar Wind Five*, ed. by M. Neugebauer, NASA **CP-2280**, 489–507, 1983.

2.206   Schwenn, R., Direct correlations between coronal transients and interplanetary disturbances, Space Sci. Rev., **34**, 85–99, 1983.

2.207   Schwenn, R., Relationship of coronal transients to interplanetary shocks: 3D aspects, Space Sci. Rev., **44**, 139–168, 1986.

2.208   Schwenn, R., Large-scale structure of the interplanetary medium, in *Physics of the Inner Heliosphere* (this volume)

2.209   Scott, S.L., B.J. Rickett, J.W. Armstrong, The velocity and the density spectrum of the solar wind from simultaneous three-frequency IPS observations, Astron. Astrophys., **123**, 191–206, 1983.

2.210   Scott, S.L., W.A. Coles, G. Bourgois, Solar wind observations near the sun using interplanetary scintillation, Astron. Astrophys., **123**, 207–215, 1983.

2.211   Sheeley, N.R., Jr., R.A. Howard, M.J. Koomen, D.J. Michels, Associations between coronal mass ejections and soft X-ray events, Ap. J., **272**, 349–354, 1983.

2.212   Sheeley, N.R., Jr., R.A. Howard, M.J. Koomen, D.J. Michels, R. Schwenn, K.H. Mühlhäuser, H. Rosenbauer, Coronal mass ejections and interplanetary shocks, J. Geophys. Res., **90**, 163–175, 1985.

2.213 Sheeley, N.R., Jr., R.A. Howard, D.J. Michels, M.J. Koomen, Solar observations with a new earth-orbiting coronagraph, in *Solar and Interplanetary Dynamics*, ed. by M. Dryer, E. Tandberg-Hanssen, 55–60, D. Reidel Publ. Co., Dordrecht, 1980.

2.214 Sheeley, N.R., Jr., D.J. Michels, R.A. Howard, M.J. Koomen, Initial observations with the Solwind coronagraph, Ap. J., **237**, L99–L101, 1980.

2.215 Sheeley, N.R., Jr., D.J. Michels, R.A. Howard, M.J. Koomen, The great solar eruption of May 24, 1979, EOS, **62**, 153, 1981.

2.216 Sheeley, N.R., Jr., A.G. Nash, Y.-M. Wang, The origin of rigidly rotating magnetic field patterns on the sun,, Ap. J., **319**, 481–502, 1987.

2.217 Sheeley, N.R., Jr., R.T. Stewart, R.D. Robinson, R.A. Howard, M.J. Koomen, D.J. Michels, Associations between coronal mass ejections and metric type II bursts, Ap. J., **279**, 839–847, 1984.

2.218 Sime, D.G., Interplanetary scintillation observations of the solar wind close to the sun and out of the ecliptic, in *Solar Wind 5*, ed. by M. Neugebauer, NASA **CP-2280**, 453–467, 1983.

2.219 Sime, D.G., The corona and interplanetary medium during the solar cycle, in *Future Missions in Solar, Heliospheric and Space Plasma Physics*, ESA **SP-235**, 23–36, 1985.

2.220 Sime, D.G., The structure and rotation of the solar corona: implications for the heliosphere, in *The Sun and Heliosphere in Three Dimensions*, ed. by R.G. Marsden, 45–51, D. Reidel Publ. Co., Dordrecht, 1986.

2.221 Sime, D.G., R.R. Fisher, M.K. McCabe, D.L. Mickey, The corona near the time of the 1983 June 11 total solar eclipse, Ap. J., **278**, L123–L126, 1984.

2.222 Sime, D.G., R.M. MacQueen, A.J. Hundhausen, Density distribution in looplike coronal transients: a comparison of observations and a theoretical model, J. Geophys. Res., **89**, 2113–2121, 1984.

2.223 Soboleva, N.S., G.M. Timofeeva, The Faraday effect in the solar supercorona during its 1977–1982 radio occultations of the Crab Nebula, Sov. Astron. Lett., **9**(4), 216–219, 1983.

2.224 Sofue, Y., K. Kawabata, F. Takahashi, N. Kawajiri, Coronal Faraday rotation of the Crab Nebula, 1971–1975, Solar Phys., **50**, 465–480, 1976.

2.225 St. Cyr, O.C., A.J. Hundhausen, On the interpretation of "Halo" coronal mass ejections, in *Proc. 6th International Solar Wind Conf.*, ed. by V.J. Pizzo et al., 235–241, NCAR/TN-306, HAO/NCAR, Boulder, CO/USA, 1988.

2.226 Steinolfson, R.S., Modeling of transient disturbances in coronal-streamer configurations, in *Solar Wind 5*, ed. by M. Neugebauer, NASA **CP-2280**, 667–673, 1983.

2.227 Steinolfson, R.S., Type II radio emission in coronal transients, Solar Phys., **94**, 193–202, 1984.

2.228 Steinolfson, R.S., S.T. Wu, M. Dryer, E. Tandberg-Hanssen, Magnetohydrodynamic models of coronal transients in the meridional plane. I. The effect of the magnetic field, Ap. J., **225**, 259–274, 1978.

2.229 Stelzried, C.T., G.S. Levy, T. Sato, W.V.T. Rusch, J.E. Ohlson, K.H. Schatten, J.M. Wilcox, The quasi-stationary coronal magnetic field and electron density as determined from a Faraday rotation experiment, Solar Phys., **14**, 440–456, 1970.

2.230 Stewart, R.T., G.A. Dulk, K.V. Sheridan, L.L. House, W.J. Wagner, C. Sawyer, R. Illing, Visible light observations of a dense plasmoid associated with a moving type IV solar radio burst, Astron. Astrophys., **116**, 217–223, 1982.

2.231 Suess, S.T., A.K. Richter, C.R. Winge, S.F. Nerney, Solar polar coronal hole – a mathematical simulation, Ap. J., **217**, 296–305, 1977.

2.232 Süss, H., Plasma-Fernerkundung durch Inversion von Messungen des Elektroneninhaltes bei den Helios-Raumsonden (Okkultationsexperiment), Ph. D. Thesis (in German), University of Bochum, 1982.

2.233 Svalgaard, L., T.L. Duvall, Jr., P.H. Scherrer, The strength of the Sun's polar fields, Solar Phys., **58**, 225–240, 1978.

2.234 Svalgaard, L., J.M. Wilcox, Long term evolution of solar sector structure, Solar Phys., **41**, 461–475, 1975.

2.235 Svalgaard, L., J.M. Wilcox, A view of solar magnetic fields, the solar corona, and the solar wind in three dimensions, Ann. Rev. Astron. Astrophys., **16**, 429–443, 1978.

2.236 Tappin, S.J., Numerical modelling of scintillation variations from interplanetary disturbances, Planet. Space Sci., **35**, 271–283, 1987.

2.237 Thieme, K.M., E. Marsch, R. Schwenn, Relationship between structures in the solar wind and their source regions in the corona, in *Proc. 6th International Solar Wind Conf.*, ed. by V.J. Pizzo et al., 317–321, NCAR/TC-306, HAO/NCAR, Boulder, CO/USA, 1988.

2.238 Tousey, R., The solar corona, Space Res., **13**, 713–730, 1973.

2.239 Tyler, G.L., J.P. Brenkle, T.A. Komarek, A.I. Zygielbaum, The Viking solar corona experiment, J. Geophys. Res., **82**, 4335–4340, 1977.
2.240 Tyler, G.L., J.F. Vesecky, M.A. Plume, H.T. Howard, A. Barnes, Radio wave scattering observations of the solar corona: first-order measurements of expansion velocity and turbulence spectrum using Viking and Mariner 10 spacecraft, Ap. J., **249**, 318–332, 1981.
2.241 van de Hulst, H.C., The electron density of the solar corona, Bull. Astron. Inst. Neth., **11**, 135–150, 1950.
2.242 Vasilyev, M.B., A.S. Vyshlov, M.A. Kolosov, A.P. Mesterton, N.A. Savich, V.A. Samoval, L.N. Samoynaev, A.Z. Sidorenko, Two-frequency radio occultation measurements with Venera-9 and Venera-10 orbiters, Acta Astron., **7**, 335–340, 1980.
2.243 Vlasov, V.I., Radio imagery of the turbulent interplanetary plasma, Sov. Astron., **23**(1), 55–59, 1979.
2.244 Volland, H., M.K. Bird, G.S. Levy, C.T. Stelzried, B.L. Seidel, Helios-1 Faraday rotation experiment: results and interpretations of the solar occultations in 1975, J. Geophys., **42**, 659–672, 1977.
2.245 Wagner, W.J., Coronal mass ejections, Ann. Rev. Astron. Astrophys., **22**, 267–289, 1984.
2.246 Wagner, W.J., E. Hildner, L.L. House, C. Sawyer, K.V. Sheridan, G.A. Dulk, Radio and visible light observations of matter ejected from the sun, Ap. J., **244**, L123–L126, 1981.
2.247 Wang, Y.-M., A.G. Nash, N.R. Sheeley, Jr., Magnetic flux transport on the sun, Science, **245**, 712–718, 1989.
2.248 Wang, Y.-M., N.R. Sheeley, Jr., A.G. Nash, L.R. Shampine, The quasi-rigid rotation of coronal magnetic fields, Ap. J., **327**, 427–450, 1988.
2.249 Ward, B.D., Interplanetary scintillation and flare-produced disturbances, Proc. Astron. Soc. Aus., **2**(6), 378–379, 1975.
2.250 Watanabe, T., T. Kakinuma, Radio scintillation observations of interplanetary disturbances, Adv. Space Res., **4**(7), 331–341, 1984.
2.251 Webb, D.F., A.J. Hundhausen, Activity associated with the solar origin of coronal mass ejections, Solar Phys., **108**, 383–401, 1987.
2.252 Weinberg, J.L., Zodiacal light and interplanetary dust, in *Properties and Interactions of Interplanetary Dust*, ed. by R.H. Giese, P. Lamy, 1–6, D. Reidel Publ. Co., Dordrecht, 1985.
2.253 Weisberg, J.M., J.M. Rankin, R.R. Payne, C.C. Counselman III, Further changes in the distribution of density and radio scattering in the solar corona in 1973, Ap. J., **209**, 252–258, 1976.
2.254 Wilcox, J.M., J.T. Hoeksema, P.H. Scherrer, Origin of the warped heliospheric current sheet, Science, **209**, 603–605, 1980.
2.255 Wilcox, J.M., A.J. Hundhausen, Comparison of heliospheric current sheet structure obtained from potential magnetic field computation and from observed polarization coronal brightness, J. Geophys. Res., **88**, 8095–8096, 1983.
2.256 Withbroe, G.L., Solar wind and coronal structure, ESA Journal, **7**, 341–356, 1983.
2.257 Withbroe, G.L., Origins of the solar wind in the corona, in *The Sun and the Heliosphere in Three Dimensions*, ed. by R.G. Marsden, 19–32, D. Reidel Publ. Co., Dordrecht, 1986.
2.258 Withbroe, G.L., The temperature structure, mass, and energy flow in the corona and inner solar wind, Ap. J., **325**, 442–467, 1988.
2.259 Withbroe, G.L., Acceleration of the solar wind as inferred from observations, in *Proc. 6th International Solar Wind Conf.*, ed. by V.J. Pizzo et al., 23–48, NCAR/TN-306, HAO/NCAR, Boulder, CO/USA, 1988.
2.260 Withbroe, G.L., J.L. Kohl, H. Weiser, Analysis of coronal H I Lyman-alpha measurements in a polar region of the sun observed in 1979, Ap. J., **307**, 381–388, 1986.
2.261 Withbroe, G.L., J.L. Kohl, H. Weiser, R.H. Munro, Probing the solar wind acceleration region using spectroscopic techniques, Space Sci. Rev., **33**, 17–52, 1982.
2.262 Withbroe, G.L., J.L. Kohl, H. Weiser, G. Noci, R.H. Munro, Analysis of coronal H I Lyman alpha measurements from a rocket flight on 1979 April 13, Ap. J., **254**, 361–370, 1982.
2.263 Wolfson, R., A coronal magnetic field model with volume and sheet currents, Ap. J., **288**, 769–778, 1985.
2.264 Woo, R., Multifrequency techniques for studying interplanetary scintillations, Ap. J., **201**, 238–248, 1975.
2.265 Woo, R., Measurements of the solar wind using spacecraft radio scattering observations, in *Study of Travelling Interplanetary Phenomena/1977*, ed. by M.A. Shea et al., 81–100, Reidel Publ. Co., Dordrecht, 1977.

2.266   Woo, R., Radial dependence of solar wind properties deduced from Helios 1/2 and Pioneer 10/11 radio scattering observations, Ap. J., **219**, 727–739, 1978. [Erratum, Ap. J., **223**, 704–705, 1978]

2.267   Woo, R., A synoptic study of Doppler scintillation transients in the solar wind, J. Geophys. Res., **93**, 3919–3926, 1988.

2.268   Woo, R., J.W. Armstrong, Spacecraft radio scattering observations of the power spectrum of electron density fluctuations in the solar wind, J. Geophys. Res., **84**, 7288–7296, 1979.

2.269   Woo, R., J.W. Armstrong, Measurements of a solar flare-generated shock wave at $13.1 R_\odot$, Nature, **292**, 608–610, 1981.

2.270   Woo, R., J.W. Armstrong, N.R. Sheeley, Jr., R.A. Howard, M.J. Koomen, D.J. Michels, R. Schwenn, Doppler scintillation observations of interplanetary shocks within 0.3 AU, J. Geophys. Res., **90**, 154–162, 1985.

2.271   Woo, R., J.W. Armstrong, N.R. Sheeley, Jr., R.A. Howard, D.J. Michels, M.J. Koomen, Simultaneous radio scattering and white light observations of a coronal transient, Nature, **300**, 157–159, 1982.

2.272   Wu, S.T., S.M. Han, M. Dryer, Two-dimensional, time-dependent MHD description of interplanetary disturbances: simulation of high speed solar wind interactions, Planet. Space Sci., **27**, 255–264, 1979.

2.273   Yakovlev, O.I., *Radio Wave Propagation in Space*, R.E. Krieger Publ. Co., Melbourne, FL/USA, 1990. (in press)

2.274   Yeh, T., A magnetohydrodynamic theory of coronal loop transients, Solar Phys., **78**, 287–316, 1982.

2.275   Zirker, J.B., *Coronal Holes and High Speed Wind Streams*, Colorado Associated University Press, Boulder, 1977.

2.276   Zirker, J.B., Progress in coronal physics, Solar Phys., **100**, 281–287, 1985.

3. Large-Scale Structure of the Interplanetary Medium

Rainer Schwenn

3.1 Introduction

3.1.1 The Solar Wind at 1 AU, Status 1974

From eclipse observations it was well known that the solar corona is highly structured and changes its shape enormously during the solar activity cycle. Hence, it was no great surprise when both these properties (spatial structure and temporal variability) were found to be reproduced in the corona's offspring, i.e. the solar wind. Even the first continuous observations of the interplanetary plasma performed on board the American Venus probe *Mariner 2* in 1962 showed a "series of long-lived, high velocity streams separated by slower moving plasma", as Neugebauer and Snyder [3.148] phrased it. (For an extensive and very informative review of the early years of solar wind research the interested reader is referred to [3.48]). There is a basic agreement between the slow solar wind parameters (velocity $v_p \approx 300 \, \text{km s}^{-1}$, proton density $n_p \approx 9 \, \text{cm}^{-3}$, proton temperature $T_p \approx 4 \times 10^4 \, \text{K}$ [3.95]) and current coronal expansion models based on Parker's theory [3.159]. This led many workers in the field to associate the slow solar wind with a "quiet state" and to regard any fast flow including that in quasistationary high-speed streams ($v \approx 600 \, \text{km s}^{-1}$, $n_p \approx 3 \, \text{cm}^{-3}$, $T_p \approx 10^5 \, \text{K}$ [3.56]) more or less as disturbances of this quiet state.

It was not until the *Skylab* era in 1973/74 when coronal holes were discovered to be the sources of the long-lived fast solar wind streams [3.115, 151, 153]. Coronal holes were found to be located over the inactive parts of the sun where "open" magnetic field lines prevail, e.g. the sun's polar caps around solar activity minima [3.229]. In contrast, the more active regions in terms of sunspot appearance and the occurrence rate of solar transients such as flares and eruptive prominences are most often associated with "closed" magnetic field structures, such as bipolar loop systems. The steady-state high-speed streams as well as their sources, the coronal holes, are representative of the inactive sun, i.e. the "quiet" sun. Consistently, the only state of the solar wind that may deserve the label "quiet state" at all was found to occur in high-speed streams rather than in the slow solar wind [3.56, 8, 59]. The slow solar wind, which apparently originates from above the more active regions on the sun, shows a much higher variability of all its properties, thus rendering its definition as a "state" difficult and probably useless.

Physics and Chemistry in Space - Space and Solar Physics, Vol. 20
Physics of the Inner Heliosphere I   Editors: R. Schwenn · E. Marsch
© Springer-Verlag Berlin Heidelberg 1990

There have been quite a number of attempts to model the solar wind expansion theoretically (for a detailed review on the work done before 1972 see the book by Hundhausen [3.95]). For all these early studies only observational data obtained around 1 AU and beyond were available as a baseline, and there was almost no information about the long stretch from the corona out to 1 AU. Thus, the modelers were left with plenty of free parameters and were able to produce more or less successfully the "right" solar wind at 1 AU, at least in the case of the slow wind. The essential differences in their assumptions showed up mainly in the radial profiles they predicted. Any measurement of these radial profiles would naturally constrain the models considerably. None of the available models has thus far been able to predict that type of "quiet" solar wind so commonly observed in high-speed streams. In particular, for a reasonable choice of boundary conditions matching the observations for coronal holes, these models fail badly in achieving the high values of the flow velocity. The *Helios* mission, and particularly the plasma instruments, were designed to provide new data over a wide range of the inner heliosphere in order to diminish this embarrassing lack of understanding.

3.1.2 The Plasma Experiment on the Helios Probes

Both *Helios* probes carried a nearly identical set of instruments dedicated to the *in situ* analysis of the solar wind plasma. Emphasis was placed on the velocity distribution functions of the different particle species, from which all important hydrodynamic quantities of the solar wind can then be deduced. Three instruments measure the positively charged particles (protons and heavier ions with energy per charge values between 0.155 and 15.32 kV), two of them allowing for an angular resolution with respect to both angles of incidence. The fourth instrument analyzes electrons at energies ranging from 0.5 to 1660 eV with a one-dimensional angular resolution. The time interval between succeeding measurements varied between 40.5 s and 43 min, depending on the telemetry bit rate. For further details on the instruments and the data evaluation the reader is referred to other papers [3.182, 200, 184, 186, 206, 171, 162].

The plasma instruments on *Helios 1* worked flawlessly from their first switch-on (12 December 1974) until the end of the mission on 2 March 1986. The increasingly severe power shortage (due to degradation of the solar cells by UV light and particle bombardment) forced a temporary switch-off of the plasma instruments and later the whole space probe during the aphelion phases, starting in 1981. The twin spacecraft *Helios 2* ceased working in March 1980, after four years of excellent operation.

Owing to the sophisticated data handling system on board the probes, a nearly uninterrupted stream of data could be obtained throughout most of the mission. Times with no ground station coverage could be bridged using the on-board memory, albeit at reduced data rates. This explains why the *Helios* plasma data set is so unique, not only in terms of its length (more than a full 11-year solar cycle), but also in terms of completeness. The plasma particle measurements were ideally complemented by instruments for diagnostics of the magnetic field [3.144, 193], its fluctuations [3.45], and plasma wave electric fields [3.77, 107, 234]. An open-minded exchange of data and ideas among the different experimenters stimulated many scientific collaborations leading to a series of multi-author papers, e.g. [3.31, 32].

3.1.3 The Basic Stream Structure

Solar sunspot cycle Nr. 20 was approaching its end at the time of the *Helios 1* launch in December 1974. The phase of minimum activity was finally reached in summer 1976, a few months after the launch of *Helios 2*. Starting in 1973, and then throughout 1974 and 1975, the occurrence of very large amplitude, broad, quasirepetitive streams of high-speed solar wind had been noted [3.6, 68, 58]. In fact, this pattern resulted from two huge, corotating, high-speed streams emanating from the corona at fixed heliospheric longitudes approximately on opposite sides of the sun. Simultaneous optical observations of the sun using *Skylab* instruments [3.115, 153, 154, 242, 15] and ground-based techniques [3.83, 84, 208] supported the, at that time rather new, hypothesis that coronal holes are the solar sources of the quasisteady high-speed streams observed in the solar wind ([3.115, 151, 154], see also the reviews in [3.243]). Such streams, one of the main subjects in this article, are generally referred to as "high-speed streams" in the following.

On approaching the sun to as close as 0.30 AU, *Helios 1* was able to confirm the validity of the coronal hole–high-speed stream association to much greater detail. The stream structure as seen from *Helios 1* is shown in Fig. 3.1 (from [3.184]). It presents the solar wind parameters proton bulk speed v_p, number density n_p and radial temperature T_p for about four solar rotations, including the first perihelion passage of *Helios 1* on 15 March 1975 (day 74). (A very similar plot showing the data from *Helios 2* for the first four solar rotations after its launch in January 1976 is given in [3.136].) The well-developed two-stream pattern mentioned above is immediately apparent. In the first two panels (covering the solar distance range between $R = 1$ and 0.84 AU) we find all the familiar signatures of this type of structure observed previously at ca. 1 AU (see, e.g., [3.95]):

- Broad high-speed streams with typical values of $v_p \approx 700 \text{ km s}^{-1}$, $n_p \approx 3 \text{ cm}^{-3}$, $T_p \approx 2 \times 10^5 \text{ K}$.
- Regions of slow solar wind with $v_p \approx 400 \text{ km s}^{-1}$, $n_p \approx 10 \text{ cm}^{-3}$ and $T_p \approx 4 \times 10^4 \text{ K}$, highly variable.
- "Leading edges" of high-speed streams extending over 15 to 30° in heliographic longitude.

Some further characteristic features not readily visible in Fig. 3.1 can be found in more detailed or other complementary data:

- The interaction of slow and fast solar wind at the leading edges of high-speed streams causes compression to high plasma densities and deflections

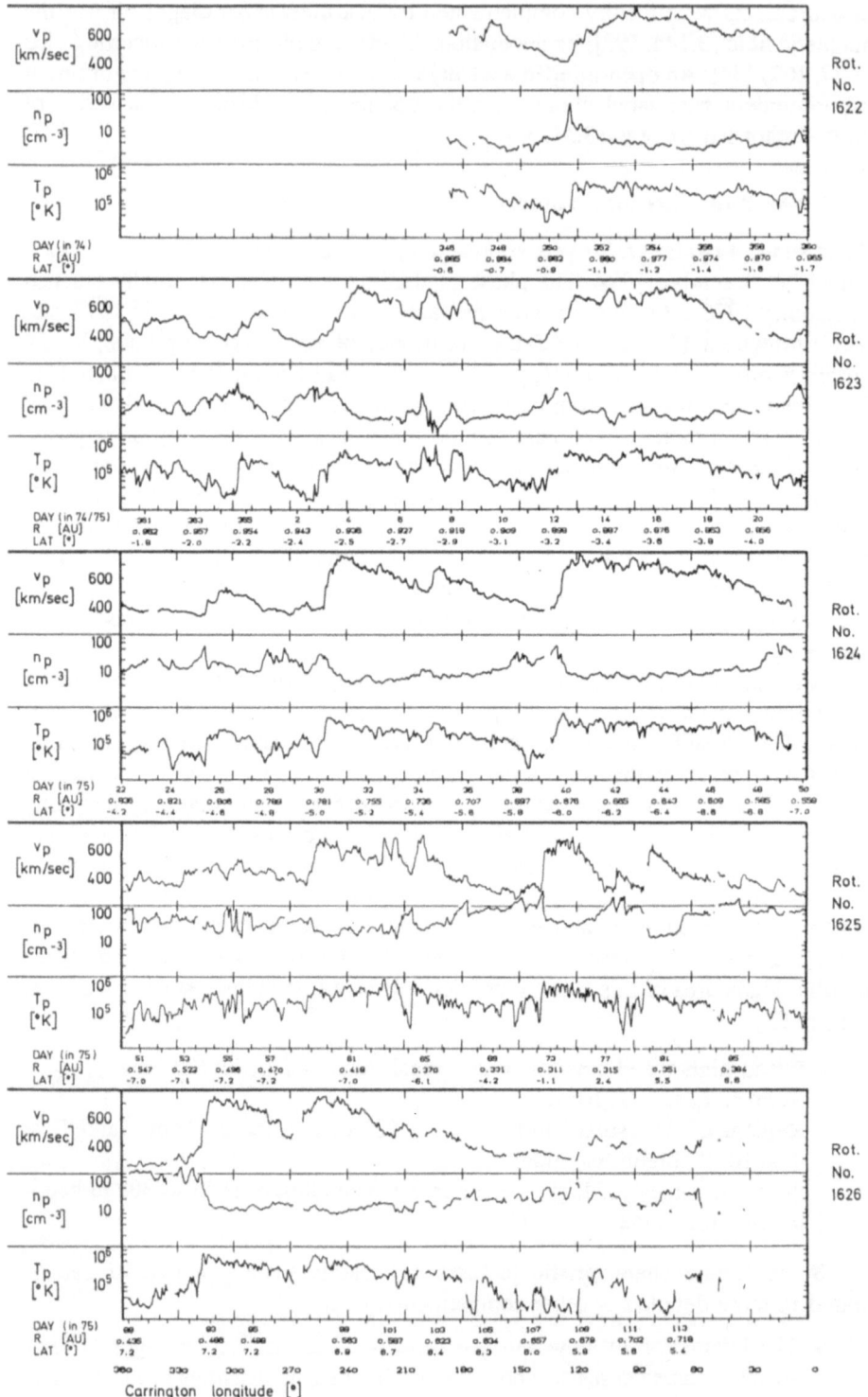

Fig. 3.1. Caption see opposite page

of the flow on both sides: westward in the slow plasma (i.e. in the sense of corotation with the sun); eastward in the fast plasma [3.216].

- Many leading edges contain a tangential discontinuity (TD) which had been termed the "stream interface" [3.26, 72].
- High-speed streams are always associated with a magnetic sector of unique polarity [3.237]. Crossing of sector boundaries often occurs slightly ahead of the leading edges of high-speed streams.

It is interesting to note the changes of these familiar signatures with decreasing distance from the sun. The most striking feature in Fig. 3.1 had already emerged from the daily *Helios* quick-look data ([3.201]; see also [3.183]). It is the dramatic steepening of the fast streams' leading edges. In fact, as stated in [3.184], "all parameter profiles tend to change toward mesa-like structures (or inverse-mesa structures respectively) with sharp leading and relatively sharp trailing edges". This result was quite unexpected since the early model calculations had to assume smooth and broad input structures near the sun in order to yield realistic stream profiles at 1 AU (see, e.g., [3.96]). The conventional view that kinematic steepening would lead to the formation of corotating shocks or shock pairs well within 1 AU turned out to contrast sharply with the *Helios* observations. We will return to this discussion later on.

Another conspicuous feature in Fig. 3.1 is the changing shape of the second stream (between Carrington longitude 120° and 90°) during the last two solar rotations. It apparently split up in Rot. 1625 and nearly disappeared in Rot. 1626. This happened right before *Helios 1* entered its telemetry blackout behind the sun, as seen from earth. The earth-orbiting satellites *IMP 7/8*, however, kept observing this particular stream without much change, both during these two rotations and even beyond [3.208]! For illustration, we show in Fig. 3.2 (adopted from [3.203]) a comparison of solar wind speeds measured from *Helios 1* and *IMP*, combined with K-corona intensity contours denoting the outlines of the coronal holes for the same solar rotation no. 1625. The plasma data have been shifted in longitude to their projected source positions at 0.1 AU by assuming that the individual plasma volume elements have traveled strictly radially from there to the spacecraft location at a constant speed equal to that measured at the *Helios* position. (This simple mapping technique, first used in [3.221], gives rise to occasional ambiguities which can be disregarded here.) *Helios 1*, which attained perihelion during this solar rotation moved quickly from a heliographic latitude of −7° to + 7°, while *IMP* remained at −7°. Obviously, *Helios 1* crossed the northern edge of the stream emanating from the southern coronal hole (at 120° longitude), while *IMP* did not. The coincidence between stream and coronal hole boundaries is not completely satisfactory in some places, probably due to the

---

Fig. 3.1. One-hour averages of solar wind proton bulk speed v_p, number density n_p and radial temperature T_p versus Carrington longitude, as measured by *Helios 1* between 12 December 1974 and 25 April 1975. The time of measurement, the radial distance, R, and heliographic latitude of the spacecraft are also indicated along the abscissa. From [3.184]

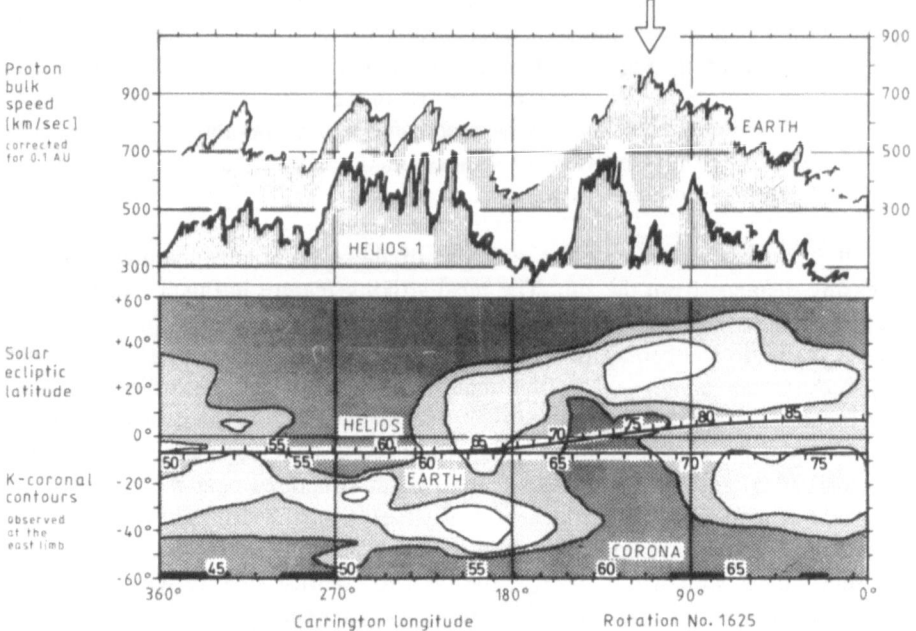

Fig. 3.2. Comparison of solar wind bulk speed data measured from *Helios 1* and *IMP* with K-corona contours for Carrington rotation 1625 in early 1975. The plasma data have been shifted in longitude to positions where the individual plasma volumes should have passed a heliocentric distance of 0.1 AU, a constant and radial flow at the measured speed being assumed. The tick marks along the source tracks in the lower panels indicate the start time at 0.1 AU. For identification of coronal holes (*dark shaded areas*) a K-corona contour of 2×10^{-8}PB (polarization times brightness at the sun's center) was chosen. The tick marks at the bottom indicate the date of corona observation at the sun's east limb. Data gaps in the corona measurements indicated as black bars at the bottom were covered by linear interpolation between adjacent intensity values. From [3.203]

13-day delay between the respective measurements. Nevertheless, as pointed out in [3.31], the general agreement between the boundaries of streams and coronal holes, as well as the sharpness of these boundaries, is remarkable.

Thanks to the long duration of the *Helios* mission, similar observations could be repeated several times, thereby allowing us to substantiate and generalize the results by applying statistical analyses. Details of these various aspects are discussed in the following sections. Furthermore, I will occasionally sketch scenarios that represent the present status of our understanding (or lack of understanding) at a point in time which is now one full solar cycle after the *Helios* launches.

3.2 Longitudinal Stream Boundaries

The "shape" of high-speed streams with respect to heliographic longitude can be easily monitored since the sun's rotation causes the stream structure pattern to sweep rapidly across any *in situ* observing instrumentation. This is why the

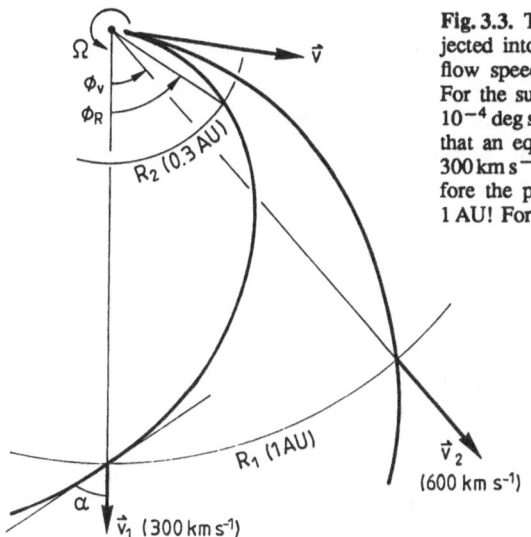

Fig. 3.3. The shape of solar wind flux tubes (projected into the sun's equatorial plane) for different flow speeds in the inner heliosphere out to 1 AU. For the sun's rotation speed Ω a value of $1.642 \times 10^{-4} \deg s^{-1}$ (sidereal value) was used. Note, e.g., that an equatorial source of slow solar wind ($v_1 = 300 \, km \, s^{-1}$) has rotated almost to the west limb before the plasma emitted from there finally reaches 1 AU! For further explanation see text

longitudinal stream structure has been one of the best-observed features ever since the start of *in situ* measurements. It is precisely this rotation, of course, that causes "flux tubes" containing plasma with different flow speeds eventually to collide and interact with each other. Structure deformation is strongest whenever a fast stream "follows" a slow stream in the sense of solar rotation, i.e. if the coronal source of the fast stream is located eastward of the slow one, as seen from earth.

The curvature of flux tubes emerging radially from a rotating source is that of an Archimedean spiral ("Parker" spiral; see, e.g., [3.95]), as illustrated in Fig. 3.3. The flux tube's angle α to the radial direction at a distance R is given by

$$\tan \alpha = \Omega R / v \ ,$$

with Ω being the sun's angular speed ($\Omega = 1.642 \times 10^{-4} \deg s^{-1}$) and v the solar wind flow speed, assumed to be radial and perpendicular to the sun's rotation axis. (For the case of nonequatorial flow, e.g. from high-latitude sources, these considerations apply to the projections of both v and R onto the equatorial plane). As a consequence, the longitudinal offset ϕ_R of that flux tube from distance R_1 to R_2 is simply

$$\phi_R = \Omega(R_1 - R_2)/v \ .$$

Furthermore, two hypothetical flux tubes containing plasma with different speeds v_1 and v_2 (again strictly radial) starting at $R = 0$ right next to each other will have an angular separation at a distance R of

$$\phi_v = \Omega R(1/v_1 - 1/v_2) \ .$$

For the case of two solar wind sources with $v_1 = 300\,\text{km s}^{-1}$ and $v_2 = 600\,\text{km s}^{-1}$ this separation would amount to as much as 41° at $R = 1\,\text{AU}$. For example, if the slow flow follows the fast one, we would expect a transition between these two streams at 1 AU that extends over 41° or more, depending on the steepness of the coronal transition. In other words, flux tubes associated with high-speed stream trailing edges expand considerably faster than according to the R^{-2} dependence of spherical symmetry. As a consequence, the particle flux density is reduced correspondingly [3.191, 66]. It is true that longitudinal pressure gradients tend to fill in these "rarefaction regions". However, interactions in this diluted medium are not very effective in causing major deflections of the large-scale flow. For the trailing edges of high-speed streams, therefore, the original profile is most appropriately reconstructed by applying the constant velocity mapping technique mentioned above. It was found that the entire flow in typical trailing edges apparently emerges from very "narrow" coronal sources (4–6° in longitude [3.154]; see also Fig. 3.2). These "dwells" [3.181] are just the eastern boundaries of high-speed streams that could be represented by almost rectangular velocity profiles at the sun.

At the front edges of high-speed streams the situation is dramatically different. In this case, the "separation" of 41° would imply that a longitudinal range of at least that size will be subject to compression and deflection. In reality, this range is substantially larger. On inspecting *in situ* observations from spacecraft at $R = 1\,\text{AU}$ we find that the total longitudinal range apparently affected by compression, the "corotating interaction region" (CIR), is typically about 30° wide (e.g., Fig. 3.1). If we map two flux tubes on each side of a CIR back to the sun (but definitely outside the CIR, i.e. unaffected by the stream interaction), it turns out that their coronal sources are separated from each other by usually more than 70° in longitude. In other words, the material involved in the CIR of a single high-speed stream at 1 AU originates from a coronal longitude range of more than 70°! The percentage of coronal longitude ranges affected by stream interactions grows quickly with increasing solar distance R and soon reaches 100%. Note that this is true only in the sun's equatorial plane. At high latitudes the situation may differ substantially!

Although not the subject of a chapter on the inner heliosphere, it is appropriate to include here at least an outline of the structural evolution in the outer solar system. Eventually, smaller streams are "entrained" [3.34], i.e. swept up by the larger ones. Their CIRs coalesce among themselves and form "merged interaction regions". At $R = 2$ to $4\,\text{AU}$ they are characterized mainly by a high total plasma pressure which exceeds that of the ambient flow by a large factor. Further out, the original stream structure will be altered to such an extent that the "memory of the source conditions has largely been erased" [3.27]. These compound stream systems may include material from the whole solar circumference and from a considerable latitude range, and this material may even have been ejected during more than one solar rotation. Spurred on by the continuing advance of spacecraft into the outer heliosphere (*Pioneer 10* at 45 AU, *Pioneer 11* at 27 AU, *Voyager 1* at 31 AU, *Voyager 2* at 28 AU, as of December 1988), the study of the dynamic solar wind stream evolution in this distant range has become exciting in its own right. The interested reader is referred to the extensive review by Burlaga [3.28] and other papers in the recent literature [3.36–38, 235, 236, 29].

On returning to the inner heliosphere, we now realize that the longitudinal range at high-speed streams' leading edges affected by stream interactions shrinks rapidly with decreasing R. The *Helios* spacecraft, which patrolled the inner heliosphere over more than 11 years, were thus well suited to revealing characteristics of the original coronal stream structure.

3.2.1 Radial Evolution of High-Speed Stream Fronts

In [3.204] the longitudinal gradients of the flow speed at high-speed stream fronts of all streams with amplitudes (i.e. speed differences with respect to the ambient slow flow) exceeding $200 \, \text{km s}^{-1}$ between December 1974 and December 1977 were analyzed. Here I report on an extension of this study including all available data up to the end of 1985. For each stream the difference in longitude was measured between the onset point of the speed increase and the point where the high-speed state was finally reached. The average speed increase per degree of heliographic longitude $\Delta v / \Delta \phi$ (in $\text{km s}^{-1} \text{deg}^{-1}$) was then calculated and plotted versus radial distance as shown in Fig. 3.4. Of course, the evaluation may suffer from some subjective judgement. However, all the data were evaluated by one person using the same criteria all the time. The process was iterated at least two times and was spot-checked by a different person. A minor bias may still exist, but it does not spoil the trends. The points at 1 AU were determined the same way from the *IMP 7/8* earth-orbiting satellites (the plasma data were kindly provided by the Los Alamos group headed by the principal investigator of the *IMP* plasma analyzer, Dr. S. Bame) and cover the time period from December 1974 to the end of 1977. Whenever one stream front was observed from two or even three spacecraft within two days and with a maximum latitudinal separation of 2°, the data points were connected by a solid line. Note that the different symbols refer to the certain presence (squares), absence (circles), or uncertain presence (dots) of a "stream interface" within the interaction region, i.e. a sharp discontinuous boundary "which separates plasma of distinctly different properties and origins" as defined in [3.72] (see the discussion in Sect. 3.2.2). The overall trend in this scatter plot is evident: the closer to the sun the higher the gradients. This trend becomes more distinct upon calculating the average values of $\Delta v / \Delta \phi$ in radial bins of 0.1 AU each (Fig. 3.5). Closer inspection of both figures reveals some more features:

- At 0.3 AU the gradients $\Delta v / \Delta \phi$ are as high as $100 \, \text{km s}^{-1} \text{deg}^{-1}$ on the average. That means that the shear between slow flow of $300 \, \text{km s}^{-1}$ and a typical high-speed stream of some $600 \, \text{km s}^{-1}$ does not extend over more than 3° in longitude.
- On the average, the steep gradients at 0.3 AU flatten out at 0.5 AU and then stay almost constant at a level of about $35 \, \text{km s}^{-1}$ per deg.

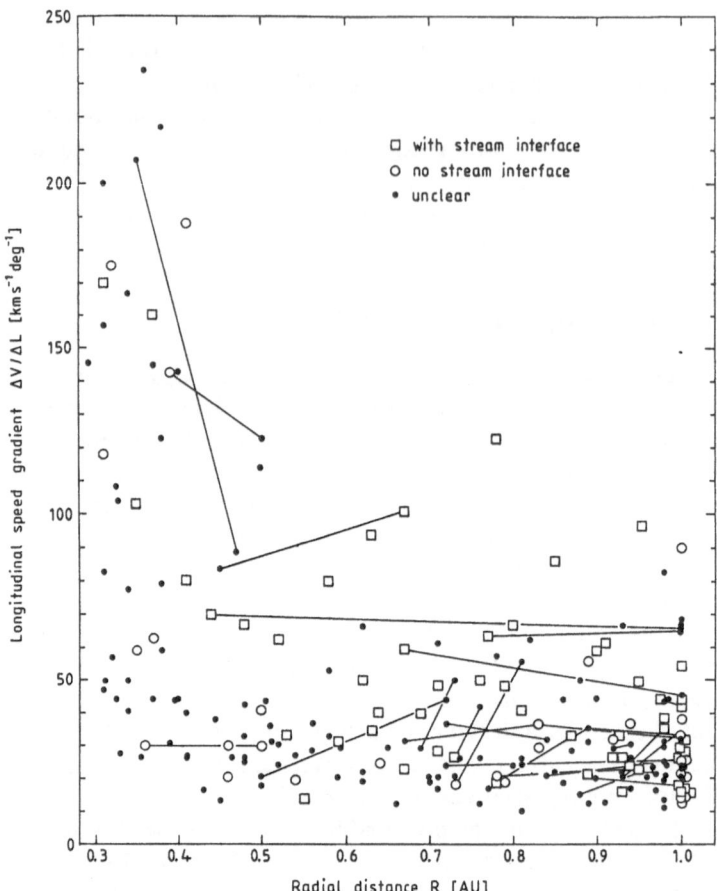

Fig. 3.4. Longitudinal bulk speed gradients at the front edges of high-speed streams between 0.29 AU and 1 AU. The data points have been connected when a stream front was observed from two spacecraft within 2 days. All streams with "amplitudes" exceeding 200 km s^{-1} (with respect to the ambient slow flow) obtained from *Helios 1* and 2 between December 1974 and December 1985 were evaluated. Data from the *IMP 7/8* satellites at 1 AU (using the Los Alamos plasma analyzers) cover the time period from December 1974 to the end of 1977

- In those cases when a CIR is observed by two almost radially aligned spacecraft we find a certain steepening tendency with increasing R. At least beyond 0.5 AU individual stream fronts appear to steepen rather than flatten. Note, however, that the statistics are poor.
- At all solar distances, one finds CIRs clearly with and others clearly without the signature of a stream interface. The number of cases in which the stream interface is clearly present or absent is about 50% at 1 AU, compared to 25% at 0.3 AU. Also, the interaction regions with clear stream interfaces generally exhibit steeper gradients than both the grand average and the clear "no stream interface" cases.

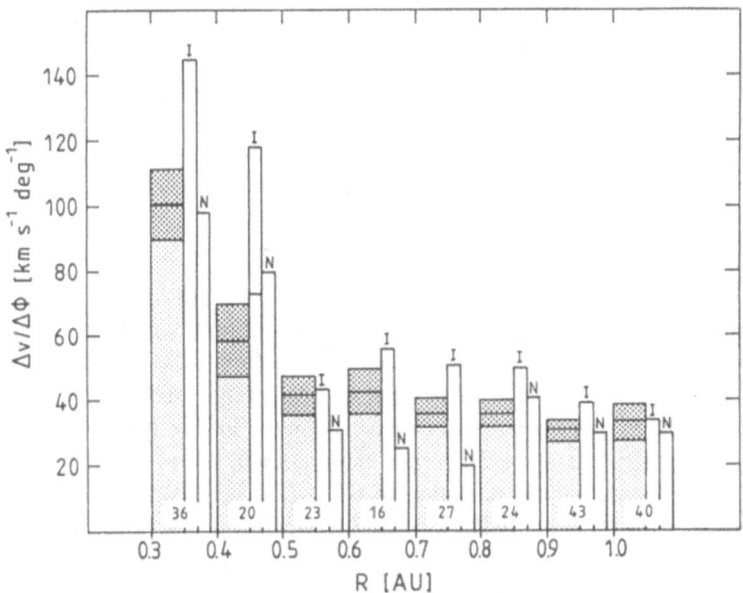

Fig. 3.5. Average longitudinal bulk speed gradients at the front edges of high-speed streams between $R = 0.29$ and 1 AU. The data from Fig. 3.4 were used as inputs. The numbers at the bottom denote the number of samples included in each 0.1 AU wide bin. The darker shaded areas indicate the statistical error margins. For the bars with "I" only those cases involving a stream interface (squares in Fig. 3.4) were evaluated. The "N" denotes cases clearly without a stream interface (circles in Fig. 3.4)

The absolute values of the speed gradients as well as the radial trend appear to contradict the results in [3.179]. Although part of the same data was used, the authors arrived at the conclusion that generally "stream steepening does indeed occur between 0.3 and 1 AU". This discrepancy, however, can be explained in terms of differences both in the definitions and in the evaluation technique. The study in [3.179] was based on a superimposed epoch analysis (as described, e.g., in [3.72]) of 16 (31) cases at 0.3–0.4 (0.9–1.0) AU. The epochs were centered at the density peak in front of the high-speed stream. After superposing all data, the resulting average profiles were analyzed and longitudinal gradients were derived. Since the density peaks show considerable variability and scatter with respect to their position relative to the stream fronts, the "mean" profile is accordingly smoothed out. Even a set of idealized rectangular profiles would be smeared out in this way. Specifically, any longitudinal "gradients" derived from such average profiles must be considered highly questionable, since they suffer most from the uncertainty in superimposing the epochs. As a result, the values for the speed gradient given in [3.179] (e.g. 9.53 km s^{-1} deg^{-1} around 0.35 AU) must be regarded as gross underestimates. In contrast, and this must be stressed again, the data points in Fig. 3.4 were obtained by simply measuring the width of the shear flow regions and thereby directly determining the individual speed gradients (some of these points can be readily verified by using Fig. 3.1 or Figs. 3.6 to 3.8 shown later). In summary, there can be no doubt that, on the grand average, originally steep speed profiles flatten out substantially with increasing solar distance. Most of that flattening occurs within $R < 0.5$ AU. Only beyond this distance do the stream fronts tend to steepen with increasing R. I will return to this issue in a slightly different context later on.

3.2.2 Stream Interfaces

The transition from a slow solar wind flow to a high-speed stream often involves "an abrupt change in density, temperature, and wave amplitude", as first noted in [3.12]. The term "stream interface" was coined by Burlaga [3.26]. It denotes "a distinct boundary in the interaction region of a stream in the solar wind, characterized by an abrupt (approximately a factor of 2 change in less than 10^6 km) drop in density, a similar increase in temperature and a small increase in speed". In some cases this is a tangential discontinuity (TD) (the occurrence of which was suggested as early as 1963, i.e. half a year after the *Mariner 2* launch [3.49]), in others it is apparently evolving into a TD. With this concept [3.26] (let us call it the "evolutional stream interface model") it is suggested that stream interfaces

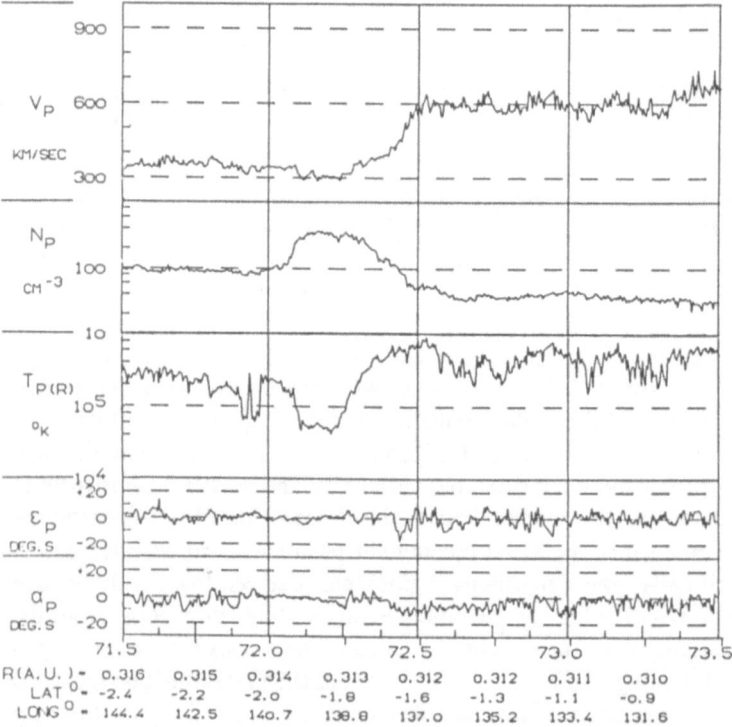

Fig. 3.6. Example of a steep speed increase at the front of a high-speed stream involving a discontinuous stream interface, observed at $R = 0.37$ AU from *Helios 1* on 5 May, 1979. ε_p and α_p denote the flow directions with respect to ecliptic latitude and longitude

form in the interplanetary medium as a consequence of the nonlinear evolution of streams generated by an increase in temperature in the solar envelope. Model calculations indicated that "the primary cause of stream interfaces appears to be a temperature variation somewhere in the corona" [3.100].

This evolutional stream interface model was questioned by Gosling et al. [3.72], who confronted it with what I would call the "discontinuous stream interface model". They stressed the fact that a stream interface is indeed a sharp discontinuous boundary ($< 4 \times 10^4$ km) and concluded that these "interfaces separate what was originally thick (i.e. dense) slow gas from what was originally thin (i.e., rare) fast gas".

They found this type of stream interface in 30% of all CIRs investigated; this percentage is, by the way, about the same as the one we found in *Helios* and *IMP* data around 1 AU. An example of this discontinuous type, observed at $R = 0.37$ AU, is shown in Fig. 3.6. The speed increases over only about 1.6° in longitude, leading to a gradient $\Delta v/\Delta\phi = 160$ km s^{-1} deg^{-1}. The size of the region containing compressed plasma on either side is hard to estimate, but it is probably not wider than 6°. For comparison, Fig. 3.7 shows an example of a similarly steep transition without a stream interface involved. Here, at 0.31 AU, $\Delta v/\Delta\phi = 175$ km s^{-1} deg^{-1}, and the total width of the CIR is only 4° at most.

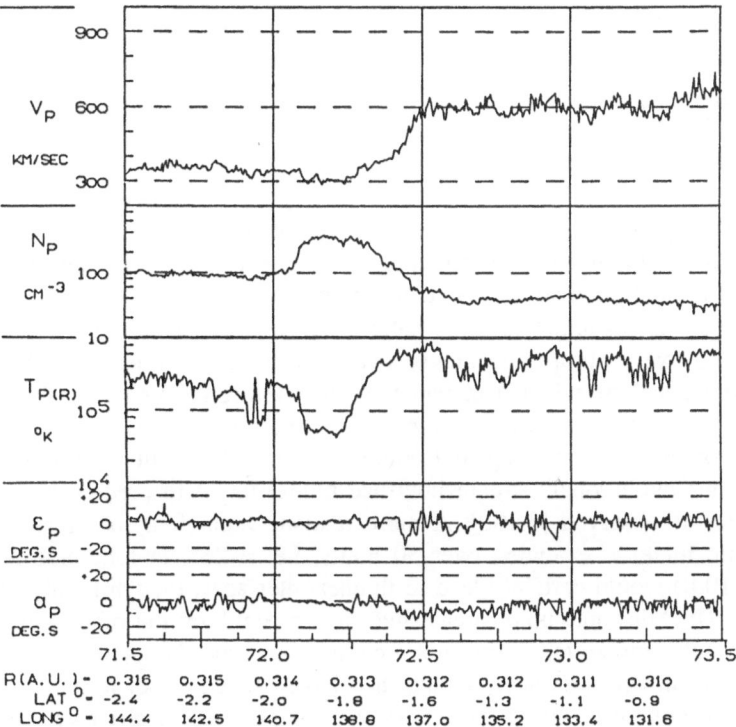

Fig. 3.7. Example of a steep speed increase at the front of a large high-speed stream without involvement of a stream interface observed at $R = 0.31$ AU from *Helios 1* on 13 March, 1975 (see also Figs. 3.1–3)

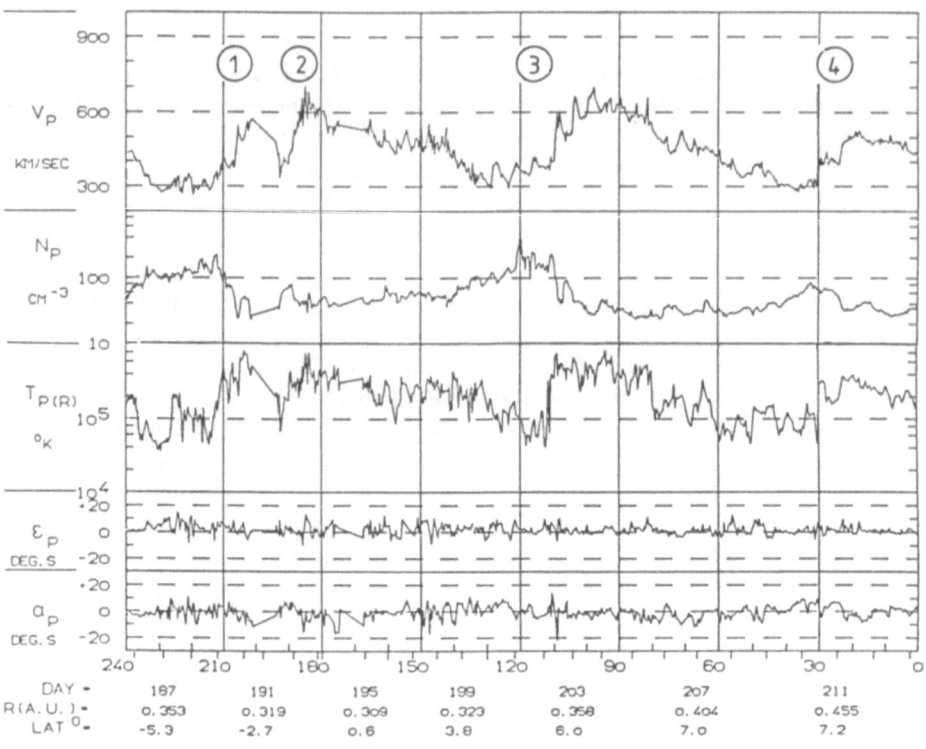

Fig. 3.8. Typical examples of fine structures of high-speed stream fronts observed around perihelion of *Helios 1* in July 1983. For explanations see text

In most cases it is not as clear whether or not a stream interface is involved. Many CIRs also exhibit much fine structure, be it due to spatial or temporal fluctuations, so that unique distinctions cannot be made.

After all, there is no doubt that the solar wind structure close to the sun is characterized by very abrupt and maybe even discontinuous transitions. We cannot at present conclude whether stream interfaces are remnants of similarly discontinuous coronal structures, or whether these stream interfaces evolve further out, as a consequence of steep, but continuous, transitions in the corona. It is true that the "evolutional" stream interface model failed to predict "thin" stream interfaces as observed [3.100]. A more realistic input function (instead of a temperature increase extended over 50 hours, i.e. = 26°, in longitude as assumed in [3.100]) would certainly lead to thinner interfaces, though hardly to discontinuities. The "discontinuous stream interface model", on the other hand, would have difficulty explaining interaction regions of the type shown in Fig. 3.7. Can discontinuous stream interfaces, should they have existed closer to the sun, have disappeared without leaving a trace upon reaching 0.3 AU? Fully aware of the fact that "with increasing distance from the sun the magnitude of the speed jump across the interface is reduced" [3.72], we would still expect some other signature of the stream interface to remain (compare, e.g., the density temperature and flow angle profiles in Figs. 3.6 and 3.7).

Note that recent evidence was found from *Helios* data for the local destruction of TDs [3.150]: there is a definite decrease in the number of interplanetary TDs with increasing R, which may be associated with the growth of the Kelvin–Helmholtz instability with decreasing Alfvén speed v_A. It is suggested in [3.150] that this instability might play an important role in the mutual diffusion of the plasmas from both sides of a TD. The possible importance of Kelvin–Helmholtz instabilities in this context was pointed out in an independent study [3.114], which indicated that this instability would cause turbulent viscosity in the shear flow region of CIRs beyond 0.1 AU and thus counteract kinematic steepening of the CIRs.

The superimposed epoch analysis centered around stream interfaces as shown in [3.72] suggests that the plasma on both sides of a stream interface may in fact emerge from different types of coronal sources. In particular, the apparent change in the helium abundance, as shown in Fig. 3 of [3.72], would be hard to understand otherwise. Even so, for interaction regions with no discernible TD, we cannot tell with certainty whether the TDs were already destroyed or whether they never existed before. We should not ignore a third possibility: there may well exist distinctly different types of stream interfaces close to the sun!

3.2.3 Fine Structure of High-Speed Stream Fronts

On inspecting Fig. 3.4 one will note quite a number of cases with comparatively small gradients even close to the sun. The plots in Fig. 3.8 give us 3 examples of this type, labeled (1), (2), and (4), while (3) denotes a case with a "steep" interface ($\Delta v / \Delta \phi = 105 \, \text{km s}^{-1} \, \text{deg}^{-1}$). In cases (1) and (4) the transition into the high-speed streams apparently takes place in two steps, each one of them with a much steeper gradient than the overall gradient: 28 and $14 \, \text{km s}^{-1} \, \text{deg}^{-1}$, respectively. In case (3) we also note several steps. Here, however, one of them happens to exceed the $200 \, \text{km s}^{-1}$ amplitude threshold, and consequently is treated as an interface in its own right. This pattern is very often observed at small solar distances: many small "rectangular" structures with a "diameter" of the order of $5°$ in longitude eventually may add up to form a high-speed stream. It appears as if the space probe has to cross several flux tubes with intermediate flow speeds before it finally reaches those flux tubes containing the high-speed solar wind. Occasionally, the transition requires only one step, i.e. "slow" and "fast" flux tubes are adjacent to each other, the speed gradient at the boundary being huge, as shown in Figs. 3.6 and 3.7.

On the way out to 1 AU the flux tubes gradually merge, probably due to increasing stream interaction at the leading edges of flux tubes with high-speed flow. Local plasma processes such as Kelvin–Helmholtz instabilities [3.114, 150] may be involved in destroying the original fine structure. Thus, the well-known pattern of large-scale high-speed streams is formed. Even these large streams often show considerable structure in their density profile at their leading edges

[3.204], thereby hindering a unique definition of the "real" compression peak (see, e.g., Fig. 3.1).

Note, by the way, that this model of gradually merging flux tubes is consistent with the concept presented recently in [3.226]. Indications were found for "fine" structures in high-speed streams with typical scale sizes of 5°, which may be interpreted as remnants of the coronal fine structure, and thus possibly even of the underlying supergranulation network.

As demonstrated in Fig. 3.5, the velocity gradients remain about constant at solar distances beyond 0.5 AU, i.e. the longitude range affected by shear flow does not change much. However, the longitudinal extent of the CIR (including the compressed gas on both sides) keeps growing with increasing R, due to the increasing amount of plasma affected by compression. Substantial heating of the plasma on both sides of the interface is a natural consequence of the compression [3.72, 60, 142]. Fine structures are gradually destroyed through the dynamic exchange of momentum. In some respects, the interplanetary medium appears to act like a "lowpass filter" [3.67, 36], with only the largest-scale stream structures surviving to great heliocentric distances [3.36].

3.2.4 Problems of Mapping Solar Wind Structures Back to the Sun

Mapping of solar wind structures back to the sun in order to locate their coronal sources and to allow correlations with detailed surface phenomena has always been important, e.g., for the solar energetic particles community. A simple mapping technique based on the assumption of constant and radial velocity [3.221] appeared to produce a unique transformation when applied to high-speed stream trailing edges [3.120, 155]. However, interaction processes render this kind of mapping totally obsolete at the streams' leading edges. According to the discussion given above, an average leading edge extending over about 10° in longitude at 1 AU would appear as a 50° ramp at the sun (look, e.g., at the many examples shown in [3.153] or [3.243]), whereas we now know that it may be only 3° wide in reality.

This knowledge, on the other hand, provides the basis of an improved mapping technique. I call it the "flux tube mapping technique". Given the measured density and speed profiles of a high-speed stream obtained at 1 AU, we want to determine the longitudinal position of the stream interface of the inferred "rectangular" stream close to the sun.

Let us assume that the profile actually observed at $R = 1$ AU results from compression and deflection in such a way that the mass fluxes $n_p v_p A$ are conserved in each fictitious flux tube with cross section A on either side of the stream interface (i.e. meridional deflections are neglected). By further assuming the flow parameters observed right outside the CIR to represent the values on both sides of the hypothetical rectangular interface (implying no acceleration or nonradial flow outside the CIR), a simple computation can be performed that finally yields the longitudinal position of the interface at the sun. This is illustrated in Fig. 3.9.

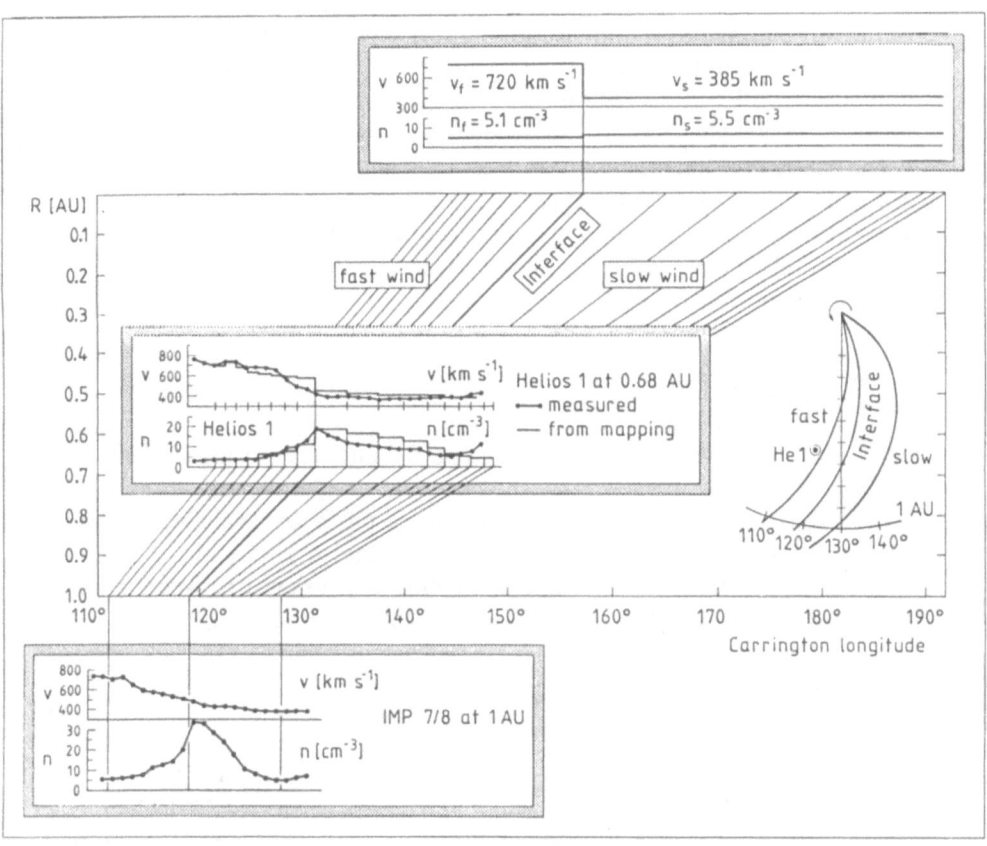

Fig. 3.9. Demonstration and test of the "flux tube technique". *IMP* data at 1 AU were taken as a baseline in order to reproduce an assumed rectangular stream profile at the sun. The diameter of each fictitious flux tube was calculated such that its mass flux $n_p v_p A$ was conserved out to 1 AU. The values of n_p were normalized to 1 AU assuming a R^{-2} dependence. It is shown that the stream interface location at the sun would be expected at 157° Carrington longitude, i.e. 38° west of its 1 AU location. The *Helios 1* data at 0.68 AU confirm the validity of this technique. Note that the material involved in this CIR at 1 AU (width: 18°) emerges from coronal regions extending over about 50° in longitude

Here I split the 1 AU data into virtual flux tubes of 1° width that are mapped back to the sun by keeping their individual flux $n_p v_p A$ = constant, and adjusting A accordingly (A goes as the square of the longitudinal width at the sun). This procedure can be started from both the high and the low speed end and thus brackets the possible position of the interface.

We were able to cross-check the validity of this procedure in the very few cases where a CIR could be observed not only at 1 AU (from *IMP 7/8*) but also almost simultaneously from *Helios* crossing the earth–sun line much closer to the sun. This is also demonstrated in Fig. 3.9. In this particular case, *Helios 1* was 8° east of the earth–sun line at $R = 0.68$ AU and thus about 25° east of the ideal line-up position with respect to the Parker spiral. This and the latitudinal

115

offset of 0.5° might explain minor differences, especially in the density profiles at the *Helios 1* position. Generally, the agreement was to within a few degrees. The remaining differences are due to the uncertainties introduced by nonunique evaluations of the pre- and post-CIR states.

The increasing longitudinal offset of a stream interface with increasing R gives it the shape of a Parker-type spiral. The curvature of this spiral can be expressed in terms of an "effective propagation speed" v_i of the stream interface. Knowing the longitudinal positions ϕ_1 and ϕ_2 of a stream interface at two distances R_1 and R_2, we find

$$v_i = \frac{\Omega(R_1 - R_2)}{(\phi_1 - \phi_2)} \; .$$

This relation may then be used to determine the interface position at any other R.

Most often, useful data from only one space probe are available. For practical evaluations in these cases, a meaningful value of v_i can also be derived by considering both radial and tangential momentum balance between the slow plasma (with v_s, n_s) and the fast one (v_f, n_f) running into it,

$$v_i = \frac{v_s n_s + v_f n_f}{n_s + n_f}$$

without further assumptions or knowledge of the detailed shape of the CIR (helium ions are neglected here). For those few cases where a CIR happened to be observed from radially aligned space probes, we determined v_i using both methods and found remarkably good agreement.

In many cases, a realistic value of v_i can simply be read from measured data, or at least a plausible range can be given. For the stream interface in Fig. 3.6, e.g., v_i would be in the range 250 to 280 km s^{-1}; our momentum balance formula would yield 275 km s^{-1}. After evaluation of 44 more stream interfaces we find the calculated values on the average to be well within the range of the measured data. In 1/3 of the cases it was slightly outside that range, in only 4 cases by more than 50 km s^{-1}. (Note, by the way, that the formula for v_i derived in [3.214] differs from ours. If applied to the same data, though, it would yield values of v_i which are larger by 5 to 10% and thus would no longer fit to the bulk of experimental data.)

The average value of all v_i is (420 ± 60) km s^{-1}. This low value is due to the usually higher particle densities in slow solar wind. In other words, the overall curvature of stream interfaces is mainly determined by the slow, dense plasma [3.72].

This whole discussion leads to an additional, much simpler recipe for practical mapping (I call it the "traveling interface technique") in order to find the longitudinal offset $\Delta\phi$ of the stream interface close to the sun with respect to an *in situ* observer:

1. Read the pre- and post-CIR values from plots of solar wind flow speed (v_s and v_f) and density (n_s and n_f) taken from spacecraft at distance R from the sun.
2. Calculate $v_i = (v_s n_s + v_f n_f) \, / \, (n_s + n_f)$.
3. Check this value of v_i with experimental value, if a stream interface is discernible.
4. Calculate $\Delta\phi = \Omega R / v_i$.
5. $\Delta\phi$ is then the longitude difference between the stream interface at the spacecraft and the same interface at the sun. In the case depicted in Fig. 3.9 we would find $v_i = 546 \, \mathrm{km\,s^{-1}}$ and $\Delta\phi = 45°$ for $R = 1\,\mathrm{AU}$ (compared to 39° in the figure).

In case the stream interface is not uniquely discernible from the *in situ* data, it can be expected to be located on the average about 1° in longitude behind the compression density peak at 1 AU, according to [3.72]. This peak, however, is not always uniquely discernible either. One may also take that longitude as a reference where the calculated v_i matches the measured value. The determination of the stream boundary position will still be accurate to within about 5° in longitude.

Note that this "traveling interface technique" does not imply any assumed stream profile at the sun or at the observer's location. It merely relates measured positions of stream interfaces for the same streams at different solar distances, and it works as well if the actual profile close to the sun is not very steep, e.g. for the cases (1) and (4) in Fig. 3.8.

Concluding this section, we now have a simple means at hand for locating the leading edges of high-speed streams in the corona with an accuracy of 5°. For many practical applications it may be sufficient to use our average value of v_i. This would mean, e.g., that the average stream interface at the sun is located 60°± 10° west of its longitude at 1 AU.

3.2.5 Mapping and Predictions

Another important application of mapping based on our present understanding of solar wind structure may become useful for prediction purposes. Based on the observations of solar features or using *in situ* measurements of the solar wind at a certain location (i.e. close to the sun), how can we deduce the solar wind behavior at some other place in the heliosphere (e.g. at the earth)? Here the arguments are essentially very similar to those discussed above. Indeed, the "traveling interface technique" works very well, provided there are detailed *in situ* solar wind data available over the same range of heliographic longitude and latitude, i.e. for sufficiently small corotation times between the spacecraft [3.185]. This was tested by using the *Helios* data of 1975/76 as a monitoring base and the *IMP* earth-orbiting satellites for checking the prediction accuracy with respect to the arrival time of stream interfaces. A total of 23 out of 25 stream observations with corotation times of less than six days were correlated, the prediction error amounting to only $1.0° \pm 4.6°$ in Carrington longitude (i.e. the actual stream fronts arrived slightly too early), regardless of *Helios'* distance from the sun. In 50 other cases with corotation times of more than 10 days, there was an apparent correlation for only half of all cases. In this group the prediction would have been "accurate" to $3.9° \pm 7.5°$. It is noteworthy that the subgroup of 11 cases with latitudinal separations of less than 3° yielded 9 examples of good correlation, despite long corotation times. We can state that trustworthy predictions of stream interface arrivals at earth can be made at times of stable solar wind structure only if a monitoring spacecraft is located within $\pm$ 3° of the earth's heliographic latitude and preferentially no further than 90° east of the earth–sun line. The monitor's distance to the sun is not of primary importance. These conclusions confirm what had been stated in a more qualitative way in [3.65]. In fact, it was this type of study that led to the discovery "that the rate of the expansion of the solar corona at any particular time is a sensitive function of latitude as well as longitude" [3.65].

Predictions of the solar wind stream structure based on coronal hole maps have not been tried yet to my knowledge. Such a procedure would depend critically on the exact definition of coronal hole properties and particularly on the practical evaluation of their boundaries. Existing simultaneous data of coronal holes [3.83, 210] and the solar wind might be used to work out and check an appropriate scheme.

3.2.6 A Scenario of Longitudinal Stream Structure

In this section I briefly summarize the conclusions on the longitudinal stream structure by sketching a scenario of the situation:

1. Bundles of flux tubes emerging from the corona containing solar wind of different speeds start interacting with each other with increasing solar distance. Close to the sun, the Parker spiral angles are small, and the characteristic speed (fast mode) c_f with which pressure signals propagate is high [3.72], keeping the compression small. The resulting flow deflections and density enhancements are thus limited to narrow longitude ranges.

2. Beyond about 0.4 AU, with now rapidly growing differences in the spiral angles of different flows, the broadening of the longitudinal extent of the shear flow becomes significant. It results mainly from lateral deflection of the flows as long as c_f is large enough. Quasiviscous effects due to Landau damping [3.52] and the development of Kelvin–Helmholtz instabilities [3.114, 150] may also contribute. These processes are probably associated with the merging of several smaller scale flux tubes into one large- scale CIR as discussed above. This widening always competes with kinematic steepening. These two processes apparently counterbalance each other so well, that the velocity gradients stay about the same, at least between 0.5 and 1 AU.

3. With increasing solar distance, the local value of c_f decreases, so that eventually the widening of the CIR due to the pressure gradient can no longer compensate the compression. Thus, the CIR will steepen again.

4. Finally, a forward–reverse shock wave pair will form [3.67, 101, 218]. At times, conditions for the formation of corotating shocks are favorable around and even within 1 AU. Several such cases were found in the *Helios* data. (We will return to this topic in Sect. 3.5).

5. The shape of stream interfaces at high-speed stream leading edges results from both the underlying coronal structure and interplanetary evolution. However, the appearance of discontinuous stream interfaces in about half of all cases and at any solar distance is still an unresolved issue. Note that there is observational evidence only for the local destruction of TDs [3.150] rather than their formation by any form of stream–stream interaction.

Figure 3.10 shows an idealized view of a stream interaction region as it may evolve between the sun and, say, 1 AU. A speed profile that was originally

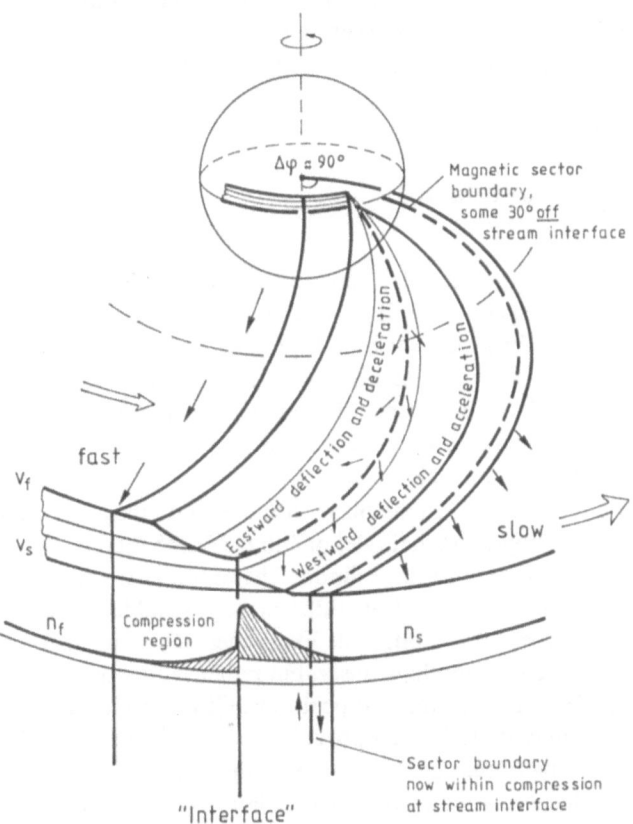

Fig. 3.10. An idealized view of a corotating interaction region and its evolution from a rectangular speed profile at the sun into a more gradual speed increase at 1 AU. Note that the angular separation between the stream interface and the preceding magnetic sector boundary decreases significantly with increasing distance to the sun. At 1 AU many sector boundaries are already engulfed by the succeeding CIR

"rectangular" is reshaped into a gradual speed increase by the action of flow compression and deflection on both sides of the interface. However, the compression region extending over some 30° in longitude at 1 AU contains plasma that has emerged from a coronal source region spanning some 70° in longitude. Thus, even "distant" features such as sector boundaries are often found well within compression regions although they were originally well separated and independent of any stream interface. This particular point will be addressed in some detail in Sect. 3.4.2.

The ridge of compressed plasma within the CIR can be visualized as a "density wave" traveling around the sun [3.70] and following its rotation. Further compression will eventually lead to formation of shock waves in the outer heliosphere which corotate as well (see Sect. 3.5).

I describe this scenario in rather qualitative terms and base it mainly on experimental data, in order to stimulate further work in checking and refining it as well as in developing appropriate theoretical models. It is worth noting here that e.g. "fluid dynamics of thin solar wind filaments" had been studied in quite some detail as early as 20 years ago (see [3.213] and references therein). This concept and maybe other early ideas deserve resuscitating and careful reevaluation in the light of our present knowledge.

3.3 Latitudinal Stream Boundaries

On studying structural details of coronal holes and the solar wind one would not expect basic differences with respect to heliographic longitude and latitude to occur close to the sun. For certain aspects it may be more convenient to study latitudinal boundaries, since they are not distorted by stream–stream interactions as strongly as their longitudinal counterparts. The analysis techniques are inevitably quite different, however, since comparatively fast scans in latitude using *in situ* instrumentation are not presently possible.

As long as spacecraft are still bound to the plane of the ecliptic they can scan only through a small band of heliographic latitude, due to the 7.25° inclination of the sun's equatorial plane with respect to the ecliptic. Therefore, two spacecraft in the ecliptic plane can actually be located at different heliographic latitudes, but only if, at the same time, they are at different longitudes. That means they cannot observe a latitudinal structure simultaneously, but rather with a time delay according to the corotation time between the two positions. Of course, spacecraft traveling at different heliographic latitudes can provide only upper limits on the thickness of those boundary layers that are thinner than the spacecraft latitudinal separation. Decreasing this separation does in fact improve resolution, but simultaneously diminishes the probability for a boundary to occur right in this gap, thus increasing the difficulty in observing it at all.

These basic contraints are the reason latitudinal stream structures in the solar wind were hardly accessible before the *Helios* twin mission. Sometimes seasonal variations of the solar wind flow were observed from earth-orbiting satellites, which follow the earth's heliographic latitude excursions of $\pm 7.25°$ through the year [3.51, 99]. The flow speed tended to be lowest near the solar equator [3.99]. If interpreted in terms of a general solar latitude effect, speed gradients of the order of 11 to $15 \, \text{km s}^{-1} \, \text{deg}^{-1}$ were implied [3.173–176]. At other times and averaged over a $11\frac{1}{2}$ year interval, this gradient was not apparent [3.51, 7]. Using the interplanetary scintillation (IPS) technique, however, the existence of a persistent and general average gradient with latitude (see, e.g. [3.43]) of the order of 1–$2 \, \text{km s}^{-1} \, \text{deg}^{-1}$ was conclusively proved and found consistent with theoretical modelling [3.168]. Soon after the launch of *Helios 1*, once plots such as Figs. 3.1 and 3.2 became available [3.201, 184, 203], the value of determining long-term average gradients in a stream structured solar wind with huge local gradients became questionable.

3.3.1 Local Latitudinal Speed Gradients

In [3.204] the issue of local latitudinal gradients was addressed. Data from both *Helios* probes were ideally suited for this study in the years 1976 and 1977, since both spacecraft were traveling fairly close to each other most of the time in terms of radial distance, latitude, and corotation time. Furthermore, in this time period around minimum solar activity the stream structure at any given latitude was almost invariant over several solar rotations. For the purpose of this study the plasma observed at the inner probe was assumed to be carried to the outer one radially at its local speed. Averages of the speed over 1° in Carrington longitude were calculated (after proper correction of some "overturnings" due to the mapping), and the speed differences at given longitudes were then plotted versus the latitude separation $\Delta\lambda$ (see Fig. 3 in [3.204]). The study can be summarized by two main results:

1. For $\Delta\lambda > 5°$, the Δv values are essentially random, i.e. two observers separated by more than 5° in latitude will have only a relatively small chance of encountering similar streams.
2. Large values of Δv (up to $250 \, \text{km s}^{-1}$) can occur even for very small $\Delta\lambda$ ($\approx 1.5°$). This applies even for corotation times of less than two days. Many of these cases are probably due to time variations between the measurements, but on several occasions considerable speed differences persisted for a day or longer, i.e. over substantial ranges in longitude and time.

One striking example from [3.204] is shown in Fig. 3.11. The two spacecraft observed very similar stream patterns, except for the leading edge (*D*) of one of the high-speed streams. Here *Helios 2* entered the stream 13° later than *Helios 1*, even though $\Delta\lambda$ was only 1.3°. Taking into account the details of this constellation, i.e. the corotation time (24 hours), travel time (12 hours for $v = 450 \, \text{km s}^{-1}$ or 8 hours for $v = 650 \, \text{km s}^{-1}$), and radial separation (0.13 AU), a purely spatial structure would appear more plausible than any other explanation, such as time effects. In fact, the coronal hole structure was very stable at that particular time

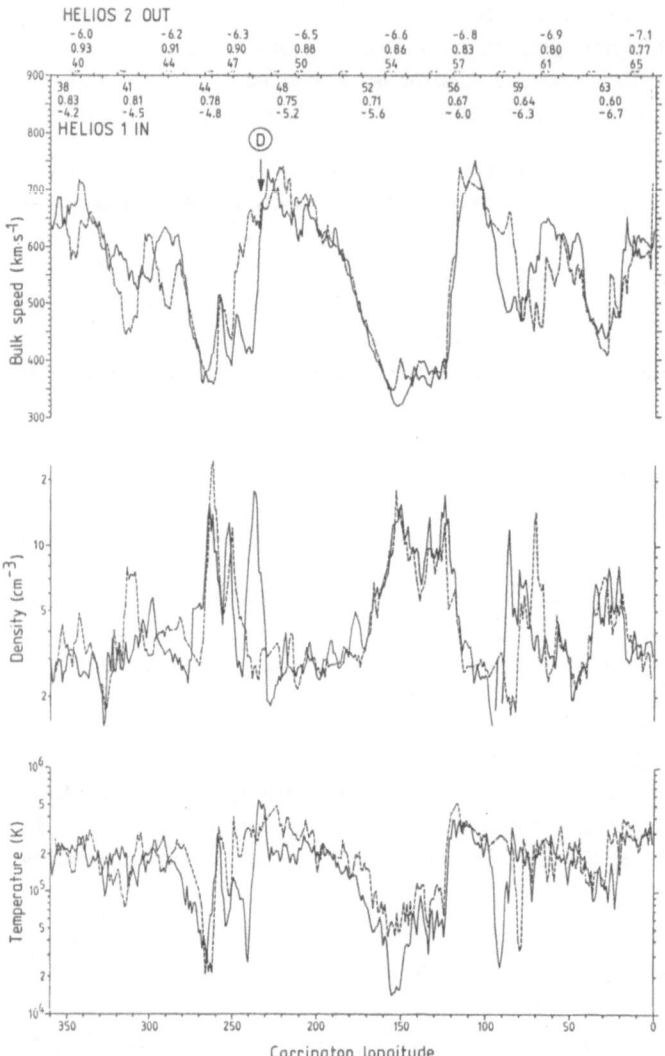

Fig. 3.11. The proton bulk speed v_p, density n_p and temperature T_p during Carrington rotation 1638 in early 1976. The data from *Helios 1* (*dashed line*) were mapped outward to the orbit of *Helios 2* (*solid line*), assuming radial outflow at the measured speed. The numbers at the tick marks on top of the figure denote heliographic latitude λ, distance to the sun R, and time (day of year 1976) of arrival of the plasma at the orbit of the outer spaceprobe. The difference of the profiles at D is discussed in the text. From [3.204]

in 1976 and was very similar to the one shown in Fig. 3.2. There was still a big "northern" coronal hole around Carrington longitude 270° with its southern edge reaching beyond the heliographic equator to about −15°. Thus, we are not surprised to find the stream to be narrower at southern latitudes (*Helios 2*) than closer to the equator (*Helios 1*). The stream boundary was observed from two points 13° and 1.3° apart in longitude and latitude, respectively, indicating that its inclination to the equator amounts to only 6°! It is interesting to note in Fig. 3.11 that not just one but two (or even three for *Helios 2*) high-density "layers" with all the typical following signatures of a high-speed stream are found at the same longitudes from both probes. Again, according to [3.204] "one is led to think of very narrow, well-separated filaments (less than 10° in angular width) of dense gas enveloping this fast stream." The final entry into the high-speed stream occurs as abruptly as usual despite the proximity of the lateral edge [3.11]. This example nicely illustrates that the whole transition between the high-speed and slow-speed state of the solar wind can take place within a layer only 1.3° wide in angle! This value 1.3° is an upper limit set by the fortuitous separation of the spacecraft; the actual width may be even less.

It is worth noting here, that sharp latitudinal stream boundaries are conserved even far beyond 1 AU. Several examples of boundaries of $\sim 5°$ latitude in width were repeatedly observed as far out as 5 AU [3.141]. This remarkable persistence of sharp latitudinal stream edges means, likewise, the persistence of substantial lateral shear flows. Following the arguments in [3.114], the development of Kelvin–Helmholtz instabilities should tend to destroy any velocity gradients between different flows. This process is predicted to be especially effective for latitudinal edges since there is no competing effect of kinematic steepening of the speed profile as for longitudinal edges. The thickness of a transition region at $R = 0.3$ AU as calculated in [3.114] amounts to 1.2×10^{12} cm (or 0.08 AU), equivalent to an angular width of 15°. This apparent discrepancy with the observations makes me suggest a thorough review of the microscopic processes possibly involved in the interaction of different plasma flows.

3.3.2 Evidence for Meridional "Squeezing" of Interaction Regions

Figure 3.11 gives an impressive example for density enhancements which apparently are not directly associated with a following high-speed stream. There are more dramatic examples of this type occurring for larger angular separations. At times, both spacecraft encounter a big compression peak, but one of them misses the succeeding high-speed stream completely. In all cases, there are strong out-of-the-ecliptic excursions of the magnetic field associated with these events. Without having reference data from a second spacecraft at hand one could possibly interpret such apparent compression regions as some sort of transients or as "noncompressive density enhancements" (NCDEs) [3.71]. We would rather conclude that compression regions may extend over larger latitudinal ranges than the high-speed streams causing them. It seems as if the plasma has been "squeezed out" meridionally, due to the strength of the compression. Similar effects have also been noted between 2 and 4 AU [3.27]. Of course, such squeezing would imply meridional flows [3.215] and magnetic field excursions within these regions. A systematic study of this phenomenon has not yet been performed with *Helios* data. So far we can only state that our observations are qualitatively consistent with model calculations of three-dimensional corotating MHD streams [3.166] which show that the "inclusion of the magnetic field and nonradial flows (including meridional ones) moderates the kinematic steepening".

In this context a very recent observation [3.121] is worth mentioning: *Voyager 2* near $R = 25$ AU detected remarkably persistent north–south flows alternating synchronously with the solar rotation rate, while the east–west flow as well as the wind speed itself were comparatively unstructured [3.122]. "The observational evidence supports stream dynamics and associated pressure gradients are responsible for driving the flow. Such a meridional flow may result in a net transport of magnetic flux from regions near the heliographic equator" [3.140]. This striking phenomenon certainly underscores the necessity of treating the heliosphere as a three-dimensional entity.

3.3.3 Problems of Associating Stream Profiles from Different Latitudes

The comparison of stream structures observed from different latitudes close to the heliographic equator has also demonstrated that two spacecraft traveling there at more than about 5° latitudinal separation have almost no chance of observing similar structures. For illustration, in Fig. 3.12, as adopted from [3.204], two solar

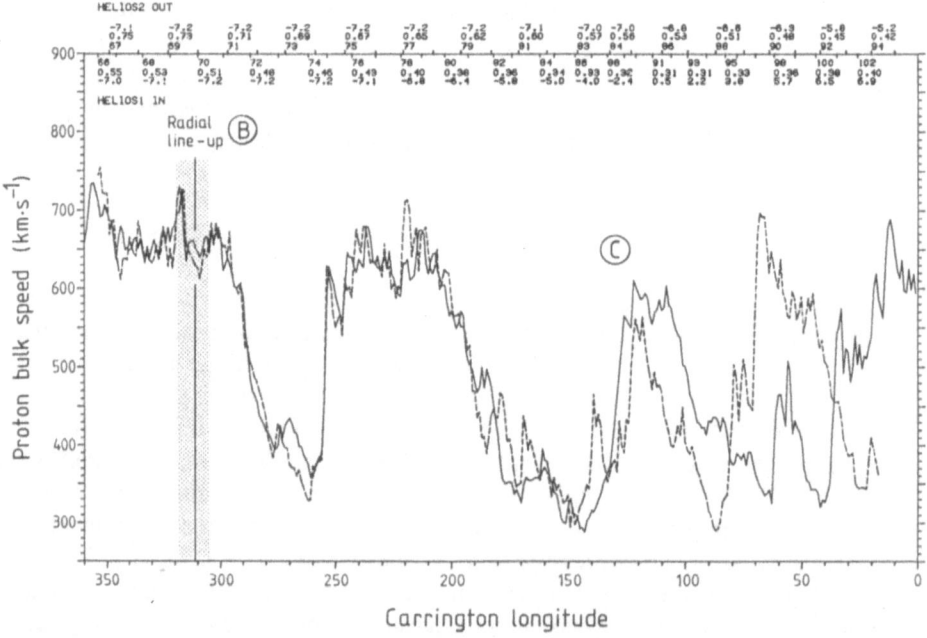

Fig. 3.12. The proton bulk speed during Carrington rotation 1639 in early 1976. The data of the inner probe *Helios 1* (*dashed line*) are mapped outward and compared to the actually measured data of the outer probe *Helios 2* (*solid line*). *B* denotes a "plasma line-up" (see text). The numbers at the tick marks on top of the figure denote heliographic latitude λ, distance to the sun R, and time (day of year 1976) of arrival of the plasma at the orbit of the outer spaceprobe. From *C* on the latitudinal separation exceeds 5°. From [3.204]

wind speed profiles are shown through a whole solar rotation in early 1976. *Helios 2* observed it between 0.75 AU and 0.42 AU, while the *Helios 1* profile was taken from 0.55 AU to perihelion passage (0.31 AU) and out again to 0.42 AU. The *Helios 1* data were mapped to the *Helios 2* orbit, and the tick marks refer to the arrival time at the *Helios 2* orbit. At B one of the few cases of a radial "plasma line-up" occurred. This term denotes a situation where a plasma volume after having passed one probe also encounters a second probe. In this case there is no latitudinal difference by definition. The speed profiles around B look very similar indeed; some minor differences may be due to a radial speed gradient [3.205, 196], or to a nonradial flow, or to processes affecting the local speed. The similarity degrades with increasing latitudinal separation, and from C on (where $\Delta\lambda$ exceeds 5°), both spacecraft are now moving through completely different stream structures. In this particular case, the two *Helios* probes even traveled on opposite sides of a magnetic sector boundary for as much as a quarter of a solar rotation [3.202, 231, 33, 21].

Such striking dissimilarities in *Helios 1* and 2 flow profiles were very commonly encountered in the years 1976 and 1977. Obviously, these differences were fostered by the conditions around solar minimum, when many coronal holes and their associated high-speed streams had their northern or southern boundaries right in the solar equatorial region.

Another consequence of this line of arguments concerns the above-mentioned significance of long-term averages of latitudinal speed profiles. Obviously, "time averages of the solar wind speed at a certain latitude must depend directly on the total breadth of the coronal holes at the same latitude" [3.203]. Thus, the long endurance of the polar coronal holes around solar minimum explains the high average speed at high latitudes measured with IPS [3.43]. Also, certain asymmetries in average latitudinal speed profiles can be understood. They fit well to asymmetries evident from K-corona maps such as in Fig. 3.2 obtained during that same time interval [3.83]. Sometimes gradient reversals are observed, i.e. the average speed right at the equator is somewhat higher than at, say, 20° latitude. The reason is that at times (see, e.g., the coronal hole map in Fig. 3.2) on scanning along the equator one will certainly encounter a coronal hole more often than in the range around ± 20° latitude. In summary, we have to be aware that "a wide class of possible spatial variations of solar wind speed are virtually undetectable in the longitude-averaged speed observed over a limited near-equatorial latitude range" [3.97] (and, of course, over any other limited latitude range).

The high values of high-latitude solar wind speeds deduced from IPS imply that the speed gradient within a high-speed stream extending from one solar pole into the equatorial regions may be very small. Indeed, the 3-D topology of solar wind high-speed streams has a remarkable resemblance to flat-topped mesas [3.184], standing out from an otherwise rugged territory with steep slopes on all sides.

3.3.4 A Scenario of Latitudinal Stream Structure

In summary of this section on latitudinal stream boundaries, let me sketch the following scenario:

1. Latitudinal boundaries of high-speed streams are often as sharp as longitudinal ones. The whole difference between being "in" or "out of" a high-speed stream may be experienced over a range of less than 1.3° in angular width.
2. There are lateral stream edges observed with very small inclinations to the heliographic equator. There is evidence that the compression region of a CIR may extend laterally into neighboring latitudes, where no succeeding high-speed stream is observed.
3. There are additional indications for a basic flux tube structure of high-speed flow, in this case deduced from the shape of lateral edges of high-speed streams.
4. Two spacecraft traveling near the ecliptic plane at latitudinal separations of more than 5° will not have more than a relatively small chance of encountering similar stream structures. This was observed during solar activity minimum when most structures were rather large and stable. The situation will be much worse at other times of the activity cycle.
5. The long-term average speed at a given latitude is determined by the fraction of time that high-speed streams occur at that latitude.
6. High-latitude solar wind streams emanating from polar coronal holes may have speeds similar to those from their equatorial extensions.

Additional issues concerning latitudinal and longitudinal stream boundaries are addressed in a wider context in the next section.

3.4 Stream Structures with Respect to the Heliomagnetic Equator

The 3-D structure of the solar wind in terms of sharply bordered slow and fast streams, of course, has to be regarded in context with the magnetic structure of both the solar corona and the interplanetary medium. It had been one of the first notions on the properties of high-speed streams that they are always fully imbedded in a single magnetic sector [3.237], and it became evident that "the magnetic field could channel the coronal expansion and strongly influence the properties of the solar wind" [3.95]. Furthermore, it is the magnetized solar wind plasma throughout the 3-D heliosphere that acts upon the propagation of high-energy particles of both galactic ("cosmic rays") and solar origin, in the latter case even affecting their acceleration (see [3.117]). This issue in particular and the 3-D heliosphere in general have been the subjects of a recent symposium dedicated to the upcoming *Ulysses* mission over the sun's poles [3.139].

As of today, *in situ* measurements have been restricted mainly to the plane of the ecliptic; in only two cases (*Pioneer 11* and *Voyager 1*) have the heavy outer planets been used to catapult spaceprobes to other heliographic latitudes. During the first excursion of *Pioneer 11* to 16° northern latitude in 1976, "an almost total absence of inward-directed fields" was observed [3.219]. This discovery explained the Rosenberg–Coleman effect of the semi-annually varying predomiment polarity of the interplanetary magnetic field as seen from earth orbit [3.187]. It also finally solved a long-lasting debate whether the 3-D magnetic sector structure can best be visualized in terms of an "orange slice" model or a "ballerina skirt" model (for details and further references see [3.219]). The latter concept of a near equatorial warped current sheet could now be firmly established. It had been hypothesized for the first time by Alfvén [3.2] and was worked out in more detail by Schulz [3.195]. Saito [3.190] arrived at a similar concept and called it the "two-hemisphere model". This whole issue is discussed in more detail in [3.131] (see also Sect. 3.6). In the context of the present chapter it is important to note that the arrival of the ballerina skirt model spurred new studies of the association of solar wind structure with *heliomagnetic* rather than *heliographic* latitude. Note that in this context the term "heliomagnetic latitude" denotes the local latitudinal separation from the warped heliomagnetic "equator", i.e. the current sheet.

3.4.1 Stream Structure with Respect to Heliomagnetic Latitude

Fortunately enough, due to the warps in the ballerina skirt a spacecraft traveling in the ecliptic plane will experience excursions to substantial heliomagnetic latitudes. Several authors have made use of this nice scanning technique and came up with profiles of solar wind speed versus heliomagnetic latitude, as shown in Fig. 3.13 from [3.23]. It displays data from several previous works

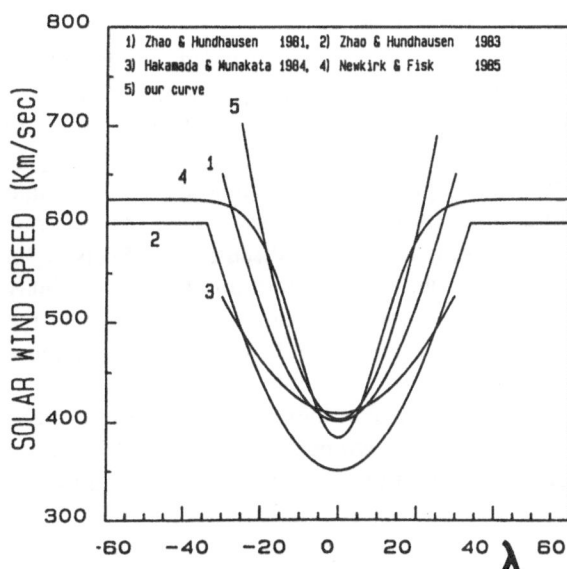

Fig. 3.13. Solar wind flow speed versus latitudinal separation from the current sheet, λ_M. The curves refer to different analyses performed by various authors as specified at the top of the figure. Curve no. 5 refers to the analysis in [3.23]

[3.240, 241, 81, 152], which all used different types of data (especially for defining the current sheet positions) and applied different evaluation criteria. However, they all show the same general trend: slow solar wind is found in a belt with a "width" of ± 20° around the heliomagnetic equator. Outside this belt there is high-speed wind only.

In a detailed study [3.23] *Helios 1* and 2 and *IMP* data taken during 1976 and 1977 were compared with current sheet positions inferred from synoptic white-light maps of the solar corona at 1.75 R_s. These, in turn, had been cross-checked with calculations from a potential field model based on photospheric magnetic field measurements [3.238, 22, 88]. The solar wind data were mapped back to 0.1 AU using the constant speed approximation. All high-speed stream leading edges were excluded, and so were all periods where the distance to the current sheet could not be measured uniquely or where other inconsistencies in the data (e.g. in field polarities or transients) were apparent.

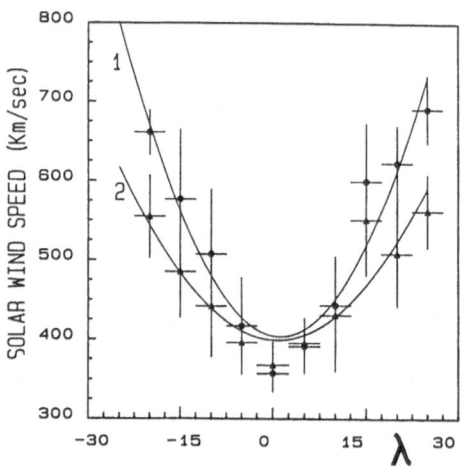

Fig. 3.14. Solar wind flow speed versus latitudinal separation from the current sheet, λ_M. Each data point represents the average value of speed in a 5° latitude interval. Vertical bars are the standard deviations associated with each interval. The horizontal bars represent the interval over which the average has been performed. Circles and triangles represent measurements from 1976 and 1977, respectively. The solid lines no. 1 and no. 2 show the best-fit curves for 1976 and 1977, respectively. From [3.23]

This extremely cautious and conservative procedure yielded, among others, the profile shown in Fig. 3.14. Here the data were split up into two time intervals ((1) from January 1976 to September 1976 and (2) from October 1976 to August 1977) and averaged over 5° latitude intervals. Curve 1, corresponding to solar minimum, clearly shows the steeper profile. Only high-speed wind is found beyond $\lambda_M = \pm 20°$. The slight asymmetry is probably associated with the one noted in coronal hole observations ([3.83], see also the discussion in Sects. 3.3.3 and 3.8.1). With increasing solar activity the profile noticeably flattens. We may suppose that this trend would continue into solar maximum [3.44]. Due to the stringent data selection criteria, however, a meaningful evaluation of the more irregular data at higher activity was no longer possible. The profiles of other solar wind parameters behave accordingly, in that they represent the low-speed state around the equator (high density, low ion temperature, higher particle flux) and the high-speed state outside. I will return to this issue in a slightly different context in Sect. 3.7.2.

3.4.2 Stream Structure with Respect to Heliomagnetic Longitude

Another point of interest is the longitudinal aspect of the current-sheet–stream-structure association. An observer crossing the current sheet will note the transition through a sector boundary [3.237]. Sector boundaries always occur in the low-speed regions between high-speed streams. However, in most cases at R = 1 AU the field reversals precede the stream interface of the next following high-speed stream by only $1\frac{1}{2}$ hours to $1\frac{1}{2}$ days [3.72]. This proximity has led to occasional confusion in the past in that the different terms "CIR", "stream interface", and "sector boundary" were not as clearly distinguished as their different nature requires. It is true that sector boundaries most often occur well within compression regions and often very close to the density peak [3.53]. However, in view of what has been discussed in Sects. 3.2.4 and 3.2.6 (see also Fig. 3.10) it is just this compression that brings flux tubes into close proximity at 1 AU, even though they may have started at the sun with substantial longitudinal separation and originally had nothing to do with each other.

In order to prove this I measured the longitude difference $\Delta\phi_{SB}$ between any stream interface and the last preceding sector boundary (from December 1974 to December 1978) and calculated averages for different solar distances (in bins of 0.1 AU each). In those cases with two streams next following each other within a single sector I accepted only the first one. The result in shown in Fig. 3.15. Clearly, $\Delta\phi_{SB}$ is largest at small distances and decreases from some 25° at 0.3 AU to 12° at 1 AU. The scatter is quite substantial, and the numbers should not be taken too seriously. The trend, however, is certainly real: close to the sun there is normally a longitudinal separation between sector boundaries and the next following high-speed stream of some 20°; this separation decreases, due to compression at the stream front, to some 10° at 1 AU.

Apparently, close to the sun the heliomagnetic current sheet is imbedded in a narrow belt of slow, dense solar wind which goes all around the sun (see also Sect. 3.6). K-corona maps like the one in Fig. 3.2 or in [3.102] probably provide a realistic impression of the shape of that belt. In most parts both the slow wind belt and the current sheet are inclined with respect to the ecliptic plane. An observer sitting at a quasifixed latitude will cut through the belt at

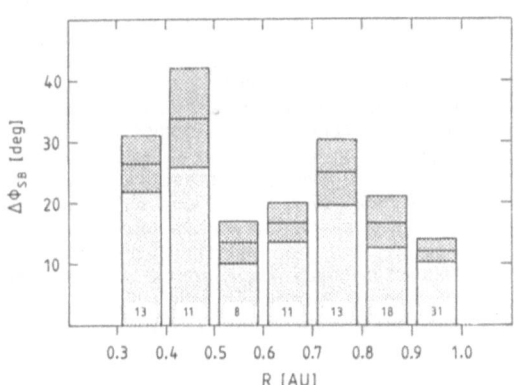

Fig. 3.15. The average longitude differences $\Delta\phi_{SB}$ between stream interfaces and the last preceding magnetic sector boundary as functions of R. At the bottom the number of cases per 0.1 AU bin is indicated. The darker shaded areas indicate the statistical error margins

many different inclination angles. This would explain the large scatter of $\Delta \phi_{SB}$ as mentioned above (see also [3.11]).

3.4.3 Solar Wind and Coronal Streamers

The properties of the "interstream flow" found in the slow solar wind belt (i.e. in between the broad high-speed stream regions) show some interesting signatures [3.16, 73, 62]. Figure 3.16 (taken from [3.73], adapted from [3.16]) was derived from a superimposed epoch analysis using a carefully selected set of 23 samples. Apparently, there is a clear depletion in the helium content centered around sector boundaries. Also, there are minima in proton temperature, in the helium ion to proton temperature ratio, and in the proton to helium ion velocity difference as well as a pronounced peak in proton density. This is true regardless of the position of the next following CIR and high-speed stream. There is a clear time lag of about one day between the two density peaks at the sector boundary and at the next following CIR. *Helios* showed that this time lag increases with decreasing R, consistent with what was said in Sect. 3.4.2 (see also Fig. 3.15). The first peak is probably a remnant of a "real" coronal structure while the second one results

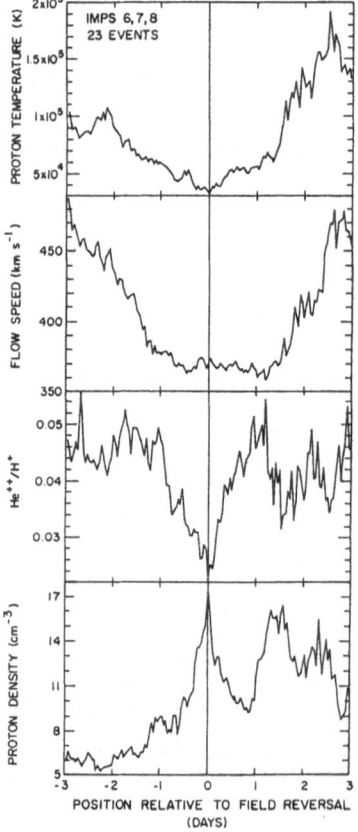

Fig. 3.16. Superposed epoch plots of the solar wind proton density, helium abundance, flow speed, and proton temperature for 23 well-defined sector boundaries more than one day removed from speed rises associated with high-speed streams. One-hour averages are employed, and sector boundary passage was used as zero epoch. The well-developed minima in abundance, flow speed, and proton temperature, and maximum in proton density are characteristic near field reversals. The data were collected at 1 AU from the *IMP 6/7/8* satellites. From [3.73]

from interplanetary stream interactions. In most individual cases of observations at 1 AU, though, these two features are found in various shapes and in such close proximity that they cannot be easily and uniquely separated. This difficulty often causes confusion and leads to strange conclusions even in recent papers, such as in [3.53]: "... both the plasma and the magnetic field in the heliospheric current sheet (HCS) do not undergo any appreciable, additional compression during an interaction with high-speed streams everywhere at $R < 215\,R_s$. Therefore, the HCS structure at the Earth's orbit is determined mainly by its properties near the solar surface." This may be partly true although the average CIR has already engulfed the preceding sector boundary, i.e. the HCS at 1 AU, (see Figs. 3.15 and 3.10). Unfortunately, the authors of [3.53] based their study on density measurements at the "wrong" peaks, i.e. the compression peaks rather than the HCS peaks!

Probably these interstream flows can be interpreted as remnants of "coronal streamers". These are large elongated structures, often helmet-like in appearance, which are brighter than the background corona and extend radially outward well beyond $2R_s$. Careful processing of eclipse photographs has revealed them reaching out to at least $10R_s$ (see, e.g. [3.239]). They emerge from above more active solar regions where a mainly closed magnetic loop topology prevails. Further out, streamers often encompass a region of magnetic field reversal, and, in fact, a "streamer belt" is thought to follow the heliomagnetic current sheet quite closely [3.92, 113]. For example, "the moderate to high plasma density and oxygen freezing-in temperatures measured at 1 AU match similar conditions within coronal streamers close to the sun" [3.62]. Although the asscociations between the interstream flow, the streamer belt, and the current sheet seem to be established now, we have to admit that the physical mechanisms connecting them are still far from understood. The low helium content may indicate that the interstream flow emerges from very high layers in a gravitationally stratified solar atmosphere [3.16].

Both the high freezing-in temperatures and the streamer location above the main activity regions indicate the crucial role that activity-related magnetic field fluctuations stirred from the photosphere probably play [3.160] in heating the corona and accelerating this type of solar wind. Somewhere in or around coronal streamers there must occur a transition from the closed coronal magnetic loops (connected to their photospheric footpoints) to the "open" field lines on both sides of the current sheet. Solar wind can emerge from here only

1. if it flows around closed structures, or
2. if it escapes in some kind of quasistationary reconnection at the loop tops, or
3. if it is released in a transient mode, following reconnection processes at the loop tops.

This discussion leaves us with a new aspect concerning the nature of NCDEs [3.71], which were originally thought to result mainly from transient solar activity. It turned out later that "at least some of these events result from spacecraft encounters with coronal streamers" [3.16]. This may no longer be contradictory, since the flow in coronal streamers may be released in a transient mode as well!

We must be aware that this concept of an interstream flow associated with coronal streamers is by no means applicable to the slow solar wind in general. There is just too much slow solar wind, in particular at high solar activity. Most of it is probably emitted from coronal regions far away from the large scale current sheet and any coronal streamers. There is, however, the possibility of similar "interstream processes" acting on smaller scales in coronal regions located above photospheric areas characterized by high activity and strong magnetic fields. This reminds us of the three-component model of the solar magnetic field suggested in [3.4]: besides long-lasting open regions (coronal holes) and long-lasting closed regions (streamer belt) there are "regions which are open but connected to sources of strong photospheric fields rather than coronal holes and also transiently open and closed regions from which coronal plasma escapes into space intermittently and where the topology of the magnetic field changes continually as a result of reconnection" [3.4]. There are in fact indications that such a three-component model might apply to the solar wind as well. This issue will be further discussed in Sect.3.8.1.

3.4.4 Interplanetary Consequences

1. Deflections of the Magnetic Field. There is no doubt that within the previously described 3-D scenario of the solar wind stream structure its radial evolution in terms of dynamical stream interactions will become even more complicated. It was suggested that CIRs can even "absorb the current sheet by overtaking the nearest fold in its wavy structure within or in close association with CIR's" [3.227]. This hypothesis offers an interesting alternative interpretation of certain latitudinal effects discussed the in context of Fig. 3.11 in Sect. 3.3.1. The deformation of the heliospheric current sheet in the presence of meridional velocity gradients has been a long-standing issue [3.224, 227, 166, 11, 225], and as of now the given explanations do not appear to be conclusive. This is discussed in more detail in [3.131].

Here, rather, I address another topic in the context of the interrelation of the current sheet and CIRs. (For illustration see Fig. 3.10). Compression and deflection of the flow has important consequences for the magnetic field. Naturally, it undergoes the same compression and participates in the deflection process too. In fact, in many cases there is an enhanced southward component of the field within the CIR which then reverses to an enhanced northward field in the rear part of that CIR [3.188]. The south-to-north turn always happened in the same way, for both toward-to-away and away-to-toward sector transitions. Before the year 1970, which coincides with the maximum activity of solar cycle 20 and its associated solar magnetic polarity reversal, the same persistent effect occurred with reversed sign.

According to [3.188], the basic reason for this effect is the fact that the magnetic field tends to remain parallel to the inclined current sheet on both sides when reasonably near it. Thus, it obtains meridional components which change sign at the sector boundary. If caught up by the following next CIR, these meridional field components will be enhanced by compression. Usually, at 1 AU the field of the "new" sector is affected most. The flow deflections around the stream interface may cause "azimuthal stress" that eventually causes a reversal of the meridional field polarity. The observed effects and their sole dependence on the basic solar dipole orientation could be consistently explained in this way.

This effect may become important for understanding the interaction of the solar wind with magnetized obstacles such as planets. A southward-pointing field, e.g., is known to favor magnetic reconnection processes on the earth's frontside magnetosphere which lead to the enhancement of geomagnetic activity. (This is a vast field of research in its own right, and the interested reader is referred to the ample literature, specifically the proceedings of a recent Chapman conference on this particular topic [3.106].) We can expect these effects to be rather different when the excursion of B_z is northward first and then southward (as during the years before 1970) or vice versa (as after 1970 to about 1980). In the first case B_z turns southward at a time when the compression is higher on the average and where, as a consequence, the ram pressure onto the magnetosphere is stronger [3.197]. The phase shift between the southward pointing of B_z and the ram pressure maximum follows the 22-year magnetic cycle of the sun (the "Hale" cycle [3.82]), and its effects should be superimposed on the well-known 11-year modulation of geomagnetic activity. There is, indeed, evidence for such 22-year periodicity in geomagnetism (e.g. [3.41, 189]).

All these relationships are very complicated and deserve further attention in the future. Significant progress appears to be possible once the "input conditions" in terms of the solar wind stream structure and its inherent features such as CIRs, stream interfaces, deflections, and sector boundaries are correctly taken into account.

2. Magnetic Clouds at Sector Boundaries? An interesting alternative hypothesis for explaining the north–to–south turn at sector boundaries deserves our attention here; it was proposed in a previous version of paper [3.188]. Further evidence supporting this hypothesis was found later on. The idea is that magnetic loops resulting from reconnecting field lines can occur at sector boundaries. Such loops would produce similar signatures: an observer cutting through a vertically oriented magnetic loop would first encounter an upward field and later on a downward field or vice versa, the order of sequence depending only on the phase of the solar magnetic cycle. In fact, a whole class of events showing a looplike magnetic topology was identified which have been called "magnetic clouds" [3.110] (for details see [3.30]). Many of these were associated with interplanetary shocks and coronal mass ejections; however, about one third of them occurred at sector boundaries, right in front of CIRs. It was suspected that some of the "thick" sector boundaries [3.109] might include magnetic clouds as well. It appears as if magnetic clouds may be common at sector boundaries. Possibly, the different types of clouds all have one feature in common: they might arise from localized transient processes involving magnetic reconnection.

Further evidence for the occurrence of magnetic clouds at sector boundaries was obtained from *Helios* measurements of the electron velocity distribution functions. Their properties are in many respects sensitive to the large-scale magnetic topology (for details see [3.131]). Within the interior of magnetic sectors, usually in high-speed streams, there is often a highly anisotropic component observed at high electron energies (above 100 eV). This strahl is well confined and points away from the sun along the IMF [3.184, 161, 164]. Apparently, scattering of these electrons is weak, and they can follow magnetic field lines from the corona far out into the heliosphere. The angular width of the strahl increases gradually upon approaching sector boundaries, indicating more effective scattering. The preferred flow direction is away from the sun and along the field. The point here is that right at the sector boundaries the distributions become nearly isotropic, with an almost vanishing net heat flux (no strahl) and much lower electron temperatures.

For illustration, Fig. 3.17 (from [3.164]) shows data from early 1976 (covering in part the same period as Fig. 3.12) including electron temperatures (T_e) and sector orientation data. Clearly, there is a distinct dip in T_e around the sector

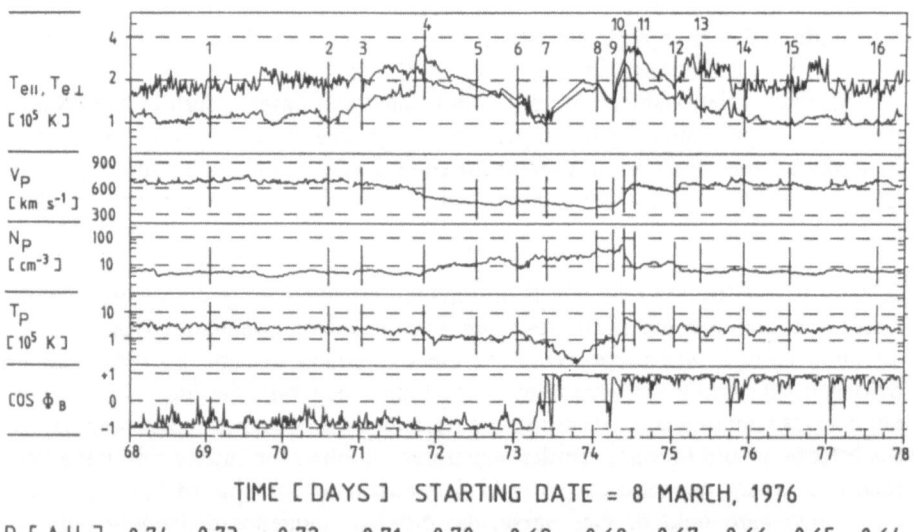

TIME [DAYS] STARTING DATE = 8 MARCH, 1976

R [A.U.] 0.74 0.73 0.72 0.71 0.70 0.69 0.68 0.67 0.66 0.65 0.64

Fig. 3.17. Twenty-minute average solar wind data observed by *Helios 2* during a time interval of 10 days, starting with 8 March 1976, when the spacecraft moved from the interior of a magnetic sector through a sector boundary into the succeeding magnetic sector. In order from top to bottom the parameters are the electron temperatures $T_{e\parallel}$, $T_{e\perp}$ ($T_{e\parallel}$ is the upper curve, $T_{e\perp}$ is the lower curve), the proton number density n_p, the proton temperature T_p, and cosine of the angle ϕ_B between the projection of the magnetic field onto the ecliptic plane and the outward directed Parker spiral. Several observation times have been selected which are indicated by vertical lines and marked with identification numbers. From [3.164]

boundary at (7) on day 73. A brief return to the "old" sector at (9) right at the stream interface caused another dip in T_e. The dips in T_e are accompanied by an almost 100% isotropization of the electron velocity distribution, (i.e. in Fig. 3.17 the components $T_{e\perp}$ and $T_{e\parallel}$ become almost equal). Only a slight bidirectional anisotropy remains. This is what we would expect if the magnetic field lines in these regions are disconnected from the sun. "The electrons should then experience stronger scattering not solely because of an increased collision frequency but also because of the larger travel time of the trapped electrons during the cloud's travel from the sun to the spacecraft" [3.164]. Of course, certain signatures of a cloud may be overridden to some extent if the next following CIR has already engulfed it and causes reheating of the plasma. It is certainly worth pursuing this kind of analysis by sorting out the different effects which may have been superimposed at 1 AU and can be more readily separated in the *Helios* data taken near perihelion.

This discussion leads us to a concept similar to that in the previous section on coronal streamers. Evidence is accumulating for the importance of transient effects in context with sector boundaries, coronal streamers, and the slow solar wind associated with them. There is, in fact, a large class of coronal mass ejections appropriately called "streamer blowouts" (see, e.g., [3.93]), which are ejected mostly at low speeds in the 100 to 400 km s^{-1} range. Very recently even slower ones (at 10 to 50 km s^{-1}) were detected. Hardly visible with present day coronagraphs, they are probably numerous enough to be considered as "the most common form of the phenomenon" [3.104]. Seemingly dissimilar phenomena such as coronal mass ejections, magnetic clouds, NCDEs, streamers, and the slow solar wind in general might be linked together more closely than we ever anticipated. Further work along these lines appears to be very promising.

3.5 Corotating Shock Waves

The main feature of CIRs on the leading edge of high-speed streams is the compression of the plasma and the IMF on both sides of the interface separating the fast and the slow flow (see Fig. 3.10). It was soon realized that "the collision of these plasmas will lead to the formation of two shock waves and a tangential velocity discontinuity between them" [3.49]. The shock at the front of the CIR would be a forward shock, and the shock on the high-speed stream side which seemingly runs back to the sun would be a reverse shock. This shock pair would corotate with the sun as does the solar wind structure that drives them.

Corotating shocks actually do form, although generally at larger heliocentric distances than originally anticipated [3.25, 218, 101, 31]. In fact, despite considerable efforts only very few candidates for corotating shocks could be found at $R = 1$ AU [3.157, 40, 26]. The real question today is why these shocks do not form much closer to the sun, especially in view of the steep velocity gradients close to the sun as discovered from *Helios*. Different MHD models of solar wind stream evolution have been developed (for references and further discussion see, e.g., [3.36]). However, "the models still show an embarrassing tendency to shock that is at odds with the observations" [3.166].

Searching through the *Helios* data set, we tried to find some experimental clues about the possible formation of shocks even within 1 AU (preliminary results are given in [3.177], see also the review [3.178]). About 25 CIRs were found in which some or all of the typical signatures of fast forward shock waves occurred. Events due to solar transients and time periods confused by transients were eliminated from consideration. In less than 10 CIRs were fast reverse shocks involved. The following analysis is mainly restricted to the forward shocks.

Each of the candidate events was checked to see whether it fulfilled the necessary conditions for qualifying as a quasistationary corotating shock (for illustration of the quantities involved, see Fig. 3.18):

1. *The "driver condition"*. In order for a corotating shock to form the fast stream driving it must ram into the ambient slow wind (v_s) fast enough. This is the case once the relative speed between the stream interface and the slow plasma along the normal to the interface exceeds the local fast magnetosonic speed (or the "characteristic speed") $c_f = \sqrt{c_A^2 + c_s^2}$ (with c_A and c_s being the Alfvén and the sound speed, respectively), i.e.

$$(v_i - v_s) \cdot \sin \alpha_i \gtrsim c_f ,$$

where α_i denotes the Parker angle between the interface (not its normal!) and the radial direction.

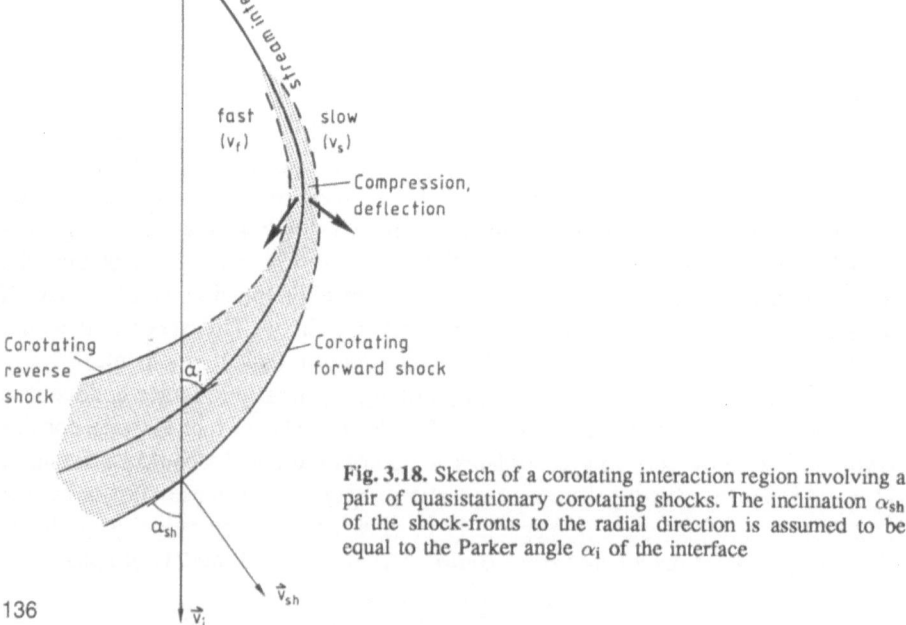

Fig. 3.18. Sketch of a corotating interaction region involving a pair of quasistationary corotating shocks. The inclination α_{sh} of the shock-fronts to the radial direction is assumed to be equal to the Parker angle α_i of the interface

136

2. *The "propagation condition".* The shock front must not propagate slower than the CIR following it:

$$v_{sh}(\text{radial}) \gtrsim v_i \ .$$

Otherwise the CIR would catch up with the shock.

3. *The "evolution condition".* In order for a corotating shock to evolve, its speed must also exceed c_f:

$$v_{sh} - v_s \sin \alpha_i \gtrsim c_f \ .$$

Otherwise there would not be a fast mode shock but just a pressure wave.

The speed v_i of the CIRs and their inclination α_i to the radial direction were calculated according to the momentum formula given in Sect. 3.2.4. Using measured values of the flow speed vectors, the local shock speeds were determined by invoking the conservation of mass through the shock front (see, e.g., [3.95]), which was taken to be oriented parallel to the CIR. This was done in order to represent better the global properties of the CIR and its associated shock rather than the local properties such as the instantaneous *in situ* shock normal. Note, by the way, that corotating shocks can be expected to be "quasiperpendicular", since their shock normal is approximately perpendicular to the magnetic field.

Under this selection procedure on the group of 25 candidates, only 11 cases were retained for further evaluation. Eight of the others, on close inspection, were not fully developed shocks. Two events were unusable because of data gaps or other shortcomings. In the remaining four cases, the above-mentioned conditions were violated so strongly that we cannot regard these events as corotating shocks. A total of 6 out of the 11 cases retained can be regarded as "safe"; the other five must be considered as "marginal" for various reasons.

In Fig. 3.19 one of the best examples of a pair of "safe" corotating shocks is shown. In this case, a discontinuous stream interface traveling at $v_i = 400 \, \text{km s}^{-1}$ was discernible (our momentum formula would yield $v_i = 393 \, \text{km s}^{-1}$). Its inclination to the radial direction would be $\alpha_i = 43.6°$. Consequently, it would travel along its normal into the ambient slow wind (with $v_s = 320 \, \text{km s}^{-1}$) with a relative speed of $(400\text{–}320) \sin \alpha_i = 55.2 \, \text{km s}^{-1}$. This speed slightly exceeds the local value of $c_f = 54.9 \, \text{km s}^{-1}$, indicating that the stream interface moves just fast enough for a shock to be driven. Furthermore, the actual shock front travels $44 \, \text{km s}^{-1}$ faster than the interface (in the radial direction) and is $31 \, \text{km s}^{-1}$ faster than the local c_f, i.e. the Mach number $M = 1.56$.

In this particular case, a similar evaluation of the reverse shock associated with that same CIR was possible, *mutatis mutandis*. We find a relative speed between the high-speed stream and the stream interface along its normal of $97.8 \, \text{km s}^{-1}$, compared with a local c_f of $97.2 \, \text{km s}^{-1}$. The shock front runs into the high-speed stream $43.5 \, \text{km s}^{-1}$ faster than the interface, and it is $32 \, \text{km s}^{-1}$ faster than the local c_f ($M = 1.32$).

The other cases of forward shocks were analyzed the same way. Their properties, displayed in the histograms of Fig. 3.20, show some interesting features:

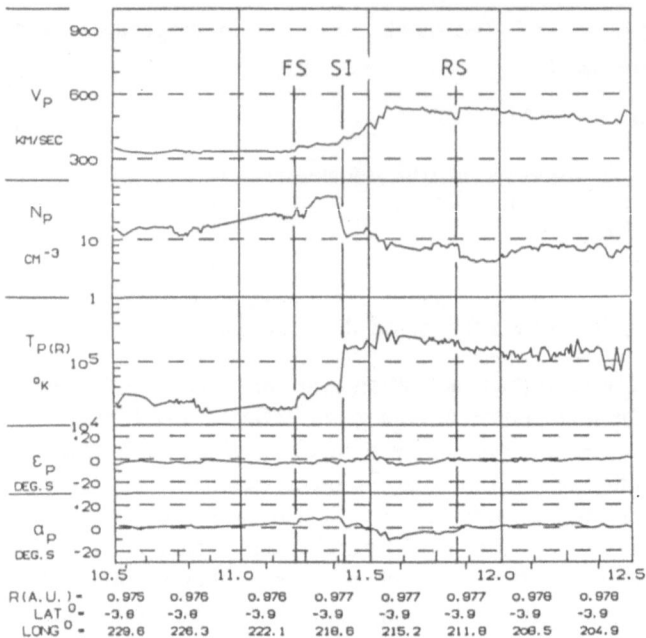

Fig. 3.19. A pair of quasistationary corotating shocks at the front edge of a high-speed stream as observed from *Helios 2* on 11 January, 1977 at $R = 0.977$ AU. Note also the discontinuous stream interface around 0900 UT

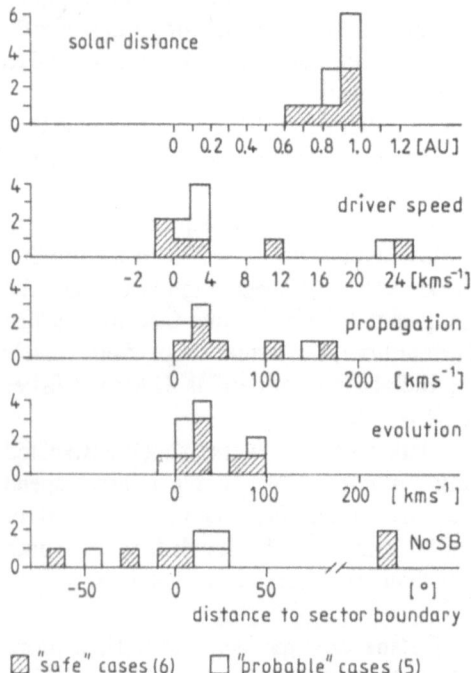

Fig. 3.20. Histograms showing properties of 11 examples ("safe" or "probable") of corotating forward shocks. The driver speed, propagation, and evolution panels indicate the margins by which the respective conditions are met (see text)

1. As expected, the chance of a corotating shock occurring grows with increasing R. The corotating shock closest to the sun was seen at 0.63 AU. The effect is to some extent overemphasized by the fact that the *Helios* probes spent much more time at large R than close to the sun. However, very many CIRs were seen at all R (see, e.g., Figs. 3.4 and 3.5). Thus, the trend is certainly real, and it will continue beyond 1 AU.

2. The "driver condition" is strikingly well met in most cases. Only rarely is the stream interface significantly faster than required for generating the corotating shock. We can expect that this will probably change beyond 1 AU, when c_f drops further. On the other hand, one is tempted to conclude that a corotating shock will form right when the "driver condition" is met, not only when it is markedly exceeded.

3. The "propagation condition" plot shows that most often the corotating shocks do not propagate much faster than the following stream interface. The relative separation will thus increase very slowly. The fate of the three very fast propagating shocks is unclear. They may eventually even detach from their driver and then run on their own in a "blast wave mode", according to the propagation conditions they encounter.

4. The "evolution condition" plot tells us that these corotating shocks are not very "strong". On the average, they move at a Mach number $M \approx 1.6$. The average local value of c_f ahead the corotating shock was $61.2\,\mathrm{km\,s^{-1}}$, with a remarkably small standard deviation of $6.8\,\mathrm{km\,s^{-1}}$.

5. There seems to be no significant correlation between the positions of sector boundaries and corotating shocks. However, shock formation in the sector boundary environment is certainly fostered by the low values of T_e and T_i (see Fig. 3.17) and sometimes low value of $|B|$, and by the high values of the density. Any of these conditions would contribute to lowering the local c_f.

It is worth noting that five of the six "safe" cases were followed by a discontinuous stream interface. Also, all five of the corotating reverse shocks uniquely identified so far were associated with discontinuous stream interfaces. The statement in [3.72] that "reverse shocks form most readily within those streams which contain abrupt interfaces" apparently applies to forward shocks as well.. Furthermore, we found abrupt stream interfaces in six out of the eight "almost shock" events. This subgroup of rejected events is interesting in itself: in six of them the driver condition was missed by an average of $42\,\mathrm{km\,s^{-1}}$. In the other two cases, despite a marginally fulfilled driver condition, the density ridge had not yet steepened into a shocklike discontinuity: the local c_f was still too high to allow a shock wave to evolve. The overall longitudinal speed gradient at a CIR appears to play a less crucial role in the shock formation process. On the average, for the 11 CIRs involving shocks $\Delta v / \Delta \phi$ amounted to $(33 \pm 14)\,\mathrm{km\,s^{-1}}$. These gradients do not differ from those for all other CIRs (see Fig. 3.5).

There is no reason to expect a corotating shock always at the outermost edge of a CIR. The point where the CIR steepens most and then shocks depends only on local conditions. The possible presence of transient magnetic clouds in this environment led me to suggest that a corotating shock will probably not be coherent on a large scale. Depending on local conditions, there will be a corotating shock in some regions, just a pressure wave in others. Furthermore, it appears that corotating shocks cannot be viewed as large-scale stationary features in front of high-speed streams. At least within and around 1 AU where their formation starts, some of them may decay again (if, e.g., they run "too fast" and reach a medium where c_f is too high) and reform later. It should be noted that the plasma driving these corotating shocks does *not* participate in this corotation. It is the shock "pattern" that corotates, while the flow itself is nearly radial everywhere. The driver gas encountered at different longitudes was emitted from the same solar source region, it is true, but at different times. Any time variations in the source will thus affect the character of the CIR and the shape of a potential corotating shock. A typical example of this effect was shown in [3.32].

3.6 Scenario in Terms of the "Ballerina Model"

In this section the basic content of the previous sections shall be recapitulated in an illustrative manner in terms of the "ballerina model" first proposed by Alfvén [3.2]. Figure 3.21 is an artist's view of the inner heliosphere as it may appear immediately before a typical solar activity minimum, e.g., in 1975 [3.197]. We find the sun's poles covered by large coronal holes. These are areas of open magnetic field lines, the northern hole being of positive (outward directed)

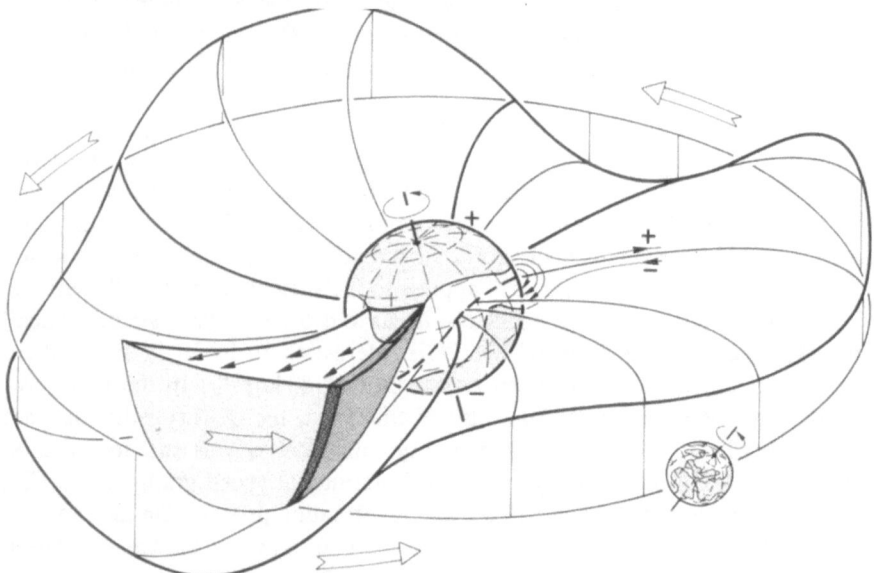

Fig. 3.21. The inner heliosphere during solar activity minimum, in terms of the "ballerina" model as proposed by Alfvén [3.2]. From [3.197]

polarity, the southern hole being negative. Some tonguelike extensions on roughly opposite sides of the sun reach well into the equatorial regions. This is due to the fact that the global solar dipole axis is tilted with respect to the rotation axis. Bright active centers (interspersed with some sunspots) and their looplike and mainly closed magnetic structures are mostly confined to the sun's equatorial region. Above this region streamers extend far out into the corona and form a bright band which meanders all around the sun according to the shape of the photospheric activity belt. What looks like the skirt of a spinning ballerina is the warped separatrix between positive and negative solar magnetic field lines (i.e. the heliomagnetic current sheet) dragged out into interplanetary space by the radially outflowing solar wind. The skirt is "attached" to the corona on top of the streamer belt, i.e. generally in the middle of the near-equatorial activity belt and in between the large coronal holes. If the spinning skirt passes an observer sitting, say, at the earth, he would notice an interplanetary sector boundary crossing. The size and number of magnetic sectors is thus closely related to the structure of the underlying corona.

The coronal holes are the sources of high-speed solar wind, while the emission of slow solar wind is sharply confined to the near-equatorial activity belt of about 40° angular width. The warps of the interplanetary separatrix (heliomagnetic current sheet) with respect to the heliographic equator allow high-speed streams to extend to low latitudes so that they become observable even in the plane of the ecliptic. The inclination of the stream boundaries to the heliographic equator causes stream–stream interactions to become effective. With increasing distance from the sun flux tubes containing plasma with higher flow speeds start jamming into the preceding slow flow and cause compression and deflection of the flow on both sides of the boundary, thus forming a CIR. In some cases there are abrupt stream interfaces separating originally fast from originally slow plasma. Local plasma processes tend to counteract the formation of a pressure ridge and thus cause gradual widening of the CIR, thereby opposing further kinematic steepening. Eventually (in rare cases even within 1 AU), the local characteristic speed c_f controlling the propagation of the developing pressure "pulse" drops below the interface speed, and a corotating shock or even a forward–reverse shock pair forms. These shocks further "erode" the originally steep flow profiles. Moreover, CIRs of adjacent high-speed streams start interacting and merging. Finally, in the outer heliosphere beyond some 10 AU, the originally spokelike radial streams of very different speeds eventually turn into huge azimuthally uniform shells of high-density plasma following each other at about the same speed. In other words (quoted from [3.39]): "A stream gives birth to an interaction region, and an interaction region leads to the death of the stream."

Close to the sun the heliomagnetic current sheet is embedded in the middle of the slow wind belt. Its angular distance to the coronal hole boundary and the high-speed flow emanating from there is typically of the order of 20°. Longitudinal stream interactions deform the original warps on a large scale. Thus, both the solar wind stream configuration and the inherent magnetic field develop into a many-layered 3-dimensional structure. This structure may become additionally complicated by transient phenomena such as magnetic clouds formed by magnetic reconnection at the sector boundaries.

Around the time of minimum solar activity all these processes will probably not affect solar ecliptic latitudes beyond $\pm 40°$, where we would rather expect a steady breeze of fast-type solar wind to prevail. The whole pattern will change dramatically during the course of the activity cycle once new magnetic flux starts emerging from fresh sunspots at high latitudes. I will address this issue in Sect. 3.8.

3.7 The Solar Wind as a Two-State Phenomenon

In the previous sections we discussed the global organization of solar wind flow in the context of its underlying coronal structure and the large-scale solar magnetic field. The distinct grouping of the solar wind into quasistationary high-speed streams and low-speed plasma and, in particular, the sharp separation between these two types leads us to think of two modes or states of the solar wind which might be due to essentially different generation mechanisms in the corona (not to speak of various types of transient flows). There have been attempts to explain the strange "switching" between these states in terms of multiple critical points. These may arise once the divergence of magnetic field lines exceeds certain limits [3.112, 90], e.g. in coronal holes, and may involve the development of "standing shocks" in the inner solar wind [3.79, 80]. Unfortunately, to this day there are no conclusive answers to such basic questions as

- What accelerates the fast solar wind to its high-speed?
- Where does the slow wind actually emerge?
- What causes the sharp boundaries, both of coronal holes and high-speed streams?

Of course, these questions are closely related to the most fundamental question in all coronal physics:

- How is the corona being heated?

Treating these issues with the care they deserve would certainly go far beyond the scope of this article. Here, rather, I restrict myself to recapitulating the basic properties of the different states of the solar wind and what we can possibly learn from them with respect to the coronal sources. It will be seen that some boundary conditions and constraints for further theoretical modeling can be derived.

From what has been said so far, it appears reasonable to categorize the solar wind according to its speed before evaluating average quantities. Such a study was done in [3.198], the main results of which are summarized in Fig. 3.22. It is based on the complete set of plasma data taken from *Helios 1* and 2 up to the end of 1976 and the data taken simultaneously by the Los Alamos plasma analyzers

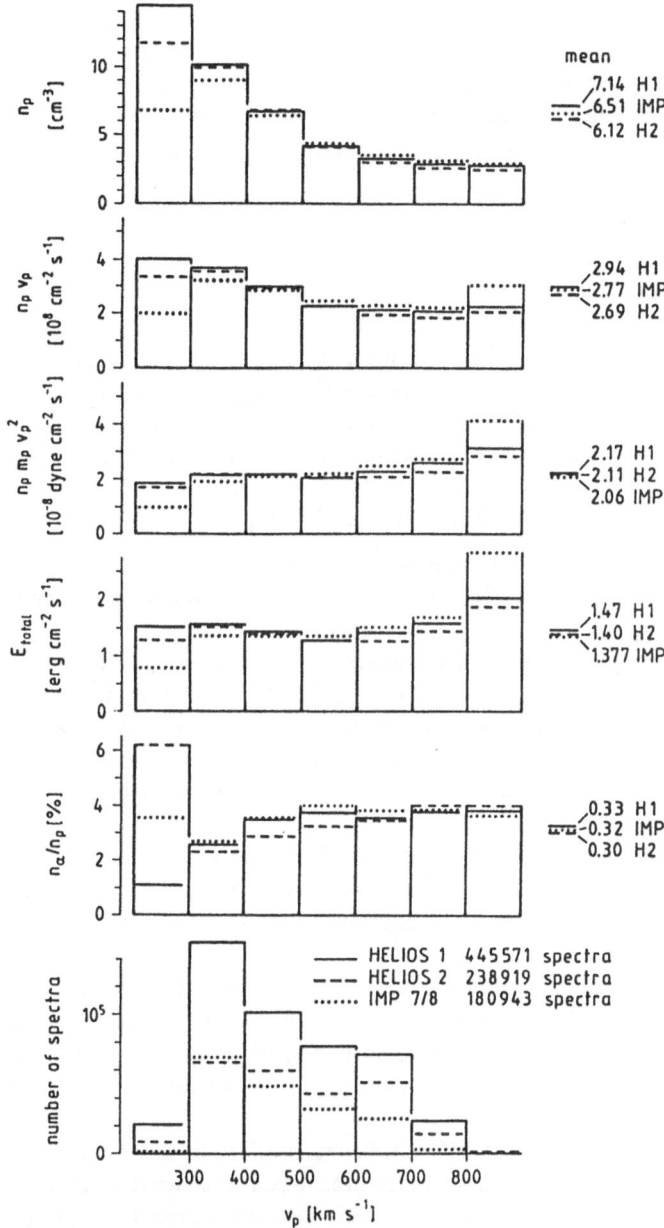

Fig. 3.22. Average values of the solar wind proton density n_p, proton flux density $F_p = n_p v_p$, momentum flux density $M_p = n_p m_p v_p^2$, total energy flux density E_t and α-particle abundance $A = n_\alpha / n_p$. The bottom panel shows the number of individual measurements in each speed class (not valid for α-particles). The data were obtained from the *Helios 1* and 2 solar probes and the *IMP 7/8* satellites and cover the time period from 12 December 1974 to 31 December 1976. All densities were normalized to 1 AU, assuming an R^{-2} dependence on solar distance. Some shock disturbed data were excluded. From [3.198]

on *IMP 7/8*. Only very few shocks occurred in this interval; the disturbed data were excluded. The data were grouped into seven speed classes, and then averages of the relevant quantities were calculated. The density values were normalized to 1 AU assuming an R^{-2} dependence. The bottom panel in Fig. 3.22 shows how the data are distributed over the different speed classes (not valid for the fifth panel).

The quantities displayed are:

- the proton density n_p,
- the proton flux density $F_p = n_p v_p$,
- the proton momentum flux density $M_p = n_p m_p v_p^2$,
- the total energy flux density E_t. It is the sum of the kinetic energy flux density $E_k = n_p m_p v_p^3/2$ and the potential energy flux density $E_g = n_p v_p G m_p M_s/R_s$ (basically the work done in moving the solar wind out of the solar gravitational potential. G is the gravitational constant. M_s and R_s denote the sun's mass and radius, respectively),
- the helium ion to proton abundance ratio $A = n_\alpha/n_p$.

All minor energy fluxes such as enthalpy transport, heat and wave energy fluxes (discussed later on) are not included in E_t, since their sum does not exceed 3% of E_t.

These average quantities are certainly affected by compressions in the CIRs. Therefore, in order to get "clean" results, one might wish to omit CIR data. However, in a dedicated study based on a representative subset of data we found that the influence of the CIRs on the overall trends is only minor. The reason is that the width of these regions is small compared to the streams themselves, in particular at decreasing R.

At a first glance we note systematic trends for all these quantities as functions of v_p (only the group with $v_p > 800 \,\mathrm{km\,s^{-1}}$ shows marked deviations; this group can be discarded, however, because of the extremely low number of samples in it).

Many parameters exhibit a significant dependence on the flow speed, such as n_p, but also F_p and A and some other parameters not shown in Fig. 3.22. In fact, they might even be used, in addition to v_p itself, as indicators of the state of a particular solar wind plasma. On the other hand, the quantities M_p and E_t are remarkably independent of v_p.

In order to evaluate both the basic differences and similarities in more detail, I assembled Table 3.1 giving average numbers for most basic solar wind parameters at 1 AU. Here a coarser binning of the data yields average values for slow (< $400 \,\mathrm{km\,s^{-1}}$) and fast (> $600 \,\mathrm{km\,s^{-1}}$) wind, plus the grand averages. Most of the values stem from [3.198], the remaining numbers (from *Helios*) were derived essentially from data taken in the same time interval around the solar activity minimum in 1976. For reference, I listed values assembled in [3.57] on the basis of *IMP 6/7/8* data at 1 AU taken during the $3\frac{1}{2}$ years before July 1974. These data are consistent with (though slightly different from) subsets published in [3.56, 8, 59].

Table 3.1. Average values of basic solar wind parameters measured from *Helios 1* and 2 between December 1974 and December 1976, and from *IMP 7/8* (at 1 AU) from 1972 to 1976. All parameters involving n_p were normalized to 1 AU assuming an R^{-2} dependence. Only those *Helios* temperature data obtained between $R = 0.9$ and 1 AU were included. All 1 AU data were taken from the data collection in [3.57] (Note that all quantities involving n_p (except for A) obtained by the Los Alamos plasma analyzers on the *IMP* satellites were multiplied by a factor 0.70 which was determined from cross-calibrations.)

Parameter		slow ($< 400\,\mathrm{km\,s^{-1}}$)		fast ($> 600\,\mathrm{km\,s^{-1}}$)		all		
		Helios	1 AU	*Helios*	1 AU	*Helios*	1 AU	Ref.
v_p	$(\mathrm{km\,s^{-1}})$	(348)	(327)	(667)	(702)	481	468	3.198
n_p	$(\mathrm{cm^{-3}})$	10.7	8.3	3.0	2.73	6.8	6.1	3.198
$F_p = n_p\,v_p$	$(10^8\mathrm{cm^{-2}s^{-1}})$	3.66	2.7	1.99	1.90	2.86	2.66	3.198
$M_p = n_p\,m_p\,v_p^2$	$(10^8\mathrm{dyn\,cm^{-2}})$	2.12		2.26		2.15	2.03	3.198
$A = n_\alpha/n_p$		0.025	0.038	0.036	0.048	0.032	0.047	3.198
$E_k = \tfrac{1}{2} n_p\,m_p\,v_p^3$	$(\mathrm{erg\,cm^{-2}s^{-1}})$	0.37	0.25	0.76	0.79	0.52	0.49	3.198
$E_g = n_p\,v_p\,G\,m_p\,M_s/R_s$	$(\mathrm{erg\,cm^{-2}s^{-1}})$	1.17	0.87	0.65	0.60	0.91	0.85	3.198
$E_{th_p} = n_p\,v_p\,\tfrac{5}{2}kT_p$	$(10^{-3}\mathrm{erg\,cm^{-2}s^{-1}})$	11	3.0	23	16	16	11	3.198
$E_{th_e} = n_p\,v_p\,\tfrac{5}{2}kT_e$	$(10^{-3}\mathrm{erg\,cm^{-2}s^{-1}})$		11		7		12.6	
Q_e	$(10^{-3}\mathrm{erg\,cm^{-2}s^{-1}})$	6	2.7	8	2.2		4.3	3.165
Q_p	$(10^{-3}\mathrm{erg\,cm^{-2}s^{-1}})$	0.08	0.029	0.3	0.16		0.13	3.138
E_A	$(10^{-3}\mathrm{erg\,cm^{-2}s^{-1}})$	–	0.7	14	6.7		5.7	3.46
$E_t = E_k + E_g$	$(\mathrm{erg\,cm^{-2}s^{-1}})$	1.55	1.18	1.43	1.505	1.45	1.46	3.198
T_p	$(10^3\,\mathrm{K})$	55	34	280	230		120	3.136
$T_{p\parallel}/T_{p\perp}$		1.7		1.2			1.5	3.136
T_e	$(10^3\,\mathrm{K})$	190	130	130	100		140	3.165
$T_{e\parallel}/T_{e\perp}$		1.2		1.6			1.18	3.165
T_α	$(10^3\,\mathrm{K})$	170	110	730	1420		580	3.135
$T_{\alpha\parallel}/T_{\alpha\perp}$		1.4		1.3				3.135
T_α/T_p		2.9	3.2	3	6.2		4.9	3.135
L_p	$(10^{30}\,\mathrm{dyn\,cm\,sr^{-1}})$	2.14		−0.56		0.2–0.3		3.167
L_M	$(10^{30}\,\mathrm{dyn\,cm\,sr^{-1}})$	0.14		0.15		0.15		3.167

In [3.57] slightly different limits were applied ($<350\,\mathrm{km\,s^{-1}}$ for slow and $> 650\,\mathrm{km\,s^{-1}}$ for fast wind, respectively); also, any data affected by CIRs or other dynamical processes were eliminated there. These procedural differences may explain some quantitative differences which are, however, of no significance in the actual context.

In the following sections I discuss various issues on the basis of Table 3.1. Error estimates are not included here, since they require very detailed discussions of the measurement and evaluation techniques applied to the individual data sets (the interested reader is referred to the original literature listed in [3.57]). In order to judge the statistical significance of the individual average quantities, I calculated in each case the standard deviation σ as a measure of the range of variations. The values of σ are not included in Table 3.1, but I sometimes refer to σ (given in percent of the average) in the discussion.

Note that all quantities involving n_p which are published by the Los Alamos group on the basis of *IMP* data should be multiplied by a factor of 0.70 before comparing them with *Helios* data [3.198]. This correction factor was determined by careful inflight cross calibrations right after both *Helios* launches. It is already incorporated both in Fig. 3.22 and in Table 3.1. With respect to the absolute density values, measurements of the local electron plasma frequency ($f_e^2 = n_e e^2 / \pi m_e$) by the plasma wave instruments on board *Helios* confirmed our calibration to within a 20% uncertainty [3.78].

3.7.1 Slow and Fast Solar Wind: Characteristic Differences

1. Flow Speed v_p. Of course, the most dramatic difference between the "fast" and "slow" state of the solar wind is the flow speed itself. This is not as trivial a statement as it may sound since it turns out that the solar wind may at times take on all properties of the "fast" type except one: the high-speed [3.134].

The numbers of the average v_p in Table 3.1 were put in parentheses since they reflect basically the somewhat arbitrary binning of the data used to define the "slow" and "fast" states.

2. Proton Density n_p and Flux Density F_p. The most prominent difference of the two states (apart from the speed) is apparently the particle density. The proton density n_p in the slow wind is a factor of 3.6 higher than in the fast wind. As a consequence of this huge factor, the proton flux density F_p in the slow wind is also higher by a factor of 1.83. The standard deviation σ of F_p is about 55% and tends to be slightly lower in fast wind.

3. Helium-to-Proton Abundance Ratio A. For determining A only those *Helios* data taken outside 0.5 AU were selected, since the evaluation of helium data inside 0.5 AU is not always possible for instrumental reasons [3.184]. There is no doubt that A in the fast wind (3.6%) is significantly higher than in the slow wind (2.5%), in general agreement with [3.57] (the differences in the actual numbers, which have been reported as 4.8% and 3.8%, respectively, may be due to measurement and evaluation techniques and to the different phases in the solar cycle). The difference in A between fast and slow wind may actually be even more pronounced since spectra with too few helium ions had to be excluded from the analysis. Such depletions occur mainly in the slow wind. The average A in slow wind is thus probably even less.

The trend toward low helium abundances in the slow solar wind had been noted earlier (see [3.8] and references therein). In particular, significant deple-

146

tions in A were found in the proximity of sector boundaries which are normally imbedded in slow solar wind [3.16], as discussed in Sect. 3.4.4. It is important to note here that this analysis is based on data taken during solar activity minimum. With increasing activity the average value A for the slow wind also increases (see Sect. 3.8.).

Another aspect should be addressed in this context: the variability of A. In high-speed streams, A is usually steady and does not vary much from stream to stream. I found a σ value of 50%. In contrast, A can be both high (sometimes up to 20%) and low in low-speed wind, despite its low average value. Here σ is 70%. This behavior is illustrated and discussed in more detail in [3.8].

The attentive reader is probably surprised to find an average solar wind helium abundance of less than half of the photospheric value (some 10%). On the other hand, theoretical models let us expect strong helium enhancements (up to some 30%) in certain coronal strata (see, e.g., [3.24]). Apparently, the value of A in any solar wind plasma volume depends on the altitude at which the solar atmosphere (thought to be stratified by the combined action of gravity, electric forces, thermal diffusion, ionization/recombination, etc.) is "tapped" to release this particular volume. This would explain the occasionally very high values of A in coronal plasma ejected from low layers in the course of large solar transients (see, e.g., [3.87, 13]). The difference in the A values for high- and low-speed solar wind would indicate that high-speed flow is generally released at lower altitudes, a conclusion consistent with other observations (see Sect. 3.7.1).

4. Energy Fluxes

- *Kinetic energy E_k*: Since v_p appears in E_k to the third power we find twice as much kinetic energy flowing in the fast wind as in the slow wind. With an average helium abundance of $A = 0.04$ the total kinetic energy flux would amount to 16% more than these numbers.

- *Potential energy E_g*: This quantity is proportional to F_p and thus falls off with increasing v_p the same way. Note that in the slow wind E_g exceeds E_k by a factor of 3. This tells us that most of the energy the sun supplies to generate the slow wind is used up for overcoming the sun's gravity. With respect to helium, the same argument holds as for E_k.

- *Thermal energy E_{th}*: The convected local proton enthalpy flux at 1 AU is less than 3% of E_k in both states. The numbers for the electron enthalpy flux are of the same order. Of course, the closer to the sun the higher are the temperatures and, consequently, the local enthalpy fluxes as well as their percentage of the total energy flux. This issue addresses both the problem of basic energy input into the solar wind and the local energy conversion processes in the evolving solar wind. For details see [3.132].

- *Electron heat flux Q_e*: Although at 1 AU the values of Q_e are about equal in slow and fast wind ($\approx 7 \times 10^{-3}$ erg cm^{-2} s^{-1}), the radial gradients differ substantially [3.165]. In the slow wind regions close to and around sector boundaries, Q_e is reduced to 3×10^{-3} erg cm^{-2} s^{-1}. This is evidence for the occurrence of closed magnetic field structures disconnected from the sun, as was discussed in Sect. 3.4.4.

- *Proton heat flux Q_p*: In high-speed streams Q_p rarely exceeds 1×10^{-3} erg cm^{-2} s^{-1} at 1 AU and drops by an order of magnitude in the slow wind [3.136, 138]. As with Q_e, Q_p is lowest around sector boundaries. Generally, there is no doubt "that the heat flux is without dynamical importance for the wind expansion within 0.3 and 1 AU" [3.138].

- *Alfvén wave energy flux E_A*: A characteristic signature of the solar wind high-speed state is the obvious occurrence of Alfvénic turbulence [3.12, 47]. Substantial fluctuations of the proton velocity vector are highly correlated (or anticorrelated) with those of the magnetic field vector. At the same time both the proton density and the field magnitude remain essentially constant. In contrast, any fluctuations within typical low-speed flows have lower amplitudes at corresponding time scales, although the variations on larger time scales are much greater. Moreover, the lack of correlation between flow and field components indicates a stronger amount of compressive components which may be due to static structures convected by the solar wind, or magneto-acoustic waves, or both [3.47].

  In high-speed streams, a typical value for E_A is 14×10^{-3} erg cm^{-2} s^{-1} [3.46] (11.6×10^{-3} erg cm^{-2} s^{-1} in [3.56]). Thus, E_A amounts to only some 3% of E_k. Although at 0.3 AU this percentage is somewhat higher (6% [3.46]), we conclude that the impact of Alfvénic fluctuations on the dynamics of solar wind evolution is only minor in this distance range. This topic is addressed again in Sect. 3.7.3 and is dealt with in more detail in [3.132, 133].

Summarizing this subsection, we note that the energy transported by the evolving solar wind is clearly dominated by the kinetic energy E_k of the protons. E_k is of the same order as the gravitational energy E_g. All other energy fluxes due to thermal motions, anisotropies, MHD fluctuations, and waves are of minor importance. This confirms our previous perception of the flow speed as one of the basic imprints in the solar wind reflecting the state of its individual coronal source. The sum of all energy fluxes E_t, however, is strikingly equal, on the average, for all flow speeds. We return to this issue in Sect. 3.7.2.

5. Temperatures of Ions and Electrons. Temperatures of all species were generally expected to be highly anisotropic in the essentially collisionless magnetized plasma of the solar wind. However, the actual degrees of anisotropy were found to be not nearly as high as predicted, e.g., by exospheric theories ([3.105]; see also the discussions in [3.95] and [3.136]).

Thanks to a variety of local microscopic processes involving plasma instabilities, turbulence, wave particle interactions, etc. (see [3.132]), the particle velocity distribution functions observed *in situ* exhibit anisotropies that rarely exceed a factor of 2 [3.136, 165, 128], despite some highly anisotropic features such as double ion beams [3.54, 136] or high-energy tails (strahl [3.184, 59, 163]). Note that the temperatures quoted here are averages, in the sense that they were derived from the physically more relevant quantities $T_\perp$ and $T_\parallel$ (referred to the magnetic field) by the simple formula

$$T = \frac{1}{3}(T_{\parallel} + 2T_{\perp}) .$$

For detailed discussions of velocity distribution functions and their implications the reader is referred to [3.132].

- *Proton temperature T_p*: Here I compare *Helios* values from [3.136] to those from *IMP* [3.57] (values in parentheses). At 1 AU we find $T_p = 55 \times 10^3$ K (34×10^3 K) in the average slow wind, and 280×10^3 K (230×10^3 K) in the fast wind. Again, the differences due to measurement and evaluation techniques as well as to observation times are apparently substantial; the trends, however, are the same.

  The usefulness of these average numbers is somewhat questionable. In the case of the slow wind the variability of T_p is large and reflects more or less the variability of spatial structures embedded in the slow wind. In that sense, T_p behaves similarly to other quantities such as n_p and A. For the fast wind, in contrast, T_p is fairly steady (apart from small-scale fluctuations due to local phenomena) and does not vary much from stream to stream.

  Additional characteristic differences between solar wind states show up in the radial temperature profiles discussed in [3.132].

- *Electron temperature T_e*: The *Helios* electron data from [3.165] (only those obtained between 0.9 and 1 AU) are compared to those from [3.57]. At 1 AU we find an average of $T_e = 1.9 \times 10^5$ K (1.3×10^5 K) in the slow to intermediate wind outside of sector boundaries and 1.3×10^5 K (1.0×10^5 K) in the fast wind. This trend first noted in [3.55] indicates that in the coronal source regions of high-speed streams T_e is probably lower than in the source regions of the low-speed wind. Furthermore, these measurements clearly indicate that T_p exceeds T_e in high-speed streams by a factor of 2 to 2.5, at 1 AU. The different radial gradients of T_e and T_p imply that this factor increases to 4.5 at 0.3 AU and to even higher values at smaller R. This apparently not widely known fact was addressed in [3.56]: "The different behaviour of protons and electrons within high-speed streams raises an interesting question concerning the transport of internal energy in the solar wind at 1 AU. It has been thought for some time (e.g. [3.94]) that heat carried by solar wind electrons provides most of the internal energy to the interplanetary plasma. The problem posed by the present data is that it is difficult to understand how a cooler gas (the electrons) can heat a hotter gas (the protons) within a region so far away from the sun, which is the ultimate source of most of the energy in interplanetary space." Although this frank statement was written 12 years ago, it is just as applicable today. This striking observational fact, of course, can only mean that ions and electrons in the high-speed solar wind are essentially decoupled from each other and obey very different radial evolutions.

6. Angular-Momentum Flow L. The sun continually loses angular momentum to the expanding corona and solar wind. This "solar spindown" has always been exciting from an astrophysical point of view. In their landmark paper [3.233] Weber and Davis set up a self-consistent theoretical model describing the magnetic coupling between stellar rotation and winds (WD model). Among other things, they predicted that the solar wind azimuthal flow component due to the estimated angular momentum flux can only be very small (1 km s^{-1} at 1 AU).

Experimenters have been struggling to measure this component ever since, yielding rather controversial results (see the summary in [3.167]). Taking advantage of the special virtues of the *Helios* double mission (identically built and calibrated instruments on identical space probes with oppositely directed spin axes, approach to the sun to 0.29 AU, exact cross calibration in orbit because of favorable orbits), the puzzle could finally be solved. In a very thorough and detailed analysis [3.167] the total angular momentum flux loss rate L (field + particles) near the solar equator was found to be $(0.2–0.3) \times 10^{30}$ dyn cm sr^{-1}. This is about one-quarter of what the WD model had predicted and much lower than previously measured values. The low value of the total flux can be reconciled with the WD model "by moving the mean Alfvén radius, R_A, in to $12 R_s$, a distance that is consistent with coronal models more realistic than the single polytrope formulation used by Weber and Davis". This statement from [3.167] tells us why the analysis of L is so relevant in the context of this section on basic properties of solar wind streams: it allows us to make inferences on the location of the Alfvén radius R_A, since, in the most simple model, we have the relation $L = \Omega \, R_A^2$.

In Fig. 3.23 (from [3.167]) we see histograms of L_p (i.e. the angular momentum transported by the particles) versus solar distance, split up into three solar

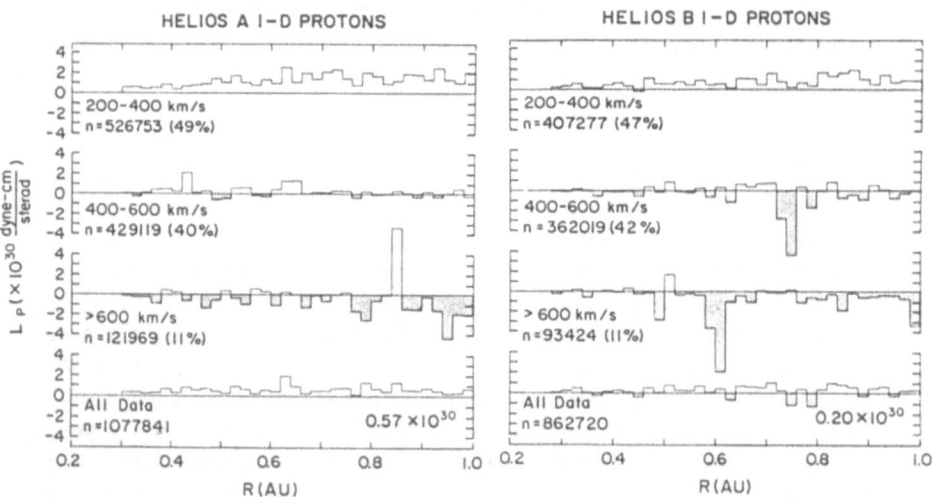

Fig. 3.23. Histograms of proton angular momentum flux L_p versus solar distance R. In the top three histograms, the data averaged in 0.02 AU bins are segregated according to flow speed, while the bottom one shows the composite. The numerals n refer to the total number of data points in each histogram, and the percentages in the parentheses list the fraction of all measurements falling within that speed regime. The mean over all values of R for the summed histograms is given at bottom as $\langle L_p \rangle$. From [3.167]

wind speed classes: "slow" (v_p from 200 to 400 km s^{-1}), "intermediate" (400–600 km s^{-1}) and "fast" (> 600 km s^{-1}), separately for *Helios 1* and *Helios 2*. As stated in [3.167], "the histograms reveal a number of interesting features. First, despite the large number of data points, the residual scatter in the histograms remains high, reflecting the inherent delicacy of the observation. Second, there is an obvious progression from slow to fast flow regimes, in that the former carries decidedly positive flux, the latter predominantly negative. Third, there is some evidence for the transfer of angular momentum from high-speed wind to low-speed wind with increasing heliocentric distance. This is most likely an artifact of the well-known east–west deflection at stream interaction fronts, whereby slow wind undergoes acceleration in the direction of corotation and fast solar wind is deflected in the opposite sense. Fourth, the composite histograms (bottom) suggest a slight positive net radial trend. This could be explained by a gradual transfer of momentum between alphas and protons".

This latter conclusion arises from the fact that the differential speed vector between heavy ions and protons is parallel to the magnetic field which, with increasing solar distance, follows the Parker spiral curvature. Thus, ions moving faster than protons would tend to be deflected "backward", i.e. to run counter to solar rotation. In fact, a negative trend for the angular momentum flux carried by helium ions in high-speed streams was observed [3.167].

The low or seemingly negative value for L_p in high-speed streams close to the sun indicates that this type of solar wind is released and thus decoupled from the rotating sun much closer in than the slow wind. Estimates for the possible differences in the Alfvén critical radii R_A were given in a complementary study [3.137] which intended to clarify in detail how the angular momentum is distributed between particles and the field. Surprisingly, the angular momentum flux L_M carried by magnetic stresses turned out to be independent of the flow speed and is constant between 0.3 and 1 AU. As stated above, L_p depends strongly on the flow speed, and thus so does the ratio L_p/L_M. One finds $R_A = 34$–$49\,R_s$ in the slow, and $R_A = 14$–$17\,R_s$ in the fast, solar wind. Along this line of arguments, other characteristics of the Alfvén critical point could be inferred which led the authors of [3.137] to conclude: "In addition, the plasma may partly corotate up to much larger heliocentric distances in low-speed solar wind, whereas in fast streams it seems to be decoupled from solar rotation already at about R_A. At this point, the corresponding radial flow speed profile has been estimated to be rather steep. If the inferred gradient of 50 km s$^{-1}/R_s$ can actually be maintained for only 10 R_s, e.g., then the typical flow speed of 680 km s^{-1} as observed at 0.3 AU might have been attained already within the first $30R_s$. Therefore, high-speed flow observed by *Helios* at $63R_s$ has essentially reached its asymptotic state. For the slow solar wind, however, it appears that there should be at least at times a fair chance to observe the wind at its stage of acceleration."

The last remark certainly catches our attention in view of the fact that on very rare occasions the solar wind was found to be sub-Alfvénic at 0.3 AU [3.198] and in one reported case even at 1 AU [3.74].

7. Different Properties of Protons and Heavier Ions. It had been noted early that the temperatures of different ion species in the solar wind are far from being equal. Instead, all ionic species tend to have the same temperature per unit mass or, in other words, all ions' thermal speeds are about equal [3.180, 194, 86]. Furthermore, helium and other heavy ions often travel markedly faster than the protons ([3.3, 75, 135]; for a survey of helium observations and interpretations see [3.147]). Both phenomena, which are apparently slightly correlated with each other [3.149, 194], are probably due to local interactions of waves with the solar wind particles. Otherwise, the extreme kinetic temperatures of the heaviest ions (up to 10^8 K for iron ions [3.14]) could not be achieved. A detailed explanation of the processes involved, however, is lacking [3.64]. Part of the problem is concerned with the generation of this highly nonthermal equilibrium state and its apparent limitation at a certain level. Another part deals with the role of damping processes that would tend to drive the nonthermal distributions back toward thermal equilibrium [3.127].

Apparently, we have now entered an area which is mainly governed by local plasma phenomena. Because of its complexity, this whole matter has been chosen to be the central topic of [3.132], to which the reader is referred for further discussion.

3.7.2 Slow and Fast Solar Wind: Similarities

1. Momentum Flux M_p. One of the most striking features in Fig. 3.22 as well as in Table 3.1 concerns the momentum flux density M_p which appears to be, on the average, an invariant of the solar wind flow state [3.222, 198]. The average value is 2.15×10^{-8} dyn cm^{-2}, compared to $(2.21 \pm 0.07) \times 10^{-8}$ dyn cm^{-2} derived in [3.222] from published *Helios 1* data. Other values derived in [3.222] and [3.223] from *Vela 3* and *Mariner 2* data differ slightly but show similar invariance with the flow speed. Between slow and fast wind M_p does not differ by more than 5% on the average. In any speed class the range of variance is about the same; σ is about 55% and compares well with σ of, e.g., the density. Apart from minor variations during the solar cycle (see Sect. 3.8.1), M_p is so constant that it may even be used for spacecraft intercalibration purposes, as first proposed in [3.222]. In fact, since v_p can easily be measured to an accuracy of less than 1%, the a priori knowledge of the invariant quantity $M_p = n_p\, m_p\, v_p^2$ can help establish absolute calibrations for n_p, which has been traditionally hard to measure.

The intrinsic invariance of M_p is immediately apparent if one inspects stream structures observed from *Helios* close to the sun (Fig. 3.24), e.g. in the time intervals preceding and following the CIR on day 72. Here the distorting action of stream–stream interactions affects only minor portions of the whole longitude range. At 1 AU these distortions can become quite large. Note that the small-scale variations of M_p can be enormous, especially in the slow wind, and have tended to conceal its long-term average invariance.

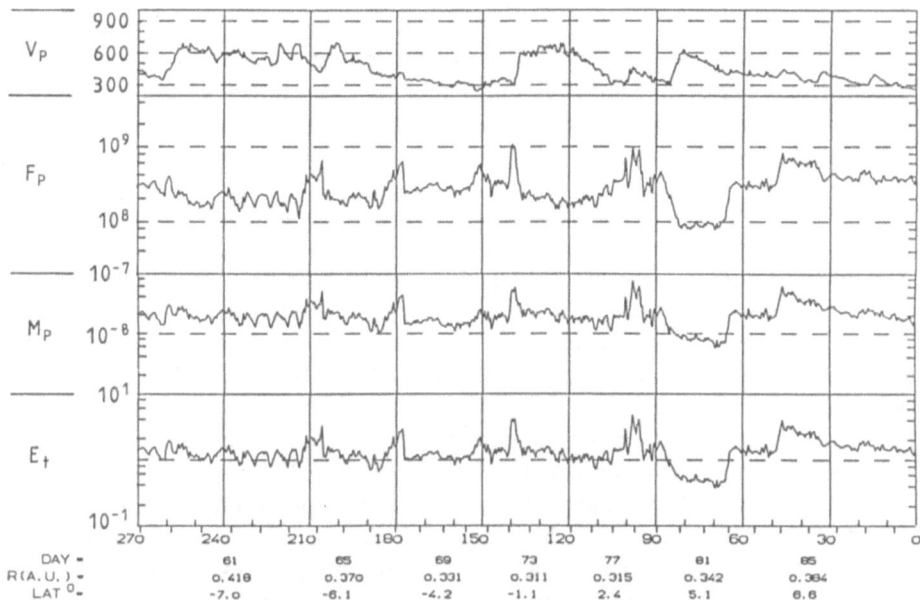

Fig. 3.24. Variations of proton bulk speed v_p (in km s^{-1}) flux density F_p (in cm^{-2}s^{-1}), momentum flux density M_p (in dyne cm^{-2}s^{-1}) and total energy (in erg cm^{-2}s^{-1}) as seen from *Helios 1* in early 1975. See also Figs. 3.1, 3.2 and 3.8

There is no unique explanation yet for the strange invariance of M_p, which may as well be regarded as the "dynamic pressure" of the solar wind plasma. I suspect that a crucial clue to the understanding of solar wind expansion may be hidden here.

One interesting trend has been found within the scatter of individual M_p data around its average [3.130, 192]. In fast wind the proton temperature T_p appears to be correlated with M_p, i.e. M_p rises with increasing T_p. In slow wind, however, there is no such correlation. This was interpreted as an indication that in fast wind additional momentum (and energy) is deposited by MHD waves above the Alfvén critical point R_A. Momentum addition below R_A would increase the particle number flux rather than speed and temperature [3.123]. The absence of correlation between M_p and T_p in slow wind might thus indicate a subsonic momentum deposition from waves, if there is any at all.

2. Total Energy Flux E_t. Although the individual components of the energy transported by the solar wind depend strongly on the solar wind flow state, their total sum does not. This is shown in Fig. 3.22, Table 3.1, and in Fig. 3.24. It may be of interest that the average value of E_t, 1.45 erg cm^{-2} s^{-1} at 1 AU, is approximately one millionth of the total energy the sun expends on radiation. The standard deviation σ of E_t is about 55%, similar to what it is for the major components of E_t.

Similarly to M_p, the striking invariance of E_t is most evident close to the sun (see Fig. 3.24), where the extent of interaction regions is only small. It appears as if the sun provides an energy density flux for driving the solar wind rather uniformly all over its surface. In view of the highly structured corona and the stream-structured solar wind this uniformity of E_t is rather striking. I do not believe that this uniformity is fortuitous, but it is certainly fair to say that we are far from understanding it.

3. Some Global Aspects. The invariance of both M_p and E_t in the solar wind stream structure also applies with respect to heliomagnetic latitude, as shown convincingly in [3.23] for heliomagnetic latitudes up to 30°. We have no reason to expect major changes at polar latitudes, since we already know the basic similarity of, e.g., polar coronal holes and their equatorial relatives. A very detailed quantitative study of a polar coronal hole [3.143] revealed that its "cross sectional area from the surface to $3R_s$ is approximately seven times greater than if the boundary were purely radial. In other words, approximately 60% of the entire solar atmosphere in the northern hemisphere above $3R_s$ is connected to about 8% at the solar surface." Accordingly, nonradial expansion of the solar wind emerging from that hole has to be expected. Thus, the invariance of M_p and E_t as encountered beyond $R = 0.3$ AU cannot hold all the way in through those nonradially shaped coronal structures and is probably severely violated down at the coronal base. Conversely, the invariance of M_p and E_t is not an intrinsic feature of the sun and the lower corona, but rather the result of mechanisms controlling the outflow of the solar wind in the more distant corona. Quantitative highly-resolved optical observations of that range up to $3R_s$ would certainly promote our understanding.

From *Helios in situ* measurements we know that at least beyond 0.29 AU there is almost no global nonradial flow, on the average. In fact, the particle flux $F_p = n_p v_p$ was found to be that quantity which varies the least with increasing R [3.198] (for details see the following section).

For times when large coronal holes cover the sun's polar caps and emit high-speed wind, the average polar mass flux can be expected to match closely the value of the typical fast wind ($F_p = 1.99 \times 10^8$ cm^{-2}s^{-1}), which is much less than the average slow wind (2.86×10^8 cm^{-2}s^{-1}) emitted simultaneously in the equatorial regions. This latitudinal dependence of F_p was actually inferred [3.116, 118] using remote sensing of Lyman-α radiation emitted from interstellar hydrogen atoms passing through interplanetary space. Charge exchange collisions with solar wind protons alter the density distribution of this neutral gas and consequently modify the Lyman-α emission according to F_p. The large-scale anisotropy of F_p became thus directly visible. The evaluation depends on the assumption of a speed profile; assuming an increase in speed from 400 to 800 km s^{-1} from equator to pole [3.44], a decrease of 30% in F_p was derived [3.118].

These observations may have important implications on our understanding of solar wind flow from coronal holes. As pointed out in [3.119] "no definitive conclusions can be drawn about the need for energy addition to the solar wind from a nonthermal energy flux in $R \lesssim 5R_s$", as had been done in the well-known study by Munro and Jackson [3.143]. It appears as if, at the present time, we are left with observational facts that "are found to be consistent with both of two extreme hypotheses: (1) that virtually all of the energy appearing as flow energy at 1 AU is added to the wind from a nonthermal energy flux in $R \lesssim 5R_s$ and (2) that no significant energy addition occurs in $R \lesssim 5R_s$. A variety of intermediate hypotheses is, of course, also consistent with these observations" [3.119].

3.7.3 Radial Gradients

It had always been anticipated that measurements of radial profiles of the basic solar wind plasma parameters would narrow down the various degrees of freedom available for modeling solar wind expansion. Some progress, in fact, has been achieved (see [3.132]), and a series of follow-up studies are on the way. In this section I very briefly summarize the observational facts as collected from papers addressing the topic of radial gradients [3.198, 205, 196]. Some minor deviations of the numbers given here to those in the original papers are due to recent recalculations based on improved algorithms and larger data sets.

As was done in [3.198], all the *Helios* plasma data were sorted into radial bins of 0.02 AU width. Within each bin the data were split up in several speed classes. I assumed linear functions of R for all parameters discussed here and performed least square fits. All quantities involving the proton density n_p were normalized assuming an R^{-2} dependence of n_p. The overall results are listed in Table 3.2 which gives the average changes in each parameter between 0.3 and 1 AU in per cent. In order to allow an estimate of the uncertainties involved, I list the percentage variations separately for *Helios 1* and *Helios 2*, and separately for both a limited data set around solar activity minimum (before 1977) and a larger data set ending in February 1980.

In spite of the huge number of data points obtained during several orbits, the scatter of the averages around ideally smooth radial profiles remained large. The main reason is the unequal occurrence rate of the different plasma "types" (in

	Helios 1		Helios 2	
	before '77	all	before '77	all
v_p	16.2	10.5	5.6	4.4
n_p	−22.6	−18.1	6.3	−10.1
F_p	−5.6	−2.8	3.3	10.0
M_p	12.6	12.7	11.8	10.6
E_k	30.9	28.7	18.8	17.6
E_g	−5.6	−2.8	3.3	10.0
E_t	5.0	4.9	7.5	5.5

Table 3.2. Radial variations of average solar wind parameters between 0.3 AU and 1 AU, in percent. The proton density n_p has been normalized to 1 AU assuming an R^{-2} dependence

terms of speed classes) at different R. For example, at certain distance ranges both *Helios* probes happened never to encounter high-speed streams, while other ranges were largely dominated by them. To compensate for this shortcoming, the values in the individual radial bins should be weighted appropriately, or, ideally, the radial profiles should be evaluated separately for the different speed classes. This kind of analysis, which requires an additional subdivision of the data and thus suffers from increased statistical scatter, will be performed in a future study. Some trends, however, are readily visible from Table 3.2, including further information from [3.198] and [3.165], and will be discussed in the following subsections. Note that the terms "slow" or "fast" refer to data taken for $v_p < 400 \text{ km s}^{-1}$ or $> 600 \text{ km s}^{-1}$, respectively. The intermediate range was omitted in order to eliminate data influenced by stream interactions.

The radial profile of the solar wind can be measured "directly", on rare occasions, by using carefully selected data sets obtained during radial line-up constellations between two space probes. This method relies on the identification of "plasma line-ups" based on appropriate mapping techniques [3.205, 196]. Uncertainties resulting from nonradial flows due to stream interactions depend on the radial separation and the actual stream profile. The two *Helios* probes, by virtue of the launch date selection, happened to attain several such favorable line-ups in 1976 and 1977.

1. Flow Speed v_p. There appears to be a general increase of v_p with increasing R. The comparatively high increase, particularly in the *Helios 1* data, is probably an artifact arising from the rather crude averaging technique discussed above. From the line-up study [3.205] we learned that the speed increase occurs mainly in the slow wind. It amounts to $52 \pm 11 \text{ km s}^{-1}$ per AU, i.e. by 12% between 0.3 and 1 AU. In fast wind, the increase is only $7 \pm 16 \text{ km s}^{-1}$ per AU, i.e. 1%.

2. Proton Density n_p. There is a marked deviation from a pure R^{-2} fall-off of the particle densities. In fact, n_p drops by some 10% faster. Using a slightly different approach, 1-hour averages of the same *Helios 1* and *Helios 2* density data were fitted by a power law as $n(R) = 6.1 \times R^{-2.1} \text{ cm}^{-3}$ [3.18]. This result was found to be consistent with an analysis of interplanetary solar radio bursts if that radiation is assumed to be emitted at the second harmonic of the local plasma frequency. Upon closer inspection, it turns out that the normalized n_p drops more steeply in slow wind than on the average, whereas in fast wind it even grows by some 15%.

3. Proton Flux Density F_p. This quantity appears to remain about constant between 0.3 and 1 AU, on the grand average. The average increase of v_p is strikingly well compensated by a simultaneous decrease in n_p. This indicates that the flux tube cross sections grow with R as R^2. In other words, there is no significant meridional flow out of or into the plane of the ecliptic.

However, we do find that F_p grows by some 15% within fast streams and tends to decrease accordingly in slow wind. We have to conclude that flux tubes carrying fast, low-density solar wind are compressed by some 15% in cross section by the much more numerous flux tubes carrying slow, high-density plasma. Remember, by the way, that the average mass flux density in high-speed streams is only half the value found in slow plasma (see also Fig. 3.22).

4. Proton Momentum Flux Density M_p. As a consequence of the average increase of v_p, M_p also rises by about 12%. This tells us that a considerable amount of momentum is still being deposited in the solar wind, even far beyond the main acceleration region in the corona.

M_p grows in the fast wind by some 15%, similarly as n_p and F_p. The simultaneous radial variations of these quantities (and E_k as well) reflect essentially nothing but the narrowing of the fast plasma flux tubes. There can only be minor "intrinsic" additions of momentum in this type of solar wind.

In the slow wind, M_p grows only slightly. Due to the widening slow flux tubes, however, the total momentum flux increases markedly, i.e. there is a continuing acceleration of the slow solar wind out to 1 AU and beyond. Both these latter conclusions are in stark contrast to what has been stated in [3.130] (see Sect. 3.7.2). Here we are left with a controversial issue for further studies.

5. Energy Flux Densities. Due to the average increase of v_p, there is also a substantial increase of the proton kinetic energy flux density E_k with increasing R. In the fast wind, the average growth of 15% corresponds well to the narrowing of the flux tubes, as discussed above. In other words, there is no or only little net gain of kinetic energy in the total high-speed solar wind flow. In slow wind, E_k clearly increases. In fact, the total kinetic energy of all slow solar wind flow may grow by more than 30% between 0.3 and 1 AU.

There may be some important implications of these observations for our understanding of solar wind acceleration. It is generally agreed that MHD waves (and in particular Alfvén waves) play a key role in this process (see Sect. 3.7.1, [3.132, 133]). Especially, it is often thought that the typically very high-speeds of the flow from coronal holes may result from some form of wave action. It therefore comes as a surprise that exactly this fast solar wind, which is characterized by the persistent presence of large amplitude Alfvén turbulence at any solar distance, does not undergo any significant postacceleration. As it turns out, the Alfvén wave energy flux at times of Alfvénic turbulence was found to be of only minor importance, dropping from 6% to 3% of the total energy flux between 0.3 and 1 AU [3.46]. On the other hand, the type of solar wind (slow) which normally does *not* exhibit Alfvénic turbulence is the one that is substantially postaccelerated even beyond 0.3 AU.

However, there is no doubt about the basic importance of wave action in the solar wind. Remember, for example, that the distinct speed difference v_D between protons and all other ion species at times of Alfvénic activity, as discussed in [3.132], is thought to be generated and controlled by some kind of resonant wave–particle interaction. The speed difference v_D is of the order of the Alfvén speed c_A, and they both decrease with increasing R. On the average, v_D drops from $70 \, \text{km s}^{-1}$ to $20 \, \text{km s}^{-1}$ between 0.5 and 1 AU. This can only mean a transfer of momentum and kinetic energy from the faster heavy ions to the protons. A complete transfer to the protons would increase v_p by some $5 \, \text{km s}^{-1}$, which is of the order of what is observed.

In slow wind, however, any speed differences between the ion species are minor at all R, due to the action of Coulomb collisions ([3.127], for details see [3.132]). Therefore, this possible energy source cannot account for the amount of kinetic energy dumped into the slow wind beyond 0.3 AU.

In this context, one may consider the effect of the interplanetary electrostatic potential Φ (caused by the global electron pressure gradient, see [3.95, 55, 165]) and its radial gradient. Φ has actually been deduced from *Helios* electron observations [3.164, 165]. It was found to decrease in typical fast wind from 60 to 30 V between 0.3 and 1 AU; in slow wind outside of sector boundaries from 130 to 50 V over the same solar distances. This potential difference would accelerate protons in slow wind of $300 \, \text{km s}^{-1}$ by $25 \, \text{km s}^{-1}$, i.e. about half of what is observed. Protons in a fast wind of $700 \, \text{km s}^{-1}$ would be accelerated by about $5 \, \text{km s}^{-1}$, a value consistent with our observations.

3.8 Variations During the Solar Activity Cycle

The long-term variations of the solar wind in the course of the solar activity cycle has always been a subject of great interest. There is no doubt that it is the solar wind which acts as the crucial agent in modulating the fluxes of galactic cosmic rays penetrating the heliosphere. The number of galactic high-energy particles reaching the earth (where they are traced using ground-based neutron monitors) varies inversely with solar activity. This well-known effect runs fairly synchronously (though with a phase lag of about 2 years) with the solar activity cycle, and characteristic signatures indicating the physical link between solar activity and cosmic ray modulation ought to be found in solar wind data [3.220]. However, it turns out that this is not at all as easy a task as initially anticipated: the long-term variation of the average plasma parameters is hidden among much larger short-term fluctuations due to the stream structure, transient shocks, and general variability. Moreover, the time coverage of reliable data sets obtained from one single instrument were rarely commensurate with a solar cycle. Combining data from different instruments always suffers from incompatibilities of the different measurement and evaluation techniques. In particular, the plasma density has traditionally been hard to measure; *in situ* cross calibrations between different instruments at times revealed differences of the order of 100% (see, e.g., [3.50, 146], and the *Interplanetary Medium Data Book* [3.108] and its recent supplements). This contrasts with the rather minor deviations obtained, for example, for the flow speed. Nevertheless, several multi-instrument studies have been attempted in order to sort out "real" solar cycle variations in the solar wind ([3.158, 50, 145, 6, 58, 61, 98] and others).

Real progress was possible thanks to the long life of the *Helios* solar probes and their instrumentation, from which we finally obtained largely continuous data in the same part of the heliosphere throughout more than a whole activity cycle. Concerning the plasma measurement calibration, special inflight-test procedures allowed for very accurate sensitivity monitoring of our two completely independent ion instruments [3.186]. For both of these the sensitivity turned out to be constant to the end of the mission within the estimated measurement accuracy of 1%. Thus, we can be sure that no drift, e.g., due to slow degradation, can be responsible for any long-term trend in the *Helios* plasma density data.

3.8.1 Variations of the "Average" Solar Wind in the Ecliptic Plane

On the basis of *Helios* plasma data from 1974 to early 1982, some trends for the basic plasma parameters were described in [3.198]. Each of the dots in Fig. 3.25 from [3.198] represent averages of the relevant parameters through one complete solar rotation. The individual data points were used as input data, the number of which per rotation is indicated in the lower panels. The plasma densities were normalized to 1 AU, assuming an R^{-2} dependence.

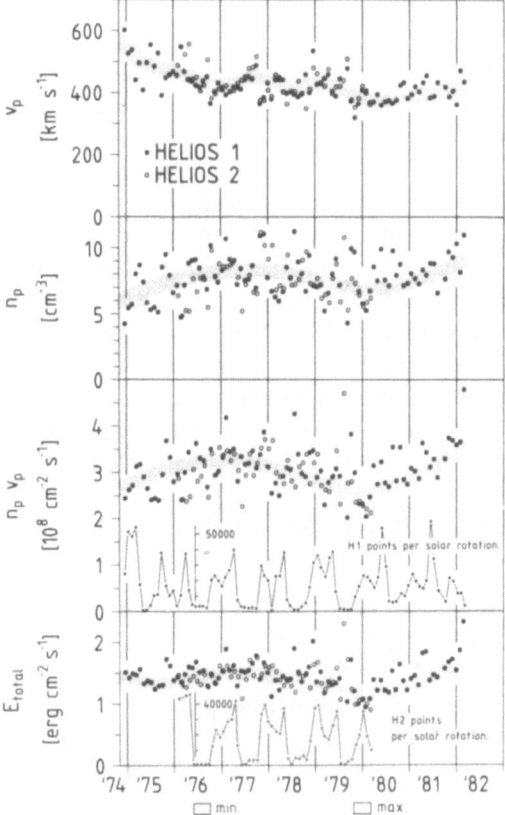

Fig. 3.25. The variation of proton flow speed v_p, density n_p, flux density $n_p v_p$ and total energy flux density from 1974 to 1982, based on all *Helios* data. Each point represents an average value during one complete solar rotation. The number of points per rotation is indicated in the two lower panels. From [3.198]

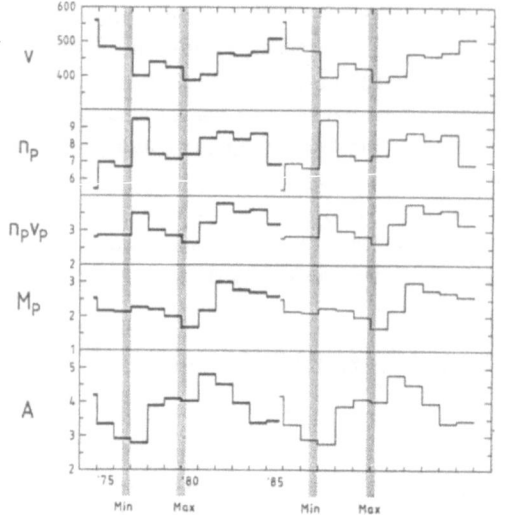

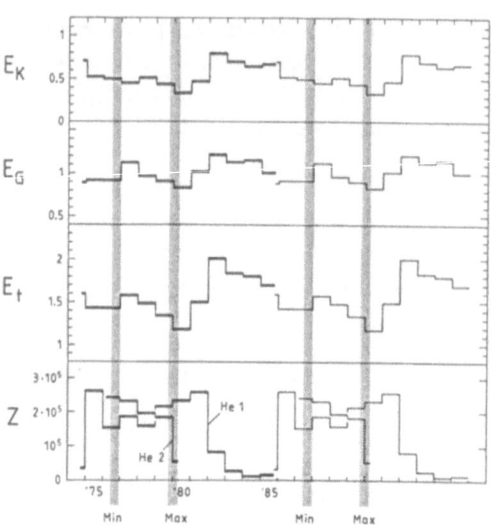

Fig. 3.26. Yearly averages of solar wind proton velocity v_p (in km s^{-1}), density (in cm^{-3}), flux density F_p (in 10^8 cm^{-2}s^{-1}), momentum flux density M_p (in 10^8 dyne cm^{-2}s^{-1}) and helium ion to proton abundance ratio A (in %), based on the combined *Helios 1* and 2 data (see also Fig. 3.27)

Fig. 3.27. Yearly averages of solar wind energy flux densities (in erg cm^{-2}s^{-1}). The bottom panel shows the number of individual *Helios 1* and 2 spectra used for the analysis here and in Fig. 3.26

Apparently, any long term variations in the parameters shown in Fig. 3.25 are minor compared with the fluctuations between individual solar rotations. In order to smooth these fluctuations I calculated yearly averages as shown in Figs. 3.26 and 3.27, including the data from 1982 to the very end of the *Helios 1* mission in early 1986 in the process. The data thus cover a time span of $11\frac{1}{4}$ years, i.e. more than an average solar activity cycle. Data from both *Helios* probes were combined; the numbers of individual data points per year are shown in the bottom panel of Fig. 3.27. The total number of data points amounts to 1,654,951 from *Helios 1* (however, the 741 points obtained in 1986 during a time interval of only a few days were excluded here) and 913,146 from *Helios 2*. In order to emphasize possible cyclic variations I plotted the yearly averaged data twice, with a 11-year shift between them. On doing so, we should be aware of the dangers inherent in this procedure: *anything* tends to look cyclic if plotted twice in succession!

With this warning in mind, we can still recognize certain cyclic long term trends:

1. There is a definite decrease in v_p from about 500 km s^{-1} in early 1975 to below 400 km s^{-1} in 1980. The average v_p is highest in 1974–1975 and 1985 which were the time periods with the largest and most persistent high-speed streams [3.6, 68]. Thus, the maximum in v_p occurs well ahead of the activity minimum as was pointed out previously (e.g., in [3.58, 61]). The minimum v_p was approximately in phase with activity maximum.

2. The density n_p is modulated by some 20% about the average. The exceptionally high density in 1977, which showed up in *IMP 8* data as well [3.61, 121], is due to the almost continuous presence of "quiet" slow solar wind. During this time, right after activity minimum, the heliospheric ballerina skirt was almost flat, and only rarely did the polar coronal holes send high-speed streams down to the ecliptic as is evident from the very low v_p value in this year.

3. The particle flux $F_p = n_p v_p$ shows a similar modulation, with an amplitude of $\pm$ 17%. Apparently, there are two minima of F_p during an 11-year cycle: one around activity maximum and a second one at the time of the large high-speed streams well before activity minimum.

4. The momentum flux M_p is modulated by $\pm$ 28%, with a clear minimum at activity maximum. The highest value of M_p is reached two years after activity maximum. This fact might be of some significance for solar terrestrial interaction processes, since M_p is a measure of the dynamic pressure the solar wind exerts on any obstacle such as the earth's magnetosphere. Similarly, the position of the heliopause would be shifted in and outward during the solar cycle, following M_p [3.121].

5. The kinetic energy flux E_k in the solar wind behaves similarly to M_p; the modulation amplitude may amount to $\pm$ 40%. The peak value in 1982 is due to the high average v_p in this year. In fact, in this phase of decreasing solar activity there still occurred several huge solar transients producing very fast mass ejections [3.212]. Furthermore, broad, long-lasting high-speed streams reappeared in that same year. The situation resembles in many respects the one in late 1972 as described in [3.58].

6. The total energy E_t the sun spends for emitting the solar wind (at least that fraction around the heliographic equator) varies by $\pm$ 27%, with a clear minimum around activity maximum and a peak value two years later.

7. The patterns of all curves are consistent with a real cyclic behavior of the parameters. There is no evidence "for an overall upward trend superimposed on a sinosoidal modulation" as had been suggested in [3.198].

8. The abundance ratio A of helium ions to protons (determined only for $R >$ 0.5 AU) shows a definite minimum ($\approx$ 2.8%) around activity minimum in 1976 and a broad maximum (some 4%) in the years around 1980. This is in good agreement with the trends reported for the previous solar cycle. Here A varied from 3% around sunspot minimum 1964) to 5% around 1969 and back to 3% in 1976 [3.158, 58, 17].

This latter point has an exciting aspect that shows up once we sort the A data into different speed classes. Figure 3.28 shows that A remains essentially constant in fast wind throughout the solar cycle, except for some excursions due to very fast helium-rich driver gases following strong shocks in the later years. Thus, the general invariability of the fast solar wind emitted from coronal holes also appears to hold for its solar cycle behavior. In contrast, A varies in slow wind from 2.5% around minimum to some 4% around maximum. This indicates that the slow solar

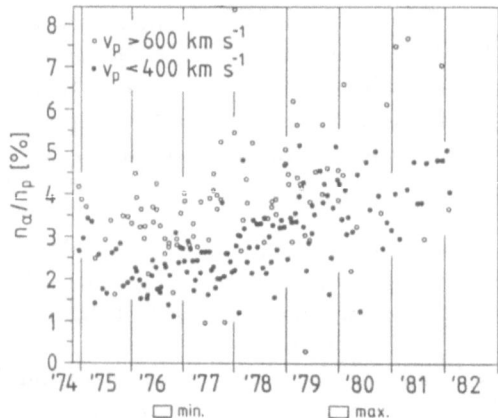

Fig. 3.28. The variation of the α-particle abundance n_α/n_p from 1974 to 1982, based on all *Helios* data taken outside 0.5 AU. Only the data for $v_p < 400\,\mathrm{km\,s}^{-1}$ and $v_p > 600\,\mathrm{km\,s}^{-1}$ were selected and averaged separately. From [3.198]

wind, as we encounter it in the ecliptic plane, changes its character dramatically during the solar cycle. A similar analysis with similar conclusions had already been described in [3.58]. At activity minimum, most of that slow solar wind we see in the ecliptic plane emerges close to the current sheet and above coronal helmet streamers where A is low due to gravitational stratification [3.16]. It appears as if that stratification is progressively disturbed with increasing activity and the disappearance of stable large-scale helmet streamers. Most of that helium-rich slow solar wind around activity maximum emerges well removed from any current sheet, however warped it may be at that time [3.89]. In terms of a three-component model as proposed in [3.4] we may conclude that the helium-rich slow solar wind is basically different from the slow flow around sector boundaries. In other words, there appear to exist two different components of slow solar wind in addition to the fast flow from coronal holes (the third component). The portions of both types of slow flow seen in the ecliptic changes dramatically during the solar cycle.

In an additional approach we sorted these data with respect to the heliographic latitude at which they were measured. The result reflects the fact that, a few years before sunspot minimum, those high-speed streams associated with the south polar coronal hole were "broader". In fact, the equatorial extension of the south polar hole was seen in K-coronagraph observations to be significantly broader than its northern counterpart [3.83] in 1973/74, and IPS measurements of the solar wind speed at high latitudes also yielded much higher yearly averages at southern latitudes in those years [3.43, 44]. This asymmetry decayed with rising solar activity and eventually disappeared altogether in all three data sets. All these results are in basic agreement with a very recent analysis [3.21] using data obtained by the MIT plasma instrument on *IMP 8* between 1974 to 1987.

Another very recent result is the apparent solar cycle invariance in the relationships between solar wind proton temperature and both the velocity v_p and the momentum flux density M_p [3.129]. The spatial and temporal variations of these parameters had been found to contain information about the possible locations of momentum addition to the different types of solar wind ([3.130, 132], see Sect. 3.7.2). These basic physical processes apparently remained unchanged throughout almost two solar cycles.

3.8.2 Structural Changes During the Solar Cycle

It might be regarded as an unfortunate coincidence that, in the only part of the heliosphere accessible so far to *in situ* solar wind measurements (i.e. in the ecliptic plane), solar cycle variations of the average solar wind are extremely subtle and hard to detect. In fact, the ecliptic plane is predominantly filled with slow solar wind at both extrema of solar activity. Coronal holes and high-speed streams are small around maximum, and slow solar wind with high helium content is emitted, on the average, at all latitudes [3.44, 111]. At minimum, the ecliptic plane almost coincides with the heliomagnetic equator and its very flat current sheet embedded in a narrow belt of slow solar wind with reduced helium content [3.155]. Simultaneously, fast wind is emitted at all latitudes beyond ± 20°. Only in the very few years right before minimum do the large polar coronal holes develop appendices that protrude down into the equatorial regions to produce the well-known long-lived high-speed streams seen in the ecliptic plane. Their number and size was found to be greatest in the years 1962 and 1964 and then again from 1973 to 1976 [3.6, 198], i.e. close to but well before activity minimum. The largest streams were found to be generally the most stable structures too [3.68]. The huge streams first seen in 1973, when *Skylab* discovered the stable coronal holes, persisted through almost three years and disappeared only at activity minimum in 1976 (see, e.g., [3.209])! Studies of the long-term evolution of coronal hole patterns, solar wind speed, the interplanetary magnetic field and the geomagnetic disturbance index C9 (see, e.g., [3.208, 209, 210] and references therein) have impressively demonstrated the very close correlation between high-speed streams and recurrent geomagnetic activity. Some 40 years after Bartels introduced the term "M-regions" for the solar sources of recurrent geomagnetic disturbances [3.9, 10], we are finally able to identify them as coronal holes. A longlasting controversy, whether or not the M-regions are tied to active regions (Bartels thought they were), was settled (see [3.151]): it is perhaps somewhat ironic that the real 'M-regions' are features typical of the quiet sun, rather than the active sun.

It is worth noting here that the long-lived high-speed streams in the past two solar cycles originated from about the same solar longitudes (assuming a synodic rotation period of the sun of 27.025 days [3.69]). A similar "preferred longitude" or "solar memory" effect was also seen in the distribution of sunspot groups [3.5]. "Some deep internal structure in the sun must ultimately be responsible for these long-lived longitudinal effects, which appear to rotate rigidly with the sun" [3.69].

In the years characterized by the large high-speed streams the average flow speed went significantly higher than in all other phases of the solar cycle [3.58,

198]. The resulting modulation of the flow speed is thus about two years out of phase with that of solar activity. This time lag is one of the arguments that makes it hard to associate this particular effect (i.e. the appearance of high-speed streams in the ecliptic plane) with the cosmic ray modulation. However, there is no doubt that the global rearrangement of the corona during the solar cycle holds the key to the problem. One can look at it in two ways: first, from the viewpoint of coronal hole evolution (the modulation of the "quiet" corona), and second, from the viewpoint of the solar activity itself, in terms of number, size, and effects of coronal transients.

1. Evolution of the "Quiet" Corona. One major manifestation of the solar activity cycle is certainly the dramatic shrinking, disappearing, and rebirth of the polar coronal holes from one solar minimum to the next [3.19, 207, 63]. The changes in shape of coronal holes are always caused by the birth and emergence of new bipolar magnetic regions which then interact with the pre-existing fields (see, e.g., [3.15, 19, 156]). In particular, the birth of new magnetic flux in the rising cycle at high latitudes shrinks the polar holes and simultaneously leads to the formation of isolated equatorial holes. The total area of coronal holes is certainly lowest around maximum. However, surprisingly enough, the total amount of magnetic flux on the sun is increased by a factor of 3 compared to minimum [3.91], while the interplanetary magnetic field is increased by only 40% or less [3.217]. This means that around maximum:

- the percentage of solar magnetic flux "leaking out" into the heliosphere is much less,
- the relatively constant interplanetary flux may have its origin in smaller areas on the sun, thus implying an even greater flux tube divergence compared to the polar hole divergence at minimum conditions [3.143, 211],
- a significant fraction of the sun's net flux (both magnetic field and solar wind) originates outside obvious coronal holes [3.85, 126].

The heliospheric current sheet changes from the flat ballerina skirt at minimum to a highly warped 3-D structure. Its evolution could be pursued even through the magnetic polarity reversal in 1980, and was found to flatten again (now "upside down") on approaching the next minimum [3.88, 89, 1]. This complete turnover is nicely illustrated in several figures in [3.89] (see also [3.13]). Of course, the polarity switch occurred during the phase of maximum activity, appropriately accompanied by a fusillade of transient coronal mass ejections. It is not a real surprise that both hemispheres of the sun may develop differently such that there appears to be a time lag between them, and that the polar coronal holes may have very different sizes and shapes [3.19].

It has been argued that this solar cycle related variation in the large-scale structure of the heliospheric current sheet might itself play a key role in the modulation of galactic cosmic rays (see, e.g., [3.228] and references therein; see also further papers in the March 1986 issue of J. Geophys. Res.). In fact, there is definitely a need for models that take into account the three-dimensional magnetic topology. In [3.44] it is suggested "that at solar minimum wide polar coronal holes connect to wide fast polar streams, which are presumably unipolar and may in turn be connected to the interstellar magnetic field, allowing easy access of cosmic rays. At solar maximum the polar holes are constricted by high latitude active regions, causing a narrowing of the polar streams, that is a smaller area of easy access which reduces the cosmic ray flux".

It is not only the magnetic topology above the poles that varies with the solar cycle. The character of the solar wind changes here as well: the steady, fast coronal hole flow at solar minimum turns into a highly variable, slow wind [3.44, 111]. One of the most characteristic signatures of this type of slow wind, its high density, could be measured optically during large parts of the *Helios* mission. The instrument dedicated to the observation of zodiacal light brightness (for details and scientific results see [3.125]) is sensitive enough, at least around perihelion, to detect and separate the Thomson scattered light from electrons along the line-of-sight at high heliographic latitudes [3.124, 125]: "The conclusion is hard to escape that for this region inside 0.4 AU the plasma density nearly doubles from minimum to maximum." Our zodiacal dust colleagues were apparently intrigued and surprised by this unexpected finding which we can now consistently explain in terms of solar cycle dependence of the high-latitude solar wind. A more quantitative study combining these optical data (plasma density n_p) with IPS data (flow speed v_p) and maybe Lyman-α measurements (particle flux F_p, [3.118]) is now on the way.

Coronal evolution during the solar cycle and the resulting modulation of the high latitude solar wind probably does weakly affect one other important quantity: the momentum flux M_p. The observed invariance of M_p with solar wind stream structure (section 7.2.1) was found to apply for heliomagnetic latitudes up to at least 30° [3.23]. As pointed out in Sect. 3.7.2, we have no reason to expect major deviations from that invariance at higher latitudes. The solar cycle modulation of M_p above the poles is not expected to exceed the one observed in the ecliptic plane ($\pm$ 25%). This conclusion may imply important consequences for the heliosphere as a whole, in particular for the three-dimensional shape, size, and the cyclic modulations of the heliospheric termination shock.

2. Modulation of the "Active" Corona. The other phenomenon that is directly related with the solar activity cycle is, quite naturally, the occurrence of coronal transients and their effects on both the coronal evolution and on the interplanetary medium. Much effort has been recently put into studying this multifaceted issue.

One aspect deals with the coronal processes, such as emergence of new magnetic flux and birth of "active centers", the formation of prominences which eventually erupt, coronal mass ejections, flares of different types, radio wave bursts of different types, particle acceleration, and coronal and interplanetary shock waves. All these processes are linked together by a complicated chain of actions and reactions which are far from understood in detail.

A second aspect concerns the transient phenomena in interplanetary space (shock waves), both macroscopically (initiation, 3-D propagation, acceleration and deceleration, magnetic topology, association with coronal ejecta, interaction with other shocks and the stream structure, etc.) and microscopically (structure of the shock front, processes affecting the particles' velocity distributions, particle acceleration, turbulence).

A third aspect concerns the associations between the other two which, unfortunately enough, have been well separated from each other, mainly for historical and instrumental reasons. In fact, coronal issues have been addressed mainly by solar physicists using remote sensing instruments such as optical telescopes, ground-based and satellite-borne white-light coronagraphs, radioheliographs, etc. On the other hand, solar wind issues were covered by the space plasma physics community, observing the interplanetary medium *in situ* from satellites and space probes. There are several recent reviews on these subjects (see, e.g., [3.172, 230, 178, 199]). Thus, I restrict myself here to a few remarks on one specific issue that is relevant in the context of this section.

It has long been realized that the fronts of interplanetary shocks launched by major solar transients may have enormous lateral extents, often exceeding 90° angular distance from their site of origin at the sun [3.212]. "Hence, we conclude that the tremor from a single solar transient may affect significantly more than half of the entire heliosphere" [3.199]. At times, particularly virulent active regions cause several transients within several days. Their resulting shock waves may eventually merge on a large scale and form a giant circumsolar, expanding "shell". It was argued that this shell, filled with highly turbulent shocked plasma, may act as an effective shield against cosmic ray particles which are denied access to the inner heliosphere as long as the shell is tightly maintained. With increasing solar activity, several such shells may follow each other and successively improve the shielding. In this manner, the distinct stepwise decrease in cosmic ray intensity with increasing activity could be consistently understood [3.35, 42].

This scenario of activity related turbulent shells is certainly an oversimplification, but it shows one possible line of reasoning in explaining the close association between solar activity and cosmic ray modulation. In reality, these shock shells will interact with the ambient solar wind in various ways. For example, close to the sun the shells may be "penetrated" by corotating high-speed streams. Further out, the interaction regions turn into almost spherical outward propagating pressure waves which eventually coalesce with the shock shells and form huge compound stream systems [3.37]. They, in turn, act upon energetic particles by effects such as channeling, trapping, scattering and accelerating. For more details on this issue see [3.117].

The different processes involved in generating shock shells and compound streams vary with solar activity in different ways, and some of these dependences are not at all uniquely understood. For example, there is still an unresolved controversy concerning the occurrence rates of different types of coronal mass ejections (see, e.g. [3.103, 93, 232]). I mention this only to illustrate the basic problems that to this day plague us in disentangling the various causes and effects determining our heliospheric environment.

3.9 Conclusions and Outlook

Parker's first theory [3.159] on the existence and evolution of the solar wind was brilliantly confirmed when the first space probes entered interplanetary space. His calculations in terms of a thermally driven continuous outflow of solar material finally yielding the supersonic solar wind predicted numbers that proved to be in generally good agreement with the observations. Meanwhile, a wealth of data on many different phenomena associated with the solar wind and the interplanetary medium has been accumulated. The inner heliosphere has been probed by several deep space missions. In particular, the *Helios* twin spacecraft approached the sun as close as 0.29 AU (i.e. $63R_s$). Other space probes have traversed the outer solar system out to 44 AU up to the present. On the other hand, powerful orbiting coronagraphs and X-ray and UV telescopes have expanded our view of the solar corona out to $10R_s$ from the sun, thereby covering the source regions of the solar wind.

In spite of the still-remaining observational gap between $10R_s$ (optical) and $63R_s$ (*in situ*), the originally large number of free parameters in designing a consistent picture of the 3D heliosphere has decreased enormously. The more observational facts we gathered and the more possible explanations we could rule out on theoretical grounds, the clearer it emerged that there are still several key issues to be resolved if we are to arrive at a sufficiently complete understanding of the solar wind's real origin. That significant progress has been achieved in the years since the first *Helios* launch in 1974 can be inferred from the quality and preciseness of the questions we can formulate today. A series of such open problems have been mentioned throughout this article. I will recapitulate some of these problems here by putting them in a somewhat broader context in order to show the reader some perspectives for further work in this field of research.

The most fundamental issue in all solar wind physics certainly deals with the solar wind's origin. This central topic has only been scratched in this chapter on the large-scale structure of the inner heliosphere. Some interrelations are worth mentioning here in more detail. Other questions already discussed will subsequently be listed by keywords only.

3.9.1 Solar Wind Origin

There is now increasing evidence that the acceleration of fast and slow solar wind might be the result of different mechanisms. High-speed solar wind can usually be traced back to coronal holes which overlie the more inactive regions of the sun, e.g., the polar caps during times of low solar activity. The high-speed streams in turn are also characterized by comparatively stable and reproducible signatures. Unfortunately, neither Parker's theory nor any other model is able to this day to explain consistently the speeds of 600 to 800 km s^{-1} commonly observed within high-speed streams.

Following the original idea of a thermal pressure gradient as the driving force, one ends in a dilemma as pointed out in [3.123]: increased heating in the lower corona would rather increase the net mass flux at an even lower average speed! Acceleration could only be achieved by adding energy and momentum to the flow *beyond* the Alfvén critical radius. We are confronted with several new questions: What kind of energy could that be? How is it transported into this distant regime? And how is it deposited there? It has been suggested that it might be the momentum flux from MHD wave fields that acts as the ultimate driver of the solar wind. The trouble with this model can be highlighted by the remark that it is very closely related to the general problem of coronal heating, which is not understood either! Furthermore, our *in situ* observations indicate that any wave energy fluxes are probably too weak to account for the acceleration of high-speed streams (Sect. 7.3.3).

An astonishing alternative for generating the fast or even all solar wind would be an acceleration in the form of numerous high-velocity jets that have been discovered on very high resolution UV observations of the sun [3.20]. Small-scale magnetic reconnection might act as a driving force to inject isolated diamagnetic bubbles or plasmoids into the ambient magnetic field [3.169, 170]. The estimated mass and energy fluxes involved in these jets or plasmoids might be sufficient to drive all types of solar wind. These observations might eventually force us to expand or even alter our views of this subject completely. Let's face it: our understanding of solar wind acceleration is not at all on firm ground yet!

The situation is no better with respect to the slow wind. Although its characteristics are, on the average, rather nicely predicted by the classical Parker model and its successors, we still do not know where and how this slow plasma originates. On tracing this type of flow back to the sun we find ourselves above active regions and their usually closed magnetic structures. For example, around activity minimum, the flow of slow solar wind is constrained to the warped activity "belt" of about 40° width in latitude which encircles the sun close to the heliographic equator. This is where the heliomagnetic current sheet is "attached", i.e. the separatrix between the open field lines of opposite polarities. At times of high solar activity, the situation is more complicated, since we now find slow wind emerging from substantially larger areas [3.111] often located far from the current sheet. This kind of slow solar wind may even be of a basically different type than the "streamer type" flow at the current sheet (see also Sects. 3.4.3, 3.7.1 and 3.8.1). The basic problem, however, is similar for both types of slow solar wind: there must be a transition from the closed loops and arches in the lower corona to the open field lines which are dragged out into the heliosphere by virtue of the outflowing (slow) solar wind.

If this transition is imbedded in the flow, the magnetic loops themselves would be dragged out and then continuously torn open, leading to a net increase of the local magnetic flux. For compensation, some process of magnetic reconnection must be active, on the average, since otherwise the total solar magnetic flux would keep building up. We do not know whether reconnection takes place right here or at any other place, whether in a more stationary way or in the form of transient events. There are a number of features such as the strong variability in density, temperature, and helium content which suggest a more transient origin of this type of solar wind.

Pictures of coronal arches, quiet prominences, and helmet streamers suggest that coronal loops are stationary or only slowly evolving. This means that the solar wind plasma would have to flow around them. Even if a few open-field lines leak out of the otherwise closed regions, as suggested in the three-component model of the corona [3.4], one still expects strongly nonradial flows. One might even think of *all* solar wind coming from open regions (i.e. coronal holes) only. However, assuming just one type of source for all types of solar wind appears to be quite unrealistic, since

1. fast and slow solar wind seem to represent two distinctly different states indicating perhaps even different acceleration mechanisms,
2. the boundary layers between these states are very "thin" and follow in much detail the shapes of the coronal holes, and
3. coronal holes change drastically during the solar cycle with respect to their total size, individual shape, position, etc. The modulation of the "average" solar wind is not at all commensurate.

On the other hand, there is this striking equality in total energy flux and momentum flux in both slow and fast type flows. This is well established from *in situ* data obtained at $R = 0.3$ AU where effects from stream–stream interactions can be ruled out. However, we still have no conclusive answer to whether this invariance is an intrinsic solar phenomenon or whether it is the result of some global rearrangement of the different flows beyond the corona. Both alternatives have their pros and cons, and whatever the ultimate solution of the puzzle may be, I believe that it may hold a key to our understanding of solar wind generation.

3.9.2 Other Open Questions on Large-Scale Phenomena

Here I will only briefly summarize the remaining questions that have come up during the discussions throughout this article. This list is by no means complete, and the reader may prefer to go back to the relevant sections for more details.

1. What is the nature of discontinuous stream interfaces? Are they remnants of similarly discontinuous coronal structures and how do they evolve or even dissolve under the effect of stream–stream interactions?
2. What does the stream fine structure at stream boundaries and within high-speed streams actually mean (some 2° to 8° in angular extent [3.226]) and how is it related with coronal structures?
3. Are there latitudinal gradients within large coherent regions, e.g., within a high-speed stream that is emitted from a coronal hole extending from one solar pole down to the ecliptic? Is there a difference at solar maximum between slow wind from the poles and that from the equatorial regions?
4. What is the effect of the three-dimensionality of the radially evolving solar wind stream structure? What happens to stream fronts inclined to the heliographic equator? How far in latitude can compression regions extend?

5. What causes that obvious difference between the slow flow around the heliomagnetic equator (reduced helium content) and the other type of slow flow (helium-rich) typical for times of high solar activity?

6. What is the nature of NCDEs, and what is their relation to coronal streamers, sector boundaries, magnetic clouds, coronal mass ejections, and the slow solar wind in general?

7. The 22-year magnetic cycle of the sun causes, on the average, a cyclically varying phase shift between the southward pointing B_z at sector boundaries and the maximum ram pressure exerted on any obstacle in the solar wind flow. Could this be the reason for certain 22-year periodicities in geomagnetism superimposed on the well-known 11-year modulation?

8. What can the characteristic features of the helium content in the solar wind (very low values around sector boundaries, low values and enormous variability in slow wind, constant value in high-speed streams, strong solar cycle modulation in slow wind, very high values in many coronal mass ejecta) possibly tell us about the coronal source regions?

9. How do we interpret the fact that the proton temperature exceeds the electron temperature in high-speed streams by at least a factor of 2, in contrast to any other solar wind flow?

10. The average slow solar wind flow expands slightly more than purely radial. This is accomplished at the expense of the fast plasma, for which the average flux tube cross section is compressed by 15% between 0.3 and 1 AU. Taking this into account, one finds no net transfer of momentum and energy to the fast solar wind beyond 0.3 AU, in distinct contrast to the slow wind. This leaves us with serious doubts about the real role of Alfvén fluctuations (mainly present in fast wind) in accelerating the solar wind.

11. The seemingly modest modulation of the solar wind during the solar activity cycle as observed *in situ* is probably an artifact caused by the thus far restricted capability of space probes to leave the plane of the ecliptic. At high latitudes, however, the modulations in such basic parameters as flow speed and flux density are considerable. This remains to be confirmed by *in situ* measurements in these remote parts of the heliosphere (Ulysses mission). The possible consequences for such "hot" topics as the cosmic ray modulation or the variability in the location or topology of the heliospheric boundary have yet to be evaluated.

12. Of course, there is a series of questions related to the evolution of the "inner" heliosphere and how it eventually becomes the "outer" heliosphere. Without further explanation, let me just throw in some key words such as: stream–stream interactions, compound streams, pressure waves, period doubling, turbulent shock shells. These topics are clearly beyond the scope of this article.

3.9.3 Outlook

The catalog of problems indicates how we should proceed in further research. The *Helios* mission has shown us the way: let's go closer to the sun. Almost all the missing answers must be sought in the solar wind source regions. Unfortunately, these very regions are the least accessible to both optical and *in situ* measurements. Therefore, concerted research efforts from different disciplines (experienced in very different measurement techniques) will be required in order to develop a new generation of diagnostic tools. In that sense, I regard the upcoming SOHO mission as the next logical step. Some twenty years after the first *Helios* launch, we hope it will be a worthy successor to take over finally the grand heritage of Helios.

Acknowledgements. I am grateful to Eckart Marsch, Pat Daly, Rolando Hérnandez, and Werner Pilipp for critical reading of the manuscript and many valuable suggestions for improvements, both in content and in style. Special thanks are due to the referees, Marcia Neugebauer and Jack T. Gosling, who ploughed through the paper in depth and requested me to clarify, specify, or even correct many details. Thanks to the generous effort of Michael Bird the English text has become much more readable than before. With great gratitude I acknowledge the untiring patience of Inge Hemmerich and Ute Spilker in typing and editing the whole piece.

References

3.1   Akasofu, S.I., C.D. Fry, Heliospheric current sheet and its solar cycle variations, J. Geophys. Res., **91**, 13679–13688, 1986.
3.2   Alfvén, H., Electric currents in cosmic plasmas, Rev. Geophys. Space Phys., **15**, 271, 1977.
3.3   Asbridge, J.R., S.J. Bame, W.C. Feldman, M.D. Montgomery, Helium and hydrogen velocity differences in the solar wind, J. Geophys. Res., **81**, 2719, 1976.
3.4   Axford, W.I., The three-dimensional structure of the interplanetary medium, in *Study of Travelling Interplanetary Phenomena*, ed. by M.A. Shea, D.F. Smart, S.T. Wu, D. Reidel Publishing Company, Dordrecht, Holland, 145–164, 1977.
3.5   Balthasar, H., M. Schüssler, Preferred longitudes of sunspot groups and high-speed solar wind streams: evidence for a 'solar memory', Solar Physics, **87**, 23–36, 1983.
3.6   Bame, S.J., J.R. Asbridge, W.C. Feldman, J.T. Gosling, Solar cycle evolution of high-speed solar wind streams, Astrophys. J., **207**, 977, 1976.
3.7   Bame, S.J., J.R. Asbridge, W.C. Feldman, H.E. Felthauser, J.T. Gosling, A search for a general gradient in the solar wind speed at low solar latitudes, J. Geophys. Res., **82**, 173–176, 1977.
3.8   Bame, S.J., J.R. Asbridge, W.C. Feldman, J.T. Gosling, Evidence for a structure-free state at high solar wind speeds, J. Geophys. Res., **82**, 1487–1492, 1977.
3.9   Bartels, J., Terrestrial magnetic activity and its relation to solar phenomena, Terr. Magn. Atmos. Elec., **37**, 1, 1932.
3.10   Bartels, J., Some problems of terrestrial magnetism and electricity, in *Terrestrial Magnetism and Electricity*, ed. by J.A. Fleming, McGraw-Hill, New York, 385–433, 1939.
3.11   Behannon, K.W., L.F. Burlaga, A.J. Hundhausen, A comparison of coronal and interplanetary current sheet inclinations, J. Geophys. Res., **88**, 7837–7842, 1983.
3.12   Belcher, J.W., L. Davis, Jr., Large amplitude Alfvén waves in the interplanetary medium, 2, J. Geophys. Res., **76**, 3534–3563, 1971.
3.13   Bird, M., P. Edenhofer, Remote sensing observations of the solar corona, (this volume).

3.14   Bochsler, P., J. Geiss, R. Joos, Kinetic temperatures of heavy ions in the solar wind, J. Geophys. Res., **90**, 10779–10789, 1985.

3.15   Bohlin, J.D., N.R. Sheeley, Jr., Extreme ultraviolet observations of coronal holes, Solar Physics, **56**, 125–151, 1978.

3.16   Borrini, G., J.T. Gosling, S.J. Bame, W.C. Feldman, J.M. Wilcox, Solar wind helium and hydrogen structure near the heliospheric current sheet: a signal of coronal streamers at 1 AU, J. Geophys. Res., **86**, 4565, 1981.

3.17   Borrini, G., J.T. Gosling, S.J. Bame, W.C. Feldman, Helium abundance variations in the solar wind, Solar Physics, **83**, 367–378, 1983.

3.18   Bougeret, J.L., J.H. King, R. Schwenn, Solar radio bursts and in situ determination of interplanetary electron density, Solar Physics, **90**, 401–412, 1984.

3.19   Broussard, R.M., N.R. Sheeley, Jr., R. Tousey, J.H. Underwood, A survey of coronal holes and their solar wind associations throughout sunspot cycle 20, Solar Physics, **56**, 161–183, 1978.

3.20   Brueckner, G.E., J.D.F. Bartoe, Observations of high energy jets in the corona above the quiet sun, the heating of the corona and the acceleration of the solar wind, Astrophys. J., **272**, 329–348, 1983.

3.21   Bruno, R., L.F. Burlaga, A.J. Hundhausen, Quadrupole distortions of the heliospheric current sheet in 1976 and 1977, J. Geophys. Res., **87**, 10339–10346, 1982.

3.22   Bruno, R., L.F. Burlaga, A.J. Hundhausen, K-coronameter observations and potential field model comparison in 1976 and 1977, J. Geophys. Res., **89**, 5381–5385, 1984.

3.23   Bruno, R., U. Villante, B. Bavassano, R. Schwenn, F. Mariani, In-situ observations of the latitudinal gradients of the solar wind parameters during 1976 and 1977, Solar Physics, **104**, 431–445, 1986.

3.24   Bürgi, A., and J. Geiss, Helium and minor ions in the corona and solar wind: dynamics and charge states, Solar Physics, **103**, 347–383, 1986.

3.25   Burlaga, L.F., A reverse hydromagnetic shock in the solar wind, Cosmic Electrodyn., **1**, 233, 1970.

3.26   Burlaga, L.F., Interplanetary stream interfaces, J. Geophys. Res., **79**, 3717–3725, 1974.

3.27   Burlaga, L.F., Corotating pressure waves without fast streams in the solar wind, J. Geophys. Res., **88**, 6085–6094, 1983.

3.28   Burlaga, L.F., MHD processes in the outer heliosphere, Space Sci. Rev., **39**, 255–316, 1984.

3.29   Burlaga, L.F., Period doubling in the outer heliosphere, J. Geophys. Res., **93**, 4103–4106, 1988.

3.30   Burlaga, L.F., Magnetic clouds, in *Physics of the Inner Heliosphere*, Vol II, ed. by R. Schwenn and E. Marsch, Springer-Verlag, Berlin, Heidelberg, New York, 1990.

3.31   Burlaga, L.F., N.F. Ness, F. Mariani, B. Bavassano, U. Villante, H. Rosenbauer, R. Schwenn, J. Harvey, Magnetic fields and flows between 1 and 0.3 AU during the primary mission of Helios-1, J. Geophys. Res., **83**, 5167, 1978.

3.32   Burlaga, L., R. Lepping, R. Weber, T. Armstrong, C. Goodrich, J. Sullivan, D. Gurnett, P. Kellogg, E. Keppler, F. Mariani, F. Neubauer, H. Rosenbauer, R. Schwenn, Interplanetary particles and fields, November 22 to December 6, 1977: Helios, Voyager, and IMP observations between 0.6 and 1.6 AU, J. Geophys. Res., **85**, 2227–2242, 1980.

3.33   Burlaga, L.F., A.J. Hundhausen, X.P. Zhao, The coronal and interplanetary current sheet in early 1976, J. Geophys. Res., **86**, 8893–8898, 1981.

3.34   Burlaga, L.F., R. Schwenn, H. Rosenbauer, Dynamical evolution of interplanetary magnetic fields and flows between 0.3 AU and 8.5 AU: entrainment, Geophys. Res. Lett., **10**, 413–416, 1983.

3.35   Burlaga, L.F., F.B. McDonald, N.F. Ness, R. Schwenn, A.J. Lazarus, F. Mariani, Interplanetary flow systems associated with cosmic ray modulation, J. Geophys. Res., **89**, 6579–6587, 1984.

3.36   Burlaga, L.F., V. Pizzo, A. Lazarus, P. Gazis, Stream dynamics between 1 AU and 2 AU: a comparison of observations and theory, J. Geophys. Res., **90**, 7377–7388, 1985.

3.37 Burlaga, L.F., F.B. McDonald, R. Schwenn, Formation of a compound stream between 0.85 AU and 6.2 AU and its effects on solar energetic particles and galactic cosmic rays, J. Geophys. Res., **91**, 13331–13340, 1986.

3.38 Burlaga, L.F., K.W. Behannon, L.W. Klein, Compound streams, magnetic clouds, and major geomagnetic storms, J. Geophys. Res., **92**, 5725–5734, 1987.

3.39 Burlaga, L.F., Interaction regions in the distant solar wind, in *Proceedings of the Sixth International Solar Wind Conference*, ed. by V.J. Pizzo, T.E. Holzer, and D.G. Sime, NCAR/TN 306+Proc, Boulder, Colorado, 547–562, 1988.

3.40 Chao, J.K., V. Formisano, P.C. Hedgecock, Shock pair observations, in *Solar Wind*, NASA SP 308, 435, 1972.

3.41 Chernosky, E.J., Double sunspot-cycle variation in terrestrial magnetic activity, 1884–1963, J. Geophys. Res., **71**, 965, 1966.

3.42 Cliver, E.W., J.D. Mihalov, N.R. Sheeley, Jr., R.A. Howard, M.J. Koomen, R. Schwenn, Solar activity and heliospheric-wide cosmic ray modulation in mid-1982, J. Geophys. Res., **92**, 8487–8501, 1987.

3.43 Coles, W.A., B.J. Rickett, IPS observations of the solar wind speed out of the ecliptic, J. Geophys. Res., **81**, 4797–4799, 1976.

3.44 Coles, W.A., B.J. Rickett, V.H. Ramsey, J.J. Kaufman, D.G. Turley, S. Ananthakrishnan, J.W. Armstrong, J.K. Harmons, LS.L. Scott, D.G. Sime, Solar cycle changes in the polar solar wind, Nature, **286**, 239–241, 1980.

3.45 Dehmel, G., F.M. Neubauer, D. Lukoschus, J. Wawretzko, E. Lammers, Das Induktionsspulen-Magnetometer-Experiment (E4), Raumfahrtforschung, **19**, 241–244, 1975.

3.46 Denskat, K.U., Untersuchung von Alfvènischen Fluktuationen im Sonnenwind zwischen 0,29 AE und 1,0 AE, Dissertation an der Naturwissenschaftlichen Fakultät der Technischen Universität Carolo-Wilhelmina zu Braunschweig, 1982.

3.47 Denskat, K.U., F.M. Neubauer, Observations of hydrodynamic turbulence in the solar wind, in *Solar Wind Five*, NASA Conf. Publ. 2280, 81–91, 1983.

3.48 Dessler, A.J., Solar wind and interplanetary magnetic field, Rev. Geophys., **5**, 1–41, 1967.

3.49 Dessler, A.J., J.A. Fejer, Interpretation of Kp index and M-region geomagnetic storms, Planet. Space Sci., **11**, 505–511, 1963.

3.50 Diodato, L., G. Moreno, C. Signorini, K.W. Ogilvie, Long-term variations of the solar wind proton parameters, J. Geophys. Res., **79**, 5095, 1974.

3.51 Diodato, L., G. Moreno, On the heliographic latitude dependence of the solar wind velocity, Astrophys. and Space Sci., **39**, 409–414, 1976.

3.52 D'Angelo, N., G. Joyce, M.E. Pesses, Landau damping effects on solar wind fast streams, in *Solar Wind Four*, Max-Planck-Inst. f. Aeronomie, Katlenburg-Lindau, MPAE-W-100-81-31, 159, 1981.

3.53 Eselevich, V.G., M.A. Filippov, An investigation of the heliospheric current sheet (HCS) structure, Planet. Space Sci., **36**, 105–115, 1988.

3.54 Feldman, W.C., J.R. Asbridge, S.J. Bame, M.D. Montgomery, Double ion streams in the solar wind, J. Geophys. Res., **78**, 2017, 1973.

3.55 Feldman, W.C., J.R. Asbridge, S.J. Bame, M.D. Montgomery, S.P. Gary, Solar wind electrons, J. Geophys. Res., **80**, 4181, 1975.

3.56 Feldman, W.C., J.R. Asbridge, S.J. Bame, J.T. Gosling, High-speed solar wind parameters at 1 AU, J. Geophys. Res., **81**, 5054–5060, 1976.

3.57 Feldman, W.C., J.R. Asbridge, S.J. Bame, J.T. Gosling, Plasma and magnetic fields from the sun, in *The Solar Output and its Variations*, ed. by O.R. White, Colorado Associated University Press, Boulder, 351–381, 1977.

3.58 Feldman, W.C., J.R. Asbridge, S.J. Bame, J.T. Gosling, Long-term variations of selected solar wind properties: Imp 6, 7, and 8 results, J. Geophys. Res., **83**, 2177–2189, 1978.

3.59 Feldman, W.C., J.R. Asbridge, S.J. Bame, J.T. Gosling, D.S. Lemons, Characteristic electron variations across simple high-speed solar wind streams, J. Geophys. Res., **83**, 5285–5295, 1978.

3.60 Feldman, W.C., J.R. Asbridge, S.J. Bame, J.T. Gosling, D.S. Lemons, Electron heating within interaction zones of simple high-speed solar wind streams, J. Geophys. Res., **83**, 5297–5303, 1978.

3.61 Feldman, W.C., J.R. Asbridge, S.J. Bame, J.T. Gosling, Long-term solar wind electron variations between 1971 and 1978, J. Geophys. Res., **84**, 7371–7377, 1979.

3.62 Feldman, W.C., J.R. Asbridge, S.J. Bame, E.E. Fenimore, J.T. Gosling, The solar origins of solar wind interstream flows: near-equatorial coronal streamers, J. Geophys. Res., **86**, 5408–5416, 1981.

3.63 Fisher, R., D.G. Sime, Solar activity cycle variation of the K corona, Astrophys. J., **285**, 354–358, 1984.

3.64 Geiss, J., P. Hirt, H. Leutwyler, On acceleration and motions of ions in corona and solar wind, Solar Physics, **12**, 458, 1970.

3.65 Gosling, J.T., Variations in the solar wind speed along the earth's orbit, Solar Physics, **17**, 499–508, 1971.

3.66 Gosling, J.T., A.J. Hundhausen, V. Pizzo, J.R. Asbridge, Compressions and rarefactions in the solar wind: Vela 3, J. Geophys. Res., **77**, 5442, 1972.

3.67 Gosling, J.T., A.J. Hundhausen, S.J. Bame, Solar wind stream evolution at large heliocentric distances: experimental demonstration and the test of a model, J. Geophys. Res., **81**, 2111–2122, 1976.

3.68 Gosling, J.T., J.R. Asbridge, S.J. Bame, W.C. Feldman, Solar wind speed variations: 1962–1974, J. Geophys. Res., **81**, 5061–5070, 1976.

3.69 Gosling, J.T., J.R. Asbridge, S.J. Bame, W.C. Feldman, Preferred solar wind emitting longitudes on the sun, J. Geophys. Res., **82**, 2371–2376, 1977.

3.70 Gosling, J.T., A.J. Hundhausen, Waves in the solar wind, Scientific American, March, 36–43, 1977.

3.71 Gosling, J.T., E. Hildner, J.R. Asbridge, S.J. Bame, W.C. Feldman, Noncompressive density enhancements in the solar wind, J. Geophys. Res., **82**, 5005, 1977.

3.72 Gosling, J.T., J.R. Asbridge, S.J. Bame, W.C. Feldman, Solar wind stream interfaces, J. Geophys. Res., **83**, 1401–1412, 1978.

3.73 Gosling, J.T., G. Borrini, J.R. Asbridge, S.J. Bame, W.C. Feldman, R.T. Hansen, Coronal streamers in the solar wind at 1 AU, J. Geophys. Res., **86**, 5438–5448, 1981.

3.74 Gosling, J.T., J.R. Asbridge, S.J. Bame, W.C. Feldman, R.D. Zwickl, G. Paschmann, N. Sckopke, C.T. Russell, A sub-alfvénic solar wind: interplanetary and magnosheath observations, J. Geophys. Res., **87**, 239–245, 1982.

3.75 Grünwaldt, H., H. Rosenbauer, Study of helium and hydrogen velocity differences as derived from HEOS-2 S-210 solar wind measurements, in *Pleins Feux sur la Physique Solaire*, editions CNRS, 377–388, 1978.

3.76 Gurnett, D.A., Waves and instabilities, in *Physics of the Inner Heliosphere*, Vol II, ed. by R. Schwenn and E. Marsch, Springer-Verlag, Berlin, Heidelberg, New York, 1990.

3.77 Gurnett, D.A., R.R. Anderson, D.L. Odem, The University of Iowa HELIOS solar wind plasma wave experiment (E5a), Raumfahrtforschung, **19**, 245–247, 1975.

3.78 Gurnett, D.A., R.R. Anderson, Plasma wave electric fields in the solar wind: initial results from Helios-1, J. Geophys. Res., **82**, 632, 1977.

3.79 Habbal, S.R., K. Tsinganos, Multiple transonic solutions with a new class of shock transitions in steady isothermal solar and stellar winds, J. Geophys. Res., **88**, 1965–1975, 1983.

3.80 Habbal, S.R., R. Rosner, Temporal evolution of the solar wind and the formation of a standing shock, J. Geophys. Res., **89**, 10645–10657, 1984.

3.81 Hakamada, K., Y. Munakata, A cause of the solar wind speed variations: an update, J. Geophys. Res., **89**, 357–361, 1984.

3.82 Hale, G.E., Preliminary results of an attempt to detect the general magnetic field of the sun, Astrophys. J., **38**, 27, 1913.

3.83 Hansen, R.T., S.F. Hansen, C. Sawyer, Long lived coronal structures and recurrent geomagnetic patterns in 1974, Planet. Space Sci., **24**, 381, 1976.

3.84 Harvey, J.W., A.S. Krieger, J.M. Davis, A.F. Timothy, G.S. Vaiana, Comparison of Skylab X ray and ground-based helium observations, Bull. Amer. Astron. Soc., 7, 358, 1975.

3.85 Harvey, K.L., N.R. Sheeley, Jr., J.W. Harvey, Magnetic measurements of coronal holes during 1975–1980, Solar Physics, 79, 149–160, 1982.

3.86 Hernandez, R., S. Livi, E. Marsch, On the He+ to H+ temperature ratio in slow solar wind, J. Geophys. Res., 92, 7723–7727, 1987.

3.87 Hirshberg, J., The transport of flare plasma from the sun to the earth, Planet. Space Sci., 16, 309, 1968.

3.88 Hoeksema, J.T., J.M. Wilcox, P.H. Scherrer, Structure of the heliospheric current sheet in the early portion of sunspot, J. Geophys. Res., 87, 10331–10338, 1982.

3.89 Hoeksema, J.T., J.M. Wilcox, P.H. Scherrer, The structure of the heliospheric current sheet: 1978–1982, J. Geophys. Res., 88, 9910–9918, 1983.

3.90 Holzer, T.E., Effects of rapidly diverging flow, heat addition, and momentum addition in the solar wind and stellar winds, J. Geophys. Res., 82, 23, 1977.

3.91 Howard, R., B.J. Labonte, Surface magnetic fields during the solar activity cycle, Solar Physics, 74, 131–145, 1981.

3.92 Howard, R.A., M.J. Koomen, Observation of sectored structure in the outer solar corona: correlation with interplanetary magnetic field, Solar Physics, 37, 469–475, 1974.

3.93 Howard, R.A., N.R. Sheeley, Jr., M.J. Koomen, D.J. Michels, Coronal mass ejections: 1979–1981, J. Geophys. Res., 90, 8173–8191, 1985.

3.94 Hundhausen, A.J., Nonthermal heating in the quiet solar wind, J. Geophys. Res., 74, 5810–5813, 1969.

3.95 Hundhausen, A.J., Coronal Expansion and Solar Wind, Springer, New York, 1972.

3.96 Hundhausen, A.J., Nonlinear model of high-speed solar wind streams, J. Geophys. Res., 78, 1528–1542, 1973.

3.97 Hundhausen, A.J., Solar wind spatial structure: the meaning of latitude gradients on observations averaged over solar longitude, J. Geophys. Res., 83, 4186–4192, 1978.

3.98 Hundhausen, A.J., Solar activity and the solar wind, Reviews of Geophysics and Space Physics, 17, 2034–2048, 1979.

3.99 Hundhausen, A.J., S.J. Bame, M.D. Montgomery, Variations of solar wind plasma properties: Vela observations of a possible heliographic latitude dependence, J. Geophys. Res., 76, 5145, 1971.

3.100 Hundhausen, A.J., L.F. Burlaga, A model for the origin of solar wind stream interfaces, J. Geophys. Res., 80, 1845–1848, 1975.

3.101 Hundhausen, A.J., J.T. Gosling, Solar wind structure at large heliocentric distances: an interpretation of Pioneer 10 observations, J. Geophys. Res., 81, 1845, 1976.

3.102 Hundhausen, A.J., R.T. Hansen, S.F. Hansen, Coronal evolution during the sunspot cycle: coronal holes observed with the Mauna Loa K-coronameters, J. Geophys. Res., 86, 2079–2094, 1981.

3.103 Hundhausen, A.J., C.B. Sawyer, L.L. House, R.M.E. Illing, W.J. Wagner, Coronal mass ejections observed during the Solar Maximum Mission: latitude distribution and rate of occurrence, J. Geophys. Res., 89, 2639, 1984.

3.104 Hundhausen, A.J., The origin and propagation of coronal mass ejections, in Proceedings of the Sixth International Solar Wind Conference, ed. by V.J. Pizzo, T.E. Holzer, and D.G. Sime, NCAR/TN 306+Proc, Boulder, Colorado, 181–214, 1988.

3.105 Jockers, K., Solar wind models based on exospheric theory, Astron. Astrophys., 6, 219–239, 1970.

3.106 Kamide, Y., J.A. Slavin (Eds.), Solar Wind-Magnetosphere Coupling, Astrophysics and Space Science Library, D. Reidel Publishing Company, Dordrecht, Boston, Lancaster, Tokyo, 126, 1986.

3.107 Kellogg, P.J., G.A. Peterson, L. Lacabanne, The electric field experiment for HELIOS (E5b), Raumfahrtforschung, 19, 248–250, 1975.

3.108 King, J.H. (Ed.), *Interplanetary Medium Data Book*, NSSDC/WDC-A-R/S, 77-04, 1977.

3.109 Klein, L., L.F. Burlaga, Interplanetary sector boundaries 1971–1973, J. Geophys. Res., **85**, 2269–2276, 1980.

3.110 Klein, L.W., L.F. Burlaga, Interplanetary magnetic clouds at 1 AU, J. Geophys. Res., **87**, 613, 1982.

3.111 Kojima, M., T. Kakinuma, Solar cycle evolution of solar wind speed structure between 1973 and 1985 observed with the interplanetary scintillation, J. Geophys. Res., **92**, 7269–7279, 1987.

3.112 Kopp, R.A., T.E. Holzer, Dynamic of coronal hole regions. I: Steady polytropic flows with multiple critical points, Solar Physics, **49**, 43, 1976.

3.113 Korzhov, N.P., Large-scale three-dimensional structure of the interplanetary magnetic field, Solar Physics, **55**, 505–517, 1977.

3.114 Korzhov, N.P., V.V. Mishin, V.M. Tomozov, On the role of plasma parameters and the Kelvin–Helmholtz instability in a viscous interaction of solar wind streams, Planet. Space Sci., **32**, 1169–1178, 1984.

3.115 Krieger, A.S., A.F. Timothy, E.C. Roelof, A coronal hole and its identification as the source of a high velocity solar wind stream, Solar Physics, **23**, 123, 1973.

3.116 Kumar, S., A.L. Broadfoot, Evidence from Mariner 10 of solar wind flux depletion at high ecliptic latitudes, Astrophys. J., **228**, 302, 1979.

3.117 Kunow, H., G. Wibberenz, G. Green, R. Müller-Mellin, and M.-B. Kallenrode, Energetic particles in the inner solar system, in *Physics of the Inner Heliosphere*, Vol II, ed. by R. Schwenn and E. Marsch, Springer-Verlag, Berlin, Heidelberg, New York, 1990.

3.118 Lallement, R., J.L. Bertaux, V.G. Kurt, Solar wind decrease at high heliographic latitudes detected from Prognoz interplanetary Lyman alpha mapping, J. Geophys. Res., **90**, 1413, 1985.

3.119 Lallement, R., T.E. Holzer, R.H. Munro, Solar wind expansion in a polar coronal hole: inferences from coronal white light and interplanetary Lyman alpha observations, J. Geophys. Res., **91**, 6751–6759, 1986.

3.120 Lazarus, A.J., Trailing edges of high speed solar wind streams, Preprint CSR-Pj-76-11, 1976.

3.121 Lazarus, A., J. Belcher, Large-scale structure of the distant solar wind, in *Proceedings of the Sixth International Solar Wind Conference*, ed. by V.J. Pizzo, T.E. Holzer, and D.G. Sime eds., NCAR/TN 306+Proc, Boulder, Colorado, 533–546, 1988.

3.122 Lazarus, A.J., B. Yedidia, L. Villanueva, R.L. McNutt, Jr., J.W. Belcher, U. Villante, L.F. Burlaga, Meridional plasma flow in the outer heliosphere, Geophys. Res. Lett., **15**, 1519–1522, 1988.

3.123 Leer, E., T.E. Holzer, Energy addition in the solar wind, J. Geophys. Res., **85**, 4681–4688, 1980.

3.124 Leinert, C., I. Richter, B. Planck, Stability of the zodiacal light from minimum to maximum of the solar cycle, Astron. Astrophys., **110**, 111–114, 1982.

3.125 Leinert, C., E. Grün, Interplanetary dust, in *Physics of the Inner Heliosphere*, (this volume).

3.126 Levine, R.H., The relation of open magnetic structures to solar wind flow, J. Geophys. Res., **83**, 4193–4199, 1978.

3.127 Livi, S., E. Marsch, H. Rosenbauer, Coulomb collisional domains in the solar wind, J. Geophys. Res., **91**, 8045–8050, 1986.

3.128 Livi, S., E. Marsch, Generation of solar wind proton tails and double beams by Coulomb collisions, J. Geophys. Res., **92**, 7255–7261, 1987.

3.129 Lopez, R.E., Solar cycle invariance in solar wind proton temperature relationships, J. Geophys. Res., **92**, 11189–11194, 1987.

3.130 Lopez, R.E., J.W. Freeman, E.C. Roelof, The relationship between proton temperature and momentum flux density in the solar wind, Geophys. Res. Lett., **13**, 640–643, 1986.

3.131 Mariani, F., F.M. Neubauer, Interplanetary magnetic field, in *Physics of the Inner Heliosphere*, (this volume).

3.132 Marsch, E., Kinetic physics of the solar wind plasma, in *Physics of the Inner Heliosphere*, Vol II, ed. by R. Schwenn and E. Marsch, Springer-Verlag, Berlin, Heidelberg, New York, 1990.

3.133 Marsch, E., MHD-turbulence in the solar wind, in *Physics of the Inner Heliosphere*, Vol II, ed. by R. Schwenn and E. Marsch, Springer-Verlag, Berlin, Heidelberg, New York, 1990.

3.134 Marsch, E., K.H. Mühlhäuser, H. Rosenbauer, R. Schwenn, K.U. Denskat, Pronounced proton core temperature anisotropy, ion differential speed, and simultaneous Alfvén wave activity in slow solar wind at 0.3 AU, J. Geophys. Res., **86**, 9199–9203, 1981.

3.135 Marsch, E., K.H. Mühlhäuser, H. Rosenbauer, R. Schwenn, F.M. Neubauer, Solar wind helium ions: observations of the Helios solar probes between 0.3 and 1 AU, J. Geophys. Res., **87**, 35–51, 1982.

3.136 Marsch, E., K.H. Mühlhäuser, R. Schwenn, H. Rosenbauer, W. Pilipp, F.M. Neubauer, Solar wind protons: three-dimensional velocity distributions and derived plasma parameters measured between o.3 and 1 AU, J. Geophys. Res., **87**, 52–72, 1982.

3.137 Marsch, E., A.K. Richter, Distribution of solar wind angular momentum between particles and magnetic field: inferences about the Alfvén critical point from Helios observations, J. Geophys. Res., **89**, 5386–5394, 1984.

3.138 Marsch, E., A.K. Richter, Helios observational constraints on solar wind expansion, J. Geophys. Res., **89**, 6599, 1984.

3.139 Marsden, R.G. (Ed.), *The Sun and the Heliosphere in Three Dimensions*, Astrophysics and Space Science Library, D. Reidel Publishing Company, Dordrecht, Boston, Lancaster, Tokyo, 123, 1986.

3.140 McNutt, R.L., Jr., Possible explanations of north–south plasma flow in the outer heliosphere and meridional transport of magnetic flux, Geophys. Res. Lett., **15**, 1523–1526, 1988.

3.141 Mitchell, D.G., E.C. Roelof, J.H. Wolfe, Latitude dependence of solar wind velocity observed $\geq 1\mathrm{AU}$, J. Geophys. Res., **86**, 165–179, 1981.

3.142 Miyake, W., T. Mukai, T. Terasawa, K. Hirao, Stream interaction as a heat source in the solar wind, Solar Physics, **117**, 171–178, 1988.

3.143 Munro, R., B. Jackson, Physical properties of a polar coronal hole from 2 to 5 Rs, Astrophys. J., **213**, 874, 1977.

3.144 Musmann, G., F.M. Neubauer, A. Maier, E. Lammers, Das Förstersonden-Magnetfeldexperiment (E2), Raumfahrtforschung, **19**, 232–237, 1975.

3.145 Neugebauer, M., Large-scale and solar-cycle variations of the solar wind, Space Sci. Rev., **17**, 221–254, 1975.

3.146 Neugebauer, M., The quiet solar wind, J. Geophys. Res., **81**, 4664–4670, 1976.

3.147 Neugebauer, M., Observations of solar-wind helium, Fundam. Cosm. Phys., **7**, 131–199, 1981.

3.148 Neugebauer, M., C.W. Snyder, Mariner 2 observations of the solar wind. 1. Average properties, J. Geophys. Res., **71**, 4469–4484, 1966.

3.149 Neugebauer, M.M., W.C. Feldman, Relation between superheating and superacceleration of helium in the solar wind, Solar Physics, **63**, 201–205, 1979.

3.150 Neugebauer, M., C.J. Alexander, R. Schwenn, A.K. Richter, Tangential discontinuities in the solar wind: correlated field and velocity changes and the Kelvin–Helmholtz instability, J. Geophys. Res., **91**, 13694–13698, 1986.

3.151 Neupert, W.M., V. Pizzo, Solar coronal holes as sources of recurrent geomagnetic disturbances, J. Geophys. Res., **79**, 3701, 1974.

3.152 Newkirk, G., Jr., L.A. Fisk, Variation of cosmic rays and solar wind properties with respect to the heliospheric current sheet. 1. Five-GeV protons and solar wind speed, J. Geophys. Res., **90**, 3391, 1985.

3.153 Nolte, J.T., A.S. Krieger, A.F. Timothy, R.E. Gold, E.C. Roelof, G. Vaiana, A.J. Lazarus, J.D. Sullivan, P.S. McIntosh, Coronal holes as sources of solar wind, Solar Physics, **46**, 303–322, 1976.

3.154 Nolte, J.T., A.S. Krieger, E.C. Roelof, R.E. Gold, High coronal structure of high velocity solar wind stream sources, Solar Physics, **51**, 459–471, 1977.

3.155 Nolte, J.T., J.M. Davis, M. Gerassimenko, A.J. Lazarus, J.D. Sullivan, A comparison of solar wind streams and coronal structure near solar minimum, Geophys. Res. Lett., **4**, 291–294, 1977.

3.156 Nolte, J.T., M. Gerassimenko, A.S. Krieger, C.V. Solodyna, Coronal hole evolution by sudden large scale changes, Solar Physics, **56**, 153–159, 1978.

3.157 Ogilvie, K.W., Corotating shock structures, in *Solar Wind*, NASA SP 308, 430–434, 1972.

3.158 Ogilvie, K.W., J. Hirshberg, The solar cycle variation of the solar wind helium abundance, J. Geophys. Res., **79**, 4595–4602, 1974.

3.159 Parker, E.N., Dynamics of the interplanetary gas and magnetic fields, Astrophys. J., **128**, 664–675, 1958.

3.160 Parker, E.N., Magnetic neutral sheets in evolving fields. II: Formation of the solar corona, Astrophys. J., **264**, 642–647, 1983.

3.161 Pilipp, W.G., R. Schwenn, E. Marsch, K.H. Mühlhäuser, H. Rosenbauer, Electron characteristics in the solar wind as deduced from Helios observations, in *Solar Wind Four*, Max-Planck-Inst. f. Aeronomie, Katlenburg-Lindau, MPAE-W-100-81-31, 241–249, 1981.

3.162 Pilipp, W.G., H. Miggenrieder, K.H. Mühlhäuser, H. Rosenbauer, R. Schwenn, Data analysis of electron measurements of the plasma experiment aboard the Helios probes, MPE report, 185, 1984.

3.163 Pilipp, W.G., H. Miggenrieder, M.D. Montgomery, K.H. Mühlhäuser, H. Rosenbauer, R. Schwenn, Characteristics of electron velocity distribution functions in the solar wind derived from the Helios plasma experiment, J. Geophys. Res., **92**, 1075–1092, 1987.

3.164 Pilipp, W.G., H. Miggenrieder, K.H. Mühlhäuser, H. Rosenbauer, R. Schwenn, F.M. Neubauer, Variations of electron distribution functions in the solar wind, J. Geophys. Res., **92**, 1103–1118, 1987.

3.165 Pilipp, W.G., H. Miggenrieder, K.H. Mühlhäuser, H. Rosenbauer, R. Schwenn, Large scale variations of thermal electron properties in the solar wind, submitted to J. Geophys. Res., 1988.

3.166 Pizzo, V.J., A three-dimensional model of corotating streams in the solar wind. 3: Magneto-hydrodynamic streams, J. Geophys. Res., **87**, 4374–4394, 1982.

3.167 Pizzo, V., R. Schwenn, E. Marsch, H. Rosenbauer, K.H. Mühlhäuser, F.M. Neubauer, Determination of the solar wind angular momentum flux from the Helios data – an observational test of the Weber and Davis theory, Astrophys. J., **271**, 335–354, 1983.

3.168 Pneuman, G.W., Latitude dependence of the solar wind speed: influence of the coronal magnetic field geometry, J. Geophys. Res., **81**, 5049–5053, 1976.

3.169 Pneuman, G.W., Ejection of magnetic fields from the sun: acceleration of a solar wind containing diamagnetic plasmoids, Astrophys. J., **265**, 468–482, 1983.

3.170 Pneumann, G.W., Driving mechanisms for the solar wind, Space Science Rev., **43**, 105–138, 1986.

3.171 Porsche, H. (Ed.), 10 Jahre HELIOS, Festschrift aus Anlaß des 10. Jahrestages des Starts der Sonnensonde Helios am 10. Dezember 1974, DFVLR Oberpfaffenhofen, 1984.

3.172 Priest, E.R. (Ed.), Solar flare magnetohydrodynamics, in *The Fluid Mechanics of Astrophysics and Geophysics*, ed. by P.H. Roberts, Gordon and Breach Science Publishers, New York, London, Paris, 1, 1981.

3.173 Rhodes, E.J., Jr., E.J. Smith, Multispacecraft study of the solar wind velocity at interplanetary sector boundaries, J. Geophys. Res., **80**, 917–928, 1975.

3.174 Rhodes, E.J., Jr., E.J. Smith, Evidence of a large-scale gradient in the solar wind velocity, J. Geophys. Res., **81**, 2123–2134, 1976.

3.175 Rhodes, E.J, Jr., E.J. Smith, Further evidence of a latitude gradient in the solar wind velocity, J. Geophys. Res., **81**, 5833, 1976.

3.176 Rhodes, E.J., Jr., E.J. Smith, Multi-spacecraft observations of heliographic latitude-longitude structure in the solar wind, J. Geophys. Res., **86**, 8877–8892, 1981.

3.177 Richter, A.K., R. Schwenn, F.M. Neubauer, Nature and origin of corotating shock waves within 1 AU, MPAE-W-79-80-38, Max-Planck-Inst. f. Aeron., Katlenburg-Lindau, 1980.

3.178 Richter, A.K., K.C. Hsieh, A.H. Luttrell, E. Marsch, R. Schwenn, Review of interplanetary shock phenomena near and within 1 AU, in *Collisionless Shocks in the Heliosphere: Reviews of Current Research*, Geophysical Monograph, **35**, 33–50, 1985.

3.179 Richter, A.K., A.H. Luttrell, Superposed epoch analysis of corotating interaction regions at 0.3 and 1.0 AU: a comparative study, J. Geophys. Res., **91**, 5873–5878, 1986.

3.180 Robbins, D.E., Helium in the Solar Wind, J. Geophys. Res., **75**, 1178–1187, 1970.

3.181 Roelof, E.C., S.M. Krimigis, Analysis and synthesis of coronal and interplanetary energetic particle, plasma, and magnetic field observations over three solar rotations, J. Geophys. Res., **78**, 5375, 1973.

3.182 Rosenbauer, H., Possible effects of photoelectron emission on a low-energy electron experiment, in *Photon and Particle Interactions with Surfaces in Space*, ed. by R.J.L. Grard, D. Reidel Publishing Company, Dordrecht, Holland, 139, 1973.

3.183 Rosenbauer, H., H. Miggenrieder, M.D. Montgomery, R. Schwenn, Preliminary results of the Helios plasma measurements, in *Physics of Solar Planetary Environments*, ed. by D.J. Williams, American Geophysical Union, 319–331, 1976.

3.184 Rosenbauer, H., R. Schwenn, E. Marsch, B. Meyer, H. Miggenrieder, M. Montgomery, K.H. Mühlhäuser, W. Pilipp, W. Voges, S.K. Zink, A survey on initial results of the Helios plasma experiment, J. Geophys., **42**, 561–580, 1977.

3.185 Rosenbauer, H., R. Schwenn, S. Bame, The prediction of fast stream front arrivals at the Earth on the basis of solar wind measurements at smaller solar distances, in *Proc. of AGARD-Symposium on 'Operation Modelling of the Aerospace Propagation Environment'*, Ottawa, Canada 24–28 April 1978, ed. by H. Soicher, AGARD-CP-238, 32-1, 1978.

3.186 Rosenbauer, H., R. Schwenn, H. Miggenrieder, B. Meyer, H. Grünwaldt, K.H. Mühlhäuser, H. Pellkofer, J.H. Wolfe, Die Instrumente des Plasmaexperiments auf den Helios-Sonnensonden, BMFT-FB-W 81-015, 1981.

3.187 Rosenberg, R.L., P.J. Coleman, Heliographic latitude dependence of the dominant polarity of the interplanetary magnetic field, J. Geophys. Res., **74**, 5611, 1969.

3.188 Rosenberg, R.L., P.J. Coleman, Jr., Solar cycle-dependent north–south field configurations observed in solar wind interaction regions, J. Geophys. Res., **85**, 3021–3032, 1980.

3.189 Russell, C.T., On the heliographic latitude dependence of the interplanetary magnetic field as deduced from the 22-year cycle of geomagnetic activity, Geophys. Res. Lett., **1**, 11–12, 1974.

3.190 Saito, T., Two-hemisphere model on the three-dimensional magnetic structure of the interplanetary space, Sci. Rep. Tohoku University, Ser. 5, Geophysics, **23**, 37–54, 1975.

3.191 Sarabhai, V., Some consequences of nonuniformity of solar wind velocity, J. Geophys. Res., **68**, 1555–1557, 1963.

3.192 Sastri, J.H., Solar wind flow associated with stream-free sector boundaries at 1 AU, Solar Physics, **111**, 429–437, 1987.

3.193 Scearce, C., S. Cantarano, N. Ness, F.R. Mariani, R. Terenzi, L. Burlaga, The Rome-GSFC magnetic field experiment for HELIOS A and B (E3), Raumfahrtforschung, **19**, 237–240, 1975.

3.194 Schmidt, W.K.H., H. Rosenbauer, E.G. Shelley, J. Geiss, On temperature and speed of He^{++} and $O\,6^+$ ions in the solar wind, Geophys. Res. Lett., **7**, 697–700, 1980.

3.195 Schulz, M., Interplanetary sector structure and the heliomagnetic equator, Astrophys. Space Sci., **24**, 371–383, 1973.

3.196 Schwartz, S.J., E. Marsch, The radial evolution of a single solar wind plasma parcel, J. Geophys. Res., **88**, 9919, 1983.

3.197 Schwenn, R., Solar wind and its interactions with the magnetosphere: measured parameters, Adv. Space Res., **1**, 3–17, 1981.

3.198 Schwenn, R., The 'average' solar wind in the inner heliosphere: structures and slow variations, in *Solar Wind 5*, NASA Conf. Publ. 2280, 489, 1983.

3.199 Schwenn, R., Relationship of coronal transients to interplanetary shocks: 3 D aspects, Space Sci. Rev., **44**, 139–168, 1986.

3.200 Schwenn, R., H. Rosenbauer, H. Miggenrieder, Das Plasmaexperiment auf Helios (E1), Raumfahrtforschung, **19**, 226–232, 1975.

3.201 Schwenn, R., H. Rosenbauer, H. Miggenrieder, B. Meyer, Preliminary results of the Helios plasma experiment, in *Proc. of the 18th Plenary Meeting of COSPAR*, Varna 1975, ed. by M.J.Rycroft, Space Research, **16**, 671, 1976.

3.202 Schwenn, R., H. Rosenbauer, K.H. Mühlhäuser, The solar wind during STIP II interval: stream structures, boundaries, shock and other features as observed on Helios-1 and Helios-2, in *Contributed Papers to the Study of Travelling Interplanetary Phenomenon/1977*, Tel Aviv, June 1977, ed. by M.A. Shea, D.F. Smart, S.T. Wu, Air Force Geophysics Laboratory Rep. Nr. 77-309, 351–361, 1977.

3.203 Schwenn, R., M. Montgomery, H. Rosenbauer, H. Miggenrieder, K.H. Mühlhäuser, S.J. Bame, W.C. Feldman, R.T. Hansen, Direct observations of the latitudinal extent of a high-speed stream in the solar wind, J. Geophys. Res., **83**, 1011, 1978.

3.204 Schwenn, R., K.H. Mühlhäuser, H. Rosenbauer, Two states of the solar wind at the time of solar activity minimum. I. Boundary layers between fast and slow streams, in *Solar Wind Four*, Max-Planck-Inst. f. Aeronomie, Katlenburg-Lindau, MPAE-W-100-81-31, 118–125, 1981.

3.205 Schwenn, R., K.H. Mühlhäuser, E. Marsch, H. Rosenbauer, Two states of the solar wind at the time of solar activity minimum. II. Radial gradients of plasma parameters in fast and slow streams, in *Solar Wind Four*, Max-Planck-Inst. f. Aeronomie, Katlenburg-Lindau, MPAE-W-100-81-31, 126–130, 1981.

3.206 Schwenn, R., H. Rosenbauer, Aufbereitung und Auswertung der Daten des Plasmaexperiments auf den HELIOS-Sonnensonden, BMFT-FB-W 82-002, 1982.

3.207 Sheeley, N.R., Jr., The evolution of the polar coronal hole, Solar Physics, **65**, 229–235, 1980.

3.208 Sheeley, N.R., Jr., J.W. Harvey, W.C. Feldman, Coronal holes, solar wind streams, and recurrent geomagnetic disturbances, 1973-1976, Solar Physics, **49**, 271, 1976.

3.209 Sheeley, N.R., Jr., J.R. Asbridge, S.J. Bame, J.W. Harvey, A pictorial comparison of interplanetary magnetic field polarity, solar wind speed, and geomagnetic disturbance index during the sunspot cycle, Solar Physics, **52**, 485–495, 1977.

3.210 Sheeley, N.R., Jr., J.W. Harvey, Coronal holes, solar wind streams, and geomagnetic disturbances during 1978 and 1979, Solar Physics, **70**, 237–249, 1981.

3.211 Sheeley N.R., Jr., R.A. Howard, M.J. Koomen, D.J. Michels, K.L. Harvey, J.W. Harvey, Observations of coronal structure during sunspot maximum, Space Sci. Rev., **33**, 219–231, 1982.

3.212 Sheeley, N.R., Jr., R.A. Howard, M.J. Koomen, D.J. Michels, R. Schwenn, K.H. Mühlhäuser, H. Rosenbauer, Coronal mass ejections and interplanetary shocks, J. Geophys. Res., **90**, 163–175, 1985.

3.213 Siscoe, G.L., Fluid dynamics of solar wind filaments, Solar Physics, **13**, 490–498, 1970.

3.214 Siscoe, G.L., Three-dimensional aspects of interplanetary shock waves, J. Geophys. Res., **81**, 6235, 1976.

3.215 Siscoe, G.L., L.T. Finley, Meridional (north–south) motions of the solar wind, Solar Physics, **9**, 452–466, 1969.

3.216 Siscoe, G.L., B. Goldstein, A.J. Lazarus, An east–west asymmetry in the solar wind velocity, J. Geophys. Res., **74**, 1759–1762, 1969.

3.217 Slavin, J.A., E.J. Smith, Solar cycle variations in the interplanetary magnetic field, in *Solar Wind 5*, NASA Conf. Publ. 2280, 323–331, 1983.

3.218 Smith, E.J., J.H. Wolfe, Observations of interaction regions and corotating shocks between one and five AU: Pioneers 10 and 11, Geophys. Res. Lett., **3**, 137, 1976.

3.219 Smith, E.J., B.T. Tsurutani, R.L. Rosenberg, Observations of the interplanetary sector structure up to the heliographic latitude of 16°; Pioneer 11, J. Geophys. Res., **83**, 717, 1978.

3.220 Snyder, C.W., M. Neugebauer, U.R. Rao, The solar wind velocity and its correlation with cosmic-ray variations and with solar and geomagnetic activity, J. Geophys. Res., **68**, 6361–6370, 1963.

3.221 Snyder, C.W., M. Neugebauer, The relation of Mariner 2 plasma data to solar phenomena, in *The Solar Wind*, ed. by R.J. Mackin and M. Neugebauer, Pergamon Press, New York, 25–34, 1966.

3.222 Steinitz, R., M. Eyni, Global properties of the solar wind. I. The invariance of the momentum flux density, Astrophys. J., **241**, 417–424, 1980.

3.223 Steinitz, R., Momentum flux invariance and solar wind sources, Solar Physics, **83**, 379–384, 1983.

3.224 Suess, S.T., J. Feynman, Sector boundary distortion in the interplanetary medium, J. Geophys. Res., **82**, 2405–2409, 1977.

3.225 Suess, S.T., E. Hildner, Deformation of the heliospheric current sheet, J. Geophys. Res., **90**, 9461–9468, 1985.

3.226 Thieme, K.M., E. Marsch, R. Schwenn, Relationship between structures in the solar wind and their source regions in the corona, in *Proceedings of the Sixth International Solar Wind Conference*, ed. by V.J. Pizzo, T.E. Holzer, D.G. Sime, NCAR/TN-306+Proc, 317–321, 1987.

3.227 Thomas, B.T., E.J. Smith, The structure and dynamics of the heliospheric current sheet, J. Geophys. Res., **86**, 11105–11110, 1981.

3.228 Thomas, B.T., B.E. Goldstein, E.J. Smith, The effect of the heliospheric current sheet on cosmic ray intensities at solar maximum: two alternative hypotheses, J. Geophys. Res., **91**, 2889–2895, 1986.

3.229 Timothy, A.F., A.S. Krieger, G.S. Vaiana, The structure and evolution of coronal holes, Solar Physics, **42**, 135–156, 1975.

3.230 Tsurutani, B.T., R.G. Stone (Eds.), *Collisionless Shocks in the Heliosphere. Reviews of Current Research*, Geophysical Monograph, **35**, 1985.

3.231 Villante, U., R. Bruno, F. Mariani, L.F. Burlaga, N.F. Ness, The shape and location of the sector boundary surface in the inner solar system, J. Geophys. Res., **84**, 6641–6648, 1979.

3.232 Webb, D.F., A.J. Hundhausen, Activity associated with the solar origin of coronal mass ejections, Solar Physics, **108**, 383–401, 1987.

3.233 Weber, E.J., Davis L., The angular momentum of the solar wind, Astrophys. J., **148**, 217–227, 1967.

3.234 Weber, R.R., The radio astronomy experiment on Helios A and B (E5c), Raumfahrtforschung, **19**, 250–252, 1975.

3.235 Whang, Y.C., L.F. Burlaga, Coalescence of the pressure waves associated with stream interactions, J. Geophys. Res., **90**, 221, 1985.

3.236 Whang, Y.C., L.F. Burlaga, The coalescence of two merged interaction regions between 6.2 and 9.5 AU: September 1979 Event, J. Geophys. Res., **91**, 13341–13348, 1986.

3.237 Wilcox, J.M., N.F. Ness, Quasi-stationary corotating structure in the interplanetary medium, J. Geophys. Res., **70**, 5793–5805, 1965.

3.238 Wilcox, J.M., A.J. Hundhausen, Comparison of heliospheric current sheet structure obtained from potential magnetic field computations and from observed polarization coronal brightness, J. Geophys. Res., **88**, 8095, 1983.

3.239 Withbroe, G.L., Origins of the solar wind in the corona, in *The Sun and the Heliosphere in Three Dimensions*, ed. by R.G. Marsden, Astrophysics and Space Science Library, D. Reidel Publishing Company, Dordrecht, Boston, Lancaster, Tokyo, 19–32, 1986.

3.240 Zhao, X.P., A.J. Hundhausen, Organization of solar wind plasma properties in a tilted, heliomagnetic coordinate system, J. Geophys. Res., **86**, 5423–5430, 1981.

3.241 Zhao, X.P., A.J. Hundhausen, Spatial structure of solar wind in 1976, J. Geophys. Res., **88**, 451, 1983.

3.242 Zirker, J.B., Coronal holes and high-speed wind streams, Reviews of Geophysics and Space Physics, **15**, 257–269, 1977.

3.243 Zirker, J. (Ed.), *Coronal Holes and High Speed Streams*, Colorado Associated University Press, Boulder, 1977.

4. The Interplanetary Magnetic Field

Franco Mariani and Fritz M. Neubauer

4.1 Introduction

4.1.1 The Magnetic Field at 1 AU, Status 1974

In 1974 a number of observations of the magnetic field at about 1 AU were already available and the concept of coronal magnetic field expansion in the heliosphere well established. On the other hand, only a few, sporadic, *in situ* field observations were available at heliocentric distances significantly inside the terrestrial orbit.

The spiral structure, more or less as outlined by Parker [4.32], was already known from the observations as was the existence of outward and inward polarities. However, the physical interpretation of magnetic sectors was still under discussion, in particular whether they are organized more like segments of an orange, i.e. essentially delimited in longitude rather than in latitude. The so-called dominant polarity effect was not yet finally understood; the same was true of the nature of and the physical properties at the sector boundaries.

Although the basic mechanism of the coronal expansion was clear, no systematic study had yet been made to trace magnetic field lines as observed inside 1 AU directly back to the solar source region.

The existence of Alfvén waves in the expanding solar wind was suggested early on by correlating solar wind velocity and magnetic field fluctuations; also, turbulence and a variety of peculiar field variations, in particular discontinuities of one or more magnetic elements (components, intensity, orientation, etc.) at different time or spatial scales were known to exist, although they were at an early stage of their physical interpretation. For example, such questions as the following were still to be satisfactorily answered: Are turbulence and/or discontinuities already present at the source close to the sun and then simply convected outward? Or are they generated *en route* out of the sun? Do they propagate in the solar wind reference system or are they frozen in it?

Helios magnetic field experiments were designed to give highly reliable and precise measurements to fill our gap in knowledge of the space between the source region at a few tens of solar radii and the earth's orbit. The wealth of *Helios* data, extending over about one full eleven-year solar cycle, give sufficient information about the inner heliosphere as deep as 0.29 AU and also allow us to combine it with the information now available on the outer heliosphere from the extended exploration of the *Pioneer* and *Voyager* spacecraft to form a comprehensive picture.

Physics and Chemistry in Space - Space and Solar Physics, Vol. 20
Physics of the Inner Heliosphere I   Editors: R. Schwenn · E. Marsch
© Springer-Verlag Berlin Heidelberg 1990

4.1.2 A Short Description of the *Helios* Magnetic Experiments

Two flux-gate triaxial magnetometers were flown on *Helios*: one [4.28] built by the Institute for Geophysics and Metereology at the Technical University of Braunschweig, FRG; the other [4.39] jointly built by the University of Rome-CNR Laboratorio Plasma Spazio, Frascati and the Laboratory for Extraterrestrial Physics of the NASA-Goddard Space Flight Center, Greenbelt, M.D. Both instruments were boom mounted and located in opposite directions on the spacecraft's equator, rotating at a spin rate of about 1 spin/second. Mechanical flipper devices made it possible to flip by command the sensor parallel to the spin axis into the spinning plane of the spacecraft to enable a determination of the zero offset of the spin component sensor to be made.

Also, the instruments were periodically calibrated. At lower bit rates built-in onboard computers were used to obtain averages over time intervals between 1 second and 64 seconds, or 1 second and 48 seconds.

The TUB magnetometer (*Förstersonde*) had automatically switchable measurement ranges of $\pm$ 100 nT and $\pm$ 400 nT with a highest resolution of $\pm$ 0.2 nT and a maximum sampling rate of 8 spin-synchronized vectors per second. The overall offsets of the components in the spin plane, composed of sensor offsets and spacecraft field, were removed on board by properly averaging over the spin rotation. The experiment also included [4.26] a so-called shock identification computer (SIC) for triggering a memory-mode to observe discontinuities and shocks characterized by positive increases of field magnitude.

The Rome-GSFC (Heliflux sensors) magnetometer [4.40] had four dynamic ranges $\pm$ 16, $\pm$ 48, $\pm$ 144, $\pm$ 432 nT with sensitivities $\pm$ 0.03 nT in the most sensitive range to $\pm$ 0.84 nT in the less sensitive range. Field determinations were time synchronized, with a maximum sampling rate of about 5 vectors/second. The instrument also had the capability of accumulating data at very high bit rates for the high temporal resolution needed in the shock mode.

4.2 Radial Variations

4.2.1 Background

The existence of a radial variation of the interplanetary magnetic field was an early discovery made after the first *in situ* measurements. The expanding solar corona model by Parker [4.32] was the theoretical background for interpreting the observations. This time-stationary spherically symmetric model predicts that under the assumption of an expanding solar wind at constant velocity v and a radial magnetic field at a heliocentric distance of a few solar radii the three field components are given by

$$B_r(r, \theta, \varphi) = B(\theta, \varphi_0) \left(\frac{b}{r}\right)^2 , \qquad (4.1)$$

$$B_\theta(r, \theta, \varphi) = 0 , \tag{4.2}$$

$$B_\varphi(r, \theta, \varphi) = B(\theta, \varphi_0) \left[\frac{\Omega b^2}{vr^2} (r - b) \sin \theta \right] , \tag{4.3}$$

where b and φ_0 are the distance and the longitude at which an inner-boundary radial field $B(\theta, \varphi_0)$ is assumed to exist; Ω is the solar angular rotation rate. The distances r and b are connected by the equation

$$\frac{r}{b} - 1 - \ln \frac{r}{b} = \frac{v}{b\Omega} (\varphi - \varphi_0) , \tag{4.4}$$

which for $r \gg b$ reduces to the familiar Archimedean spiral equation. The radial component does not depend upon the solar wind velocity, so it is expected to be more "regular" than the azimuthal (transverse) component which is also a function of v.

A more elaborate model [4.14] has since been worked out to take into account longitudinal and temporal variations, the explicit results of which could only be derived by numerical integration of specific sets of inner heliospheric boundary conditions.

Early observational results on the field parameters on board *Helios 1*, which traveled well inside the inner heliosphere [4.24], have shown a high degree of variability, up to a factor of 2 or so in a few hours; this is in contrast to an expected variation in the order of 40% at 0.3 AU in a time interval of 2–3 days relative to a fixed point, as predicted [4.12] by extrapolation of observations at 1 AU using a model of a corotating stream. This means that either spatial nonuniformities in B near the sun are simply convected outward, retaining their individuality, and/or nonlinear dynamical processes strongly affect the short-term variability of the interplanetary magnetic field at heliocentric distances less than 0.3 AU.

4.2.2 The Radial Component B_r

Before the *Helios* measurements inside 1 AU, the only available information was from individual radial excursions of *Pioneer 6*, *Mariner 5*, and *Mariner 10* (see [4.5]). Data were in the first case solar rotation averages, in the second and third cases smoothed 3-day running averages or daily averages. Linear best fits of log $\langle |B_r| \rangle$ versus log r indicated a slope s close to the theoretically expected $s = -2$, although with non-negligible statistical uncertainties, with the exception of *Mariner 5* which gave $s = -1.78 \pm 0.02$.

Helios 1 and 2 had the unique possibility of exploring repeatedly and quickly the inner heliospheric range of distances from 1 to 0.3 AU. Long-term variations do not generally affect the results and thus a good time resolution is possible.

As concerns the specific variation law of the field components, the absolute values of daily averages $|\langle B_r \rangle|$ were shown [4.27] to scale with r as Ar^s with $s = -2.00$ during the first radial excusion of *Helios 1* (10 December 1974– 15 March 1975), Fig. 4.1a shows the large scatter of individual averages which is

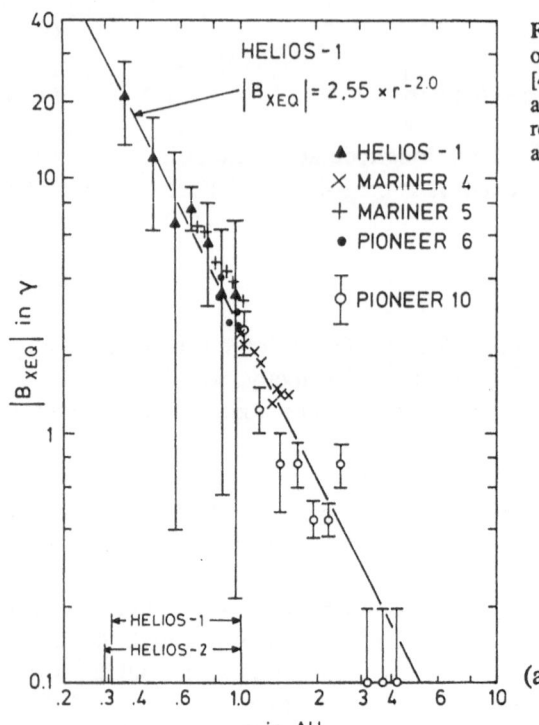

Fig. 4.1. Heliocentric distance variations of daily averages $\langle|B_r|\rangle$ from *Helios 1* [4.27] (Fig. 4.1a) and twelve-hour averages of $\langle|B_\varphi|\rangle$ for two different velocity regimes [4.23] (Fig. 4.1b). Best linear fits are also shown

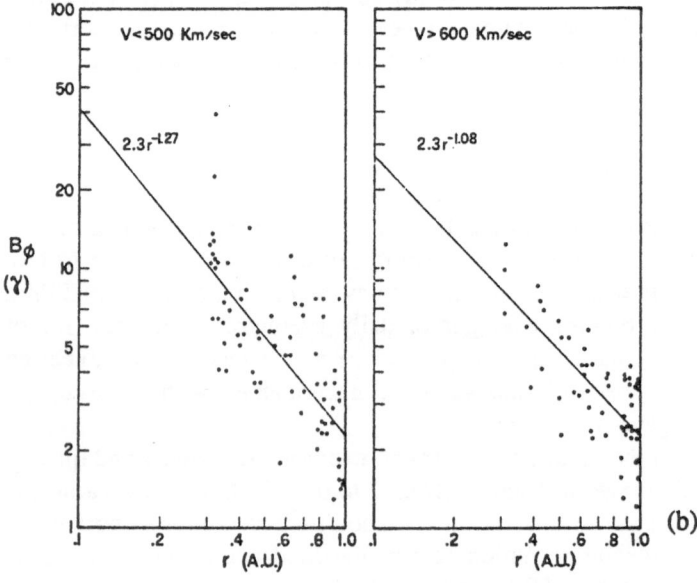

at least partially an effect of the variations of the heliographic longitude and to a lesser extent of the latitude and time. The observed B_r variation appears to be in good agreement with the model predictions. Actually, data by other spacecraft show some differences which are probably due to the use of different time intervals of different solar cycle conditions and, not insignificantly, due to the use of different definitions of the displayed "radial" component.

An analysis of data with higher time resolution (and hence also higher spatial) resolution – 3 to 12 hours – where two regimes of low and high solar wind velocity were also considered, has led [4.23, 25] to a substantial agreement with the expected power law with $s = -2$, for the B_r component (Table 4.1).

A comprehensive overall study has been recently performed [4.31] using 12-hours averages from *Helios 1* (covering 10 December 1974 to 17 August 1979) and *Helios 2* (covering 15 January 1976 to 7 August 1979). Average values for A and s for these intervals turned out to be $A = 2.02 + 1.13\,\text{nT}$, $s = -2.21 + 0.26$ for *Helios 1* and $A = 2.23 + 1.18\,\text{nT}$, $s = -2.10 + 0.16$ for *Helios 2*.

Table 4.1.

Helios 1

All Data
$0.31 < r < 1.0\,\text{AU}$

	A	s
$\langle |B_r| \rangle$	2.36 ± 0.16	-2.06 ± 0.09
$\langle |B_\varphi| \rangle$	2.42 ± 0.17	-1.13 ± 0.10
$\langle |B_\theta| \rangle$	1.41 ± 0.10	-1.35 ± 0.10
$\langle B \rangle$	3.85 ± 0.24	-1.56 ± 0.09
σ_c^2	9.46 ± 0.94	-3.24 ± 0.14

Helios 2

	All Data $0.29 < r < 0.98\,\text{AU}$		High $v\,(v > 550\,\text{km/s})$ $0.29 < r < 0.98\,\text{AU}$		Low $v\,(v < 450\,\text{km/s})$ $0.31 < r < 0.98\,\text{AU}$	
	A	s	A	s	A	s
$\langle |B_r| \rangle$	2.77 ± 0.10	-1.97 ± 0.06	2.77 ± 0.10	-1.97 ± 0.05	2.47 ± 0.23	-2.02 ± 0.16
$\langle |B_\varphi| \rangle$	2.43 ± 0.11	-1.14 ± 0.08	2.19 ± 0.10	-1.13 ± 0.07	2.96 ± 0.34	-1.07 ± 0.20
$\langle |B_\theta| \rangle$	1.59 ± 0.09	-1.32 ± 0.09	1.44 ± 0.07	-1.34 ± 0.08	1.95 ± 0.29	-1.32 ± 0.25
$\langle B \rangle$	3.29 ± 0.17	-1.84 ± 0.08	3.28 ± 0.11	-1.86 ± 0.05	3.45 ± 0.48	-1.64 ± 0.24
σ_c^2	14.80 ± 1.00	-3.07 ± 0.12	14.10 ± 0.73	-3.20 ± 0.08	16.31 ± 3.22	-2.97 ± 0.34

4.2.3 The Azimuthal Component B_φ

It was evident from the early studies that in general observational data using $\langle |B_\varphi| \rangle$ do not exactly fit the asymptotic power law r^{-1} expressed by (4.3). Values $s = -1.29$ and $s = -1.85$ were derived in the inner heliosphere by *Mariner 10* (0.46 to 1 AU) and *Mariner 5* (0.66 to 1 AU) data.

An early best fit of *Helios 1* 24-hour averages [4.27] on a three-month time span has given an exponent $s = -1.0$. A number of following studies which all used $\langle |B_\varphi| \rangle$ have consistently led to somewhat larger values of $|s|$, either inside or outside the terrestrial orbit. Higher-resolution *Helios* data [4.23] lead (Fig. 4.1b) to an exponent $s = -1.27$ for solar wind velocity $v < 500$ km/s and $s = -1.08$ for $v > 600$ km/s. Table 4.1 shows a homogeneous set of results for B_φ computed on both *Helios* probes, essentially in the same time interval, corresponding to the primary mission of *Helios 2*. A recent study on the full set of *Helios* data [4.31] leads to best-fit values $s = -1.12 \pm 0.40$ for *Helios 1* and $s = -1.09 \pm 0.26$ for *Helios 2*. So, in conclusion, a value $s \approx -1.10$ can be taken as representative for the inner heliosphere, independently of the solar wind velocity regime, down to temporal scales as low as few hours. And the same s value is also suggested by the field observations far beyond 1 AU [4.5].

The discrepancy between the predicted $s = -1$ and the observed $s = -1.1$ may be attributed to several different mechanisms. One possibility is that the rectification of B_φ in $\langle |B_\varphi| \rangle$ enhances the influence of fluctuations on the results. In contrast, using absolute values of B_φ averaged over fairly long time intervals (of the order of one day) would lead to a physical description less dependent upon fluctuations and then, possibly, more representative of magnetic flux elements.

It was suggested early on [4.18] that an additional contribution to the radial variation may be due to correlated fluctuations of B_φ and the solar wind velocity. Under the conditions that $B_\theta = 0$ and v is independent of the distance, the radial variation of B_φ is given by

$$\langle B_\varphi \rangle = \frac{c}{r} - \frac{\langle \delta B_\varphi \delta v \rangle}{v} \, , \qquad\qquad (4.5)$$

where any correlation between B_φ and v fluctuations implies a nonzero $\langle \delta B_\varphi \delta v \rangle$ and hence a decrease steeper than r^{-1} if $(\partial/\partial r)(r \langle \delta B_\varphi \delta v \rangle) > 0$. Further theoretical considerations and model calculations [4.14] for an idealized situation of stream-associated fluctuations inside the propagating solar wind indicate that particular choices of the boundary conditions in the source region of the solar wind may lead to an exponent $|s|$ slightly larger than 1, with a predicted decrease of the product $r B_\varphi$ up to 10% between 1 and 6 AU; but, due to its nonlinearity, the effect, although still observable, might be significantly smaller, and then not easily identifiable over shorter distance intervals.

Other mechanisms have been suggested [4.12, 30, 52] to justify discrepancies from (4.3). One of these [4.30, 52] is based on the fact that, as a consequence of a meridional magnetic pressure gradient, a meridional flow is established which tends to decrease the azimuthal Maxwell stress in the equatorial plane where azimuthal magnetic flux is accumulated compared with the polar region. The B_φ decrease, close to r^{-1} up to 10 solar radii, beyond that becomes steeper so that it is on the order of several percent at 1 AU, and on the order of 25% between 1 and 5 AU. The mechanism also predicts nearly similar deviations for the radial field component B_r, which implies that the angular azimuthal orientation is not appreciably affected. The importance of this effect is questionable in the inner heliosphere, where the experimental data do not support a faster-than-r^{-2} variation of B_r. However, at distances beyond 1 AU there is some evidence for the overall structure of the field topology being consistent with meridional transport of magnetic flux to higher latitudes [4.53].

Another mechanism [4.12] is based on the consideration of the kinematic effects which are expected because of the presence of streams, which tend to modify both B_r and B_φ. Two parameters are potentially important, the azimuthal angle at the source region, which may be different from the exact radial orientation ($\varphi_0 = 90°$ or $270°$), and the solar wind velocity azimuthal profile (in particular due to the presence of streams), which although it may be stationary in a corotating frame appears temporally variable to a fixed observer. A model computation with four identical streams, simulating in some way the typical four sectors around the sun, shows that while the averaged B_r over the time profile

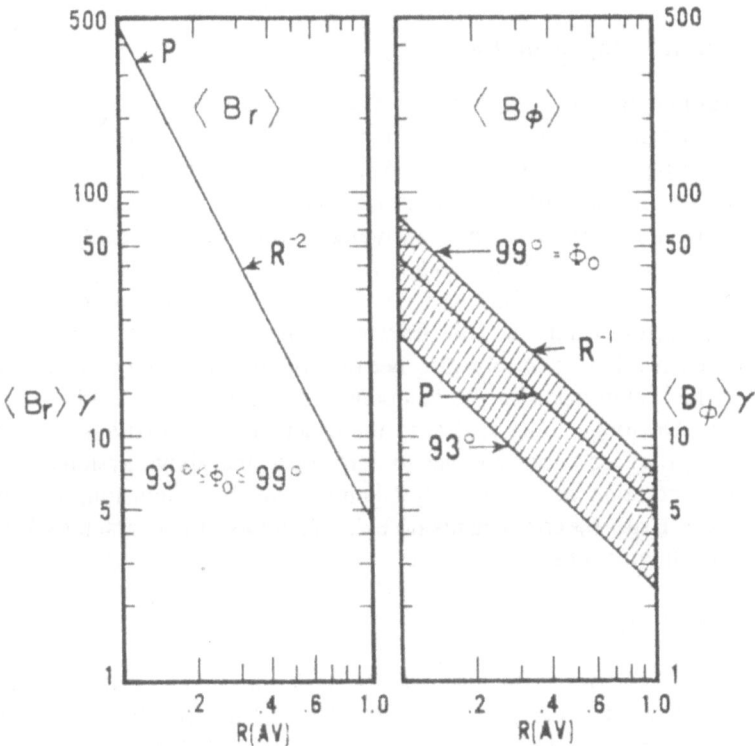

Fig. 4.2. Heliocentric distance variation of $\langle B_r \rangle$ and $\langle B_\varphi \rangle$ averaged over the time profile of a "standard stream" [4.12]. The straight lines indicated by P are the Parker's variations

of an individual stream still has an r^{-2} variation law, the average B_φ exhibits (Fig. 4.2) a strong dependence upon the deviation of the field orientation from the ideal Parker model and on a shorter time scale its actual values depend very much upon the averaging time interval. This may justify the usually large scatter of individual points by *Helios* (Fig. 4.1b), as well as a significant discrepancy from the r^{-1} law, in particular when averages of B_φ are computed over time intervals during which streams are present.

On a much longer time scale, averages from *Pioneer 10* and *11* data computed over distance intervals of 0.5 AU (corresponding to temporal intervals of several months) show [4.47] that the product rB_φ is practically independent of r; however, using annual averages of the same data a best fit value $s = -1.12 \pm 0.04$ has been found [4.46] in the distance range 1 to 12 AU.

In conclusion, a variety of effects can affect the B_φ radial variation at any distance from the sun (the rectification effect, latitudinal and temporal effects, solar wind velocity fluctuations, presence of streams, kinematic effects, etc.). The high temporal resolution of *Helios* data in the inner heliosphere, summarized in Table 4.1, consistently suggests that a value $s \approx -1.1$ can be taken as a good representation of the actual radial variation of $\langle |B_\varphi| \rangle$ inside the terrestrial orbit; and this same value can be safely adopted even at much larger heliocentric distances and/or at much lower time resolutions.

4.2.4 The Normal Component B_θ

Despite the fact that $B_\theta = 0$ is predicted by Parker's model, a significant component B_θ was observed on relatively short time scales. Best fits on *Helios 1* and *2* using 12-hour averages $\langle |B_\theta| \rangle$ suggest [4.25] a radial power law with an average value $B_{\theta,0}$ at 1 AU in the range 1.4 to 1.9 nT and a slope $s = -1.24$ to -1.35, with no appreciable difference at different velocity regimes. The analysis made in [4.27] led to $s = -1.29 + 0.34$ for *Helios 1* and $s = -1.25 \pm 0.33$ for *Helios 2*. The values of s are consistent with those from *Mariner 4* and *10* beyond or inside the terrestrial orbit; the values of $B_{\theta,0}$ from different spacecraft are remarkably different from each other, probably also because of the rather large systematic uncertainties which may occur in the component of the observed fields parallel to the spin axis. On longer time scales, in the order of solar rotations or more, the averages $\langle B_\theta \rangle$ are essentially equal to zero, once the above systematic effects on the spin component are properly taken care of. In conclusion, although occasionally even large B_θ components do exist, they have the characteristics of "short" temporal fluctuations.

4.2.5 The Field Magnitude B

Some early rough fits by a simple power law have produced values of s less than 2: actual values range from -1.56 for *Helios 1* to -1.86 for *Helios 2*, which compare with values $s = -1.65$ for *Mariner 10* and -1.37 for *Pioneer 10* (in this case at distances up to 3 AU).

However, the appropriate approach, as derived by (4.1) and (4.3), predicts a field strength

$$B(r) = B_r(r)\sqrt{1 + \left(\frac{\Omega r}{v}\right)^2} \,, \tag{4.6}$$

where r is the heliocentric distance in AU, B_r is the radial field at the distance r and the other constants are such that $c = \Omega r/v$ is typically ≈ 1 at $r = 1$ AU.

A study of combined data from *Helios 1* using heliocentric distance bins of 0.1 AU [4.27] and from *Pioneer 10* up to 3 AU has shown that the radial variation of B is consistent with the predicted expression with $B_r = 5$ nT at 1 AU. A similar result has been found by *Mariner 10*, in the range $0.46 < r < 1$ AU; in this case, however, a value $B_r = 4$ nT was found.

An extended study on the *Voyager* data between 1 and 10 AU has shown that the radial variation of the field strength is in good agreement with Parker's model if appropriate account is taken of temporal and bulk velocity variations [4.19]. As concerns the latitudinal variation (in the limited range accessible to the *Voyagers*) there is agreement with the model only to first approximation.

4.2.6 The Field Orientation

The magnetic field vector orientation, expressed by the elevation θ from the equator and the azimuth φ, exhibits significant fluctuations from the heliocentric expected variation, expecially at lower time scales.

According to the *Helios* observations the elevation θ is centered [4.58] about a value close to $0°$ (with more than 2/3 of the available data having $|\theta| < 20°$), which implies that (4.2) is indeed satisfied on the average over long time intervals.

As concerns the azimuth φ, it is useful to represent these fluctuations by the difference $\varphi' = \varphi - \varphi_0$ between the observed and the predicted angles φ and φ_0. By definition one should expect $\varphi' = 0°$ (or $180°$) only if the dependence of (4.3) upon the solar wind velocity is neglected (a nominal value of $v = 430$ km/s was indeed used by the authors). As a result of the study, extended over 21 solar rotations centered on the *Helios* perihelia (the spacecraft maximum distance from the sun, although variable, never exceeded 0.68 AU; the time duration of each solar rotation was close to 39 days), it has been shown that the peaks of the azimuthal frequency distribution in the inner heliosphere are indeed nearly coincident with $\varphi' = 0°$ and $\varphi' = 180°$ with a minimum in between corresponding to about 4% of the peak value. By comparison, a similar result was found in the distance range 1 to 8.5 AU [4.53], with a minimum corresponding to about 17% of the peak value, if quiet regions are considered, and significantly less inside interaction regions.

The narrower angular distributions observed closer to the sun might be the simple consequence of the still-small effects of stream interactions. Also, the included angle between sunward and antisunward peaks is $180°$, not only on the average during the seven-year total interval, but also for individual years; some small differences in the last case are only latitudinal effects, i.e. a consequence of the different percentages of inward and outward field lines observed at different latitudes.

4.2.7 The Field Variability at Low Frequencies

A variable nature of the observed interplanetary magnetic field elements was evident from the very first *in situ* measurements. Low-frequency fluctuations superimposed on the average behavior discussed in the previous sections are an important parameter for understanding the physical properties of the interplanetary medium. The need to separate "ordered" from random fluctuations was soon realized. If pure compressive-mode waves are present in the interplanetary medium, a variation of the field magnitude B is expected with no direction change; if pure Alfvén waves are present, only directional variations are expected. In the case of fast-mode waves, oscillations of both magnitude and direction are expected.

Use of field variances proved to be very appropriate for the above purpose. Magnitude fluctuations are estimated by $\sigma_B^2 = 1/N \sum_i (B_i - \bar{B})^2$, while directional variations are better estimated by means of the so-called Pythagorean variance $\sigma_c^2 = \sigma_x^2 + \sigma_y^2 + \sigma_z^2$. Although directional and magnitude variations can contribute somewhat to σ_B^2 and σ_c^2, respectively, there is enough evidence and confidence that σ_c^2 represents directional fluctuations because the power of directional variations in the interplanetary medium is on the average a factor of 10 or so higher than that of the field intensity. In some cases, variances along the average field direction or perpendicular to it ("parallel and perpendicular variances") have also been used.

Helios data first of all confirmed [4.23] down to 0.3 AU the predominance of directional over magnitude fluctuations, at a power level about an order of magnitude larger than that of the field strength, indicative of the presence of transversal fluctuations in the inner heliosphere. A power law r^s describes well (Fig. 4.3) the decrease of the Pythagorean variance σ_c^2, as the distance r increases. A value $s \sim -3$ is computed when averages over 3 or 12 hours are used as input data [4.23]. A behavior of just this kind is predicted for propagating Alfvén waves, so that the observations give support to the idea that no local generation of Alfvén waves occurs in the distance range explored by *Helios*. The total power level appears consistently independent of the solar wind velocity (see Table 4.1), although some small differences were found in 1975 *Helios 1* data. These results mean essentially that at frequencies below $\approx 10^{-4}$ Hz, a necessary, although not sufficient, condition for the existence of very low frequency Alfvénic fluctuations is satisfied. Due to the known fact that the latter are associated with the trailing edge of the solar wind streams, special time intervals of *Helios 1* and *2* data were selected when the two spacecraft were almost aligned on the same line of force [4.60]. The study was done in different frequency ranges up to $\approx 3 \times 10^{-3}$ Hz (corresponding to a period of a few minutes). The exponent s (which is ≈ -3.2 for periods of 6 hours) consistently increases in value as the frequency increases

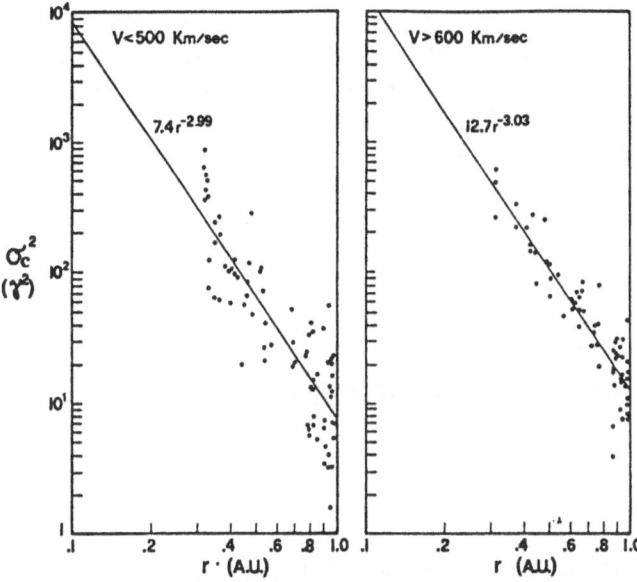

Fig. 4.3. Heliocentric distance variation of the variance σ_c^2, at two different solar wind regimes. Best linear fits are also shown [4.23]

(up to $s \approx -4$ for periods of 5 min.). On the other hand, the field strength variance σ_B^2 increases with respect to the Pythagorean variance σ_c^2, as the distance r increases. This suggests the possibility that wave modes other than stable Alfvén waves also contribute to the total power in an increasing relative proportion as the frequency increases. Assuming that unstable waves "saturated" at a certain level (defined by a given constant ratio σ_c/B) were being observed, it has been possible to match the observed [4.60] radial variation of σ_c^2. A somewhat different approach [4.4] was that of using observations at three different distances of the same stream by *Helios 2* during three consecutive solar rotations. While for the ratios σ_c/B and σ_B/σ_c similar results were obviously found, a more complete study of the statistical properties of the fluctuations was possible. Using the eigenvalues λ_i (defined such that $\lambda_1 > \lambda_2 > \lambda_3$) and the eigenvectors of the variance matrix computed at the three distances and five different time scales it was shown that the total power (normalized to the square of the average field) tends to increase approaching the sun up to periods of around 1 hour, above which an inversion of trend is observable.

Towards the sun, the increasing ratio λ_2/λ_1 of the eigenvalues implies a decreasing anisotropy in the plane perpendicular to the minimum variance direction.

The degree of compression, i.e. the ratio σ_B^2/B^2, also decreases especially at lower frequencies. The ratio σ_c^2/B^2 decreases with increasing distance and this is just the opposite of what is expected in the frame of a WKB propagation theory in the range of distances explored by *Helios*. This means that a damping mechanism is present and it is more effective at higher frequencies.

More recent studies [4.3, 36] using *Helios* and *Voyager* data allow a better interpretation of the low-frequency field fluctuations. The Alfvénic character, which is a predominant feature closer to the sun, becomes less pronounced as the distance increases, as well as when larger spatial scales are considered. The overall picture is that of outward-propagating Alfvénic fluctuations generated near the sun, and whose dynamical evolution, already taking place in the inner heliosphere, becomes well evident beyond the earth's orbit.

4.3 The Solar Magnetic Field and the Interplanetary Magnetic Field

4.3.1 Background Observations

The existence of internal organization of the interplanetary magnetic field with a well-defined gross structure on the scale of hourly averages, was the result of the early observations of *Explorer 18* [4.62]. The so-called magnetic sector structure was found, i.e., a spiral field line configuration, as implied by Parker's solar wind model, with alternatively outward (positive) or sunward (negative) magnetic polarities. At the time of discovery four sectors per solar rotation were present. Beyond a few degrees above or below the ecliptic plane, due to the limited latitudinal heliographic excursion of the presently possible orbital planes, no direct information was then available on the three-dimensional structure. There was an early feeling that the separation line of positive and negative sectors might be more or less meridional [4.62]. It was found, however, in subsequent studies that a characteristic magnetic polarity signature existed, the so-called dominant polarity effect [4.37], i.e. that there was a preferential polarity of the field when the observing spacecraft was north of the ecliptic plane, and an opposite preferential polarity with the spacecraft in the southern hemisphere. It was also observed that, as the sector polarity changed its sign, the preferred polarity in each hemisphere was also changing.

Some years later it was suggested [4.22, 43] that the surface separating positive and negative sectors around the sun might be a warped (nonplanar) surface at some variable inclination to the ecliptic plane, at least much closer to the ecliptic than to a meridian. Observations by *Pioneer 11*, all around the sun, during a complete solar rotation at solar minimum, indeed showed an almost total absence of inward field lines at 16° above the solar equatorial plane [4.49].

All observational evidence was interpretable in terms of a rotation of the warped surface past the observing spacecraft with its angular velocity relative to the sun. Alvén [4.2] gave this type of field configuration the name "ballerina effect". If the warped surface does not extend too far from the equator, the same polarity is observed above a certain heliospheric latitude during a full solar rotation; otherwise, one or two pairs of polarities are seen.

4.3.2 The Heliospheric Current Sheet

The *Helios* observations allow significant improvements in the knowledge of the latitudinal dependence of the IMF polarity, also because of the availability of simultaneous plasma data. Actually, precise knowledge of the solar wind radial velocity allows the projection of the field polarities from the observing spacecraft onto the actual coronal source point of each field line through the craft. So, the azimuthal width at the location of *Helios* was used [4.56] as input to determine the corresponding width on the corona. Data for early 1976 relative to 4 solar rotations show that the azimuthal widths of the unipolar region were actually independent of the heliocentric distance; the positive and negative polarity distributions and the projected sector boundary locations were consistent with an average quasiplanar configuration at an inclination $\alpha_S \approx 10°$ with respect to the heliographic equator. In addition, a distortion of the boundary surface above the equator was noticed such that a four-sector structure could be observed within a certain latitudinal interval, but only a two-sector structure in another latitudinal interval south of the equator (Fig. 4.4). Local triangulation of a sector boundary using *Helios 1* data [4.29] also showed significant deviations from the picture of a rigidly rotating, flat current sheet.

One study [4.59] extended over the solar minimum, based on data from the two *Helios* probes, has shown that the angle defining the least-squares fit of the

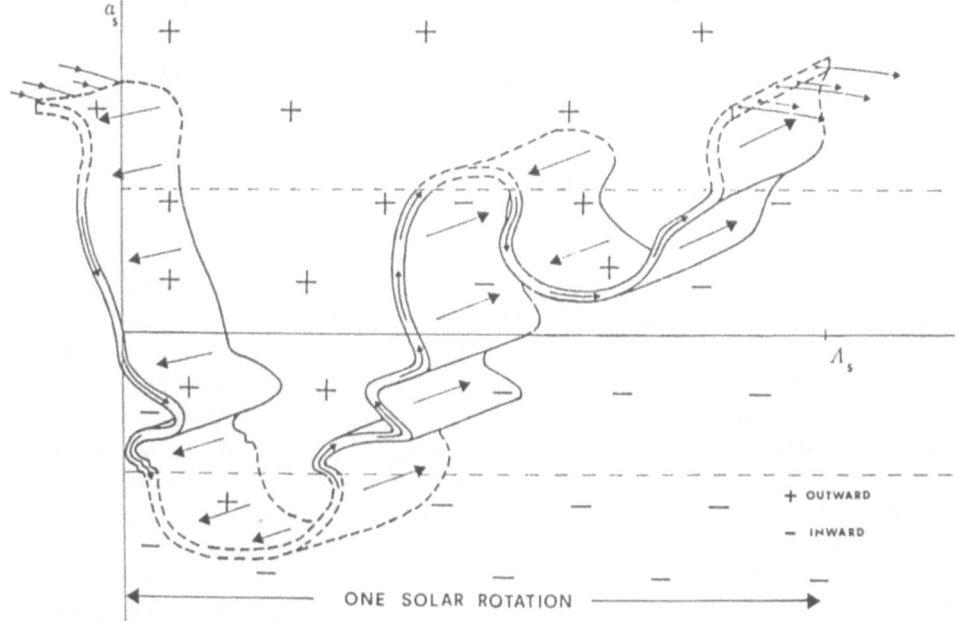

Fig. 4.4. The shape of the current sheet as extrapolated from the location of the boundary crossings; the orientation of the local normals of the current sheet and the longitudinal extension of the unipolar regions of the interplanetary magnetic field. Solid lines correspond to the region of direct knowledge of the current sheet by *Helios* observations [4.56]

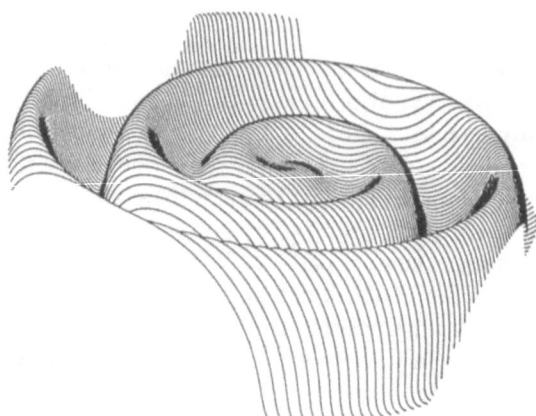

Fig. 4.5. A schematic simple representation of the spiral wave heliospheric current sheet [4.54]

inclination of the boundary surface to the equatorial plane in 1975 and early 1976 was close to 19°, with the dominant two-sector configuration which was also observed in the nearly equatorial region on the *Helios* data since 1972. The change of α_S between 1975 and 1976 can be considered a direct indication of a temporal, i.e. solar cycle, variation of the inclination of the boundary surface. Both values of α_S are compatible with the results obtained at larger distances by *Pioneer 11*.

The overall picture for the years 1972–75 is then that of an approximately flat current sheet inclined to the sun's equatorial plane, from which, due to the solar rotation, a spiral wave [4.54] is produced propagating outward (Fig. 4.5); this configuration is obviously only representative of what the sheet looks like in simple cases, at low solar activity.

Just before and after the solar maximum in 1978, the dominant polarity effect was no longer observable [4.57], a feature which is thought to be related to the change of configuration of coronal holes and consequently of the warped boundary surface, as well as to solar wind streams [4.45]: these features are the consequence of opposite magnetic polarities extending from the poles, even beyond the equator, in the opposite hemispheres, and of a highly erratic behavior, as expected for a highly irregular and rapidly evolving IMF organization. In other words, as the solar activity increased, the current sheet progressively changed its shape reaching high latitudes; and changes are also expected in connection with the disappearance of the sun's polar field near the maximum of the solar activity.

The appearance of a four-sector configuration at *Helios* perihelion implies a heliospheric current sheet that is not planar in the external solar corona, which, mathematically speaking, means that significant higher-order terms are superimposed on the first-order dipole-field terms. Working in this frame, the quadrupole contribution to the magnetic field potential was estimated [4.10] to be on the order of 17% of the dipole for the one year time interval May 1976 to May 1977. The picture emerging from previous observational evidence and confirmed by more recent longer-term studies [4.1, 38] is that of a nonplanar heliospheric current sheet, whose "average inclination" with respect to the solar equatorial plane is in most cases only a crude representation. Actually, depending upon the phase of the solar cycle, the overall magnetic field topology varies as a consequence of the changing geometry of the large solar scale features, in particular of the coronal holes, in the solar source region. Interesting examples of complex tridimensional sheet configurations have indeed been produced [4.1] for different stages of solar activity.

In addition to the large-scale effects, smaller-scale warping can also be expected: a kinematic analysis of the effects of solar wind velocity inhomogeneities on the heliospheric current sheet has shown [4.51] that even small velocity gradients may produce significant distortions. *Helios* data have shown that even much closer to the sun than 1 AU local appreciable warping does exist (Fig. 4.4).

4.3.3 The Source Region of the Heliospheric Current Sheet

The very existence of a heliospheric current sheet immediately raises the problem of connecting the observed interplanetary field of solar origin to its source on the sun.

After the early attempts [4.41, 42] to derive the interplanetary field by a magnetic potential external to a spherical source surface about 0.6 solar radii above the photosphere, on which a radial field was supposed to exist with no currents in between, a number of improved models have been worked out.

The basic concept of these models is that of constructing a solution of Laplace's equation for coronal fields under some simplifying assumptions. The problem is typically a Neumann boundary problem, where in the familiar spherical harmonics expansion of a current-free magnetic field the normal, i.e. the radial, field is assumed known. The available observational data is indeed the line-of-sight field component obtained by magnetograph observations. If the true field were exactly radial, a simple relationship could be used to derive its value from the line-of-sight field. However, first, the field is certainly not exactly radial; second, and more importantly, its value at high latitudes can be only poorly derived by the observations.

In general, the line-of-sight component (lying on the meridian plane on the sun) can be expressed as a linear combination of the radial and north–south field components; so, least-mean-square-fit procedures have been used to treat the problem. Even taking advantage of the orthogonality relations of the harmonic functions, this procedure leads to prohibitively long computing times when the order n of the expansion increases. (Actually matrices of $[n(n+2)]$ components are theoretically to be inverted! But just by using some symmetry properties the difficulty can be partially overcome.) More recently, by using some classical recursion formulae, significant progress has been made [4.35]: the coefficients with the highest order are computed first and then all lower-order ones are computed by recursion formulae.

Field models derived by solution of the Laplace equation, usually called potential field (PF) models, have been used extensively by several authors to construct the projection of the heliospheric current sheet onto a spherical source surface and the photosphere. The photospheric field map, used as a boundary field configuration, is based on field observations made at central meridian passage. A usual assumption of the PF models is that everywhere there is zero electric current density. Possible effects of photospheric currents have been studied [4.20] in two cases (force-free magnetic field and curl-free current distribution) to show that the overall PF topology is not significantly affected by the currents unless they are really high. Also, the effects of a nonspherical surface [4.21, 44], although they tend to improve the physical description, are not such as to affect the overall structure of the field greatly.

Also effective in determining some features of the interplanetary field configuration is the magnetic field in the solar polar regions, whose relevance was first realized by Pneuman et al. [4.33].

A different approach [4.15] was to locate the projection of the sector boundary on the source surface at the bright features observed in the white-light corona. In a study of the coronal and interplanetary current sheet, the sector pattern from the *Helios 1* and 2 data in early 1976 was compared [4.13] with the maximum brightness curve (MBC) obtained from plots of the K-coronameter brightness contours as the line encircling the sun that connects the latitudes of maximum brightness.

The projected polarities have been compared with the neutral line as computed by the PF model during four consecutive solar rotations. Interestingly enough, although the neutral lines derived from the two methods are somewhat different from each other, they also show considerable qualitative similarities. Apparently, the excursion of the neutral line about the equator was a little larger for the PF model than for the MBC model. The observed sector boundaries extrapolated back to the sun match the crossings of this line with the warped neutral sheet inferred by either technique to within the expected accuracy of the sector boundary extrapolation and the neutral sheet (in the order of $\pm 10°$). In the case shown in Fig. 4.6a the separation of the *Helios 1* and 2 orbital paths by about 10° in latitude made possible a precise latitudinal determination of the neutral line which looks better represented by the MBC curve than the PF-determined contour. The time resolution of the two neutral line determinations was good, under the hypothesis of little structural change during 27 days (for the MBC) or 6 months (for the PF model). But this is certainly not true, since short-term transients do exist and they were not properly taken care of in the computations. Apparently, in the case of Fig. 4.6a the PF neutral line should be displaced southward by about 20° to agree with the pattern observed by *Helios*.

The combined results obtained by different methods have been instrumental in emphasizing that, while the basic features are correct, a more detailed approach was needed. One particular aspect to inquire about was the distortion, due to the poorly known strength of the polar magnetic field, of the shape of the source region, which might well be far from spherical, and of the field orientation at the source, which is not necessarily radial as generally assumed. Introducing a reasonable solar polar field, although largely unknown and therefore arbitrary,

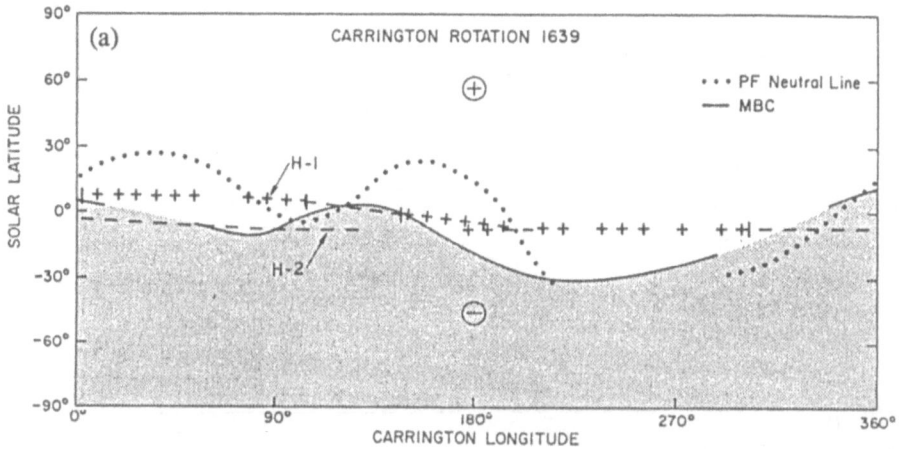

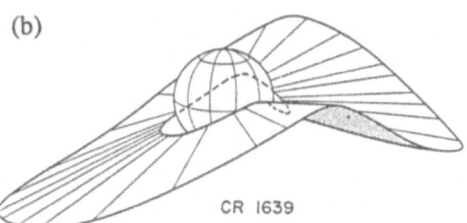

CR 1639

Fig. 4.6. (a) The projected magnetic field polarities observed by *Helios 1* and *2* and the maximum brightness curves for Carrington rotation 1639; together with the calculated potential field neutral line [4.13]. (b) The sector boundary surface between 1.5 and 5 R_s (solar radii) for the same rotation, obtained by using the white light maximum brightness curves at 1.5 R_s and a radial projection out to 5 R_s [4.13]

in the PF model [4.16], has the important effect of reducing the latitudinal excursion of the computed sheet. The study over the 18 months following the last sunspot minimum has indeed shown a slow but steady increase of the latitudinal excursion from about 15° near the start of the interval to about 45° near the end. An overall survey [4.17] of the structure of the heliospheric current sheet in the years 1978–1982 further substantiates the picture of an evolving structure during the solar cycle up to and beyond the reversal of the solar polar field orientation. A remarkable feature is the continuous evolution from a four-sector boundary topology to a complex situation just before the solar maximum when the heliographic latitude of the sheet approaches the poles. Also, disconnected current sheets are predicted and they do not necessarily intersect the terrestrial orbit (Fig. 4.7).

The improved neutral lines derived from the PF and the MBC models have been shown [4.61] to be closer to each other. Comparison over six solar rotations show that only during one 6-day period did the PF and MBC curves fail to be close to each other, and this occurred with the appearance of an unusually large photospheric region of unbalanced "toward" polarity which was present in neither the previous nor the following solar rotations. A study of data from a constellation of spacecraft (*IMP 8, Helios 1* and *2, Voyager 2*) throughout 18 Carrington rotations gives further support to above results [4.11]. Data from *Voyager 2* during its initial phase, when it was close to the earth at a heliographic latitude 1° higher than *Helios 2*, were also used to test the geometry inferred by *Helios 1* and *2* and *IMP 8*.

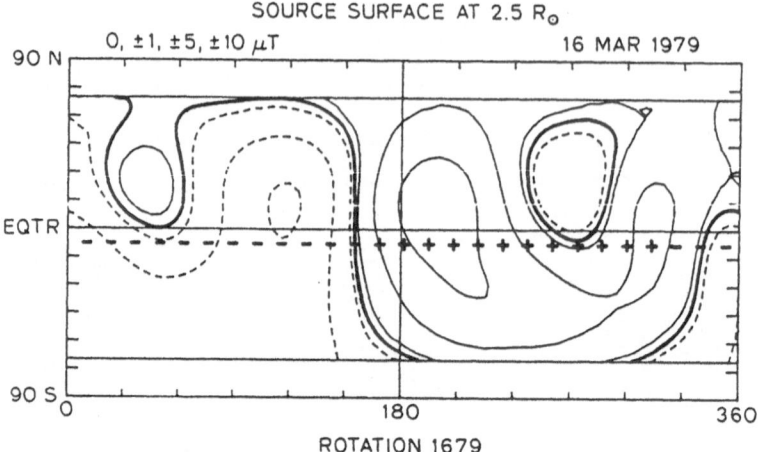

Fig. 4.7. An example of the complex configurations of the current sheet during the period near the sunspot maximum [4.17]. There is a disconnected current sheet near 270° longitude in the northern hemisphere which does not intersect the latitude of the Earth. A two-sector pattern is observed at the earth

It is worth remarking that extrapolation of the heliospheric shape predicted by the MBC method to *Voyager 1* and *2*, at distances of 1.4 and 2.8 AU, fits [4.7, 50] reasonably well the local shape estimates. Another interesting point is that convincing indirect evidence of changes in the shape of the heliospheric sheet also comes from temporal changes of galactic cosmic rays [4.48].

The computed heliospheric current sheets from both PF and MBC methods consistently indicate a four-sector pattern. Generally speaking, the two methods are nearly equivalent in predicting positive and negative polarity signatures, with a slightly better success for the MBC method. The MBC method works better when large bipolar magnetic regions suddenly appear on the photosphere; the PF method works better when fast-evolving deformations appear. This means that the PF method is preferred just before solar maximum, when the coronal structure becomes much more complex, while the MBC method is preferred in more stable situations.

The spatial variation and evolution of the sector boundary configurations have been studied combining data from the *Pioneer Venus* orbiter inside and the *Voyager* spacecraft outside the terrestrial orbit [4.6]. It has been shown that the sector shape evolves with heliocentric distance: while below 1 AU the shape follows well enough the neutral line on the solar source region, significant and increasing distortion occurs at typical *Voyager* distances (up to 25 AU).

The fact that no direct measurements of the magnetic field are available between the photosphere and 0.29 AU is a rather severe limitation on our knowledge. In this respect, it is interesting that for a limited number of cases Faraday rotation measurements were obtained when the spacecraft–earth line passed through the corona, so that some significant information at distances much lower than 0.3 AU has been at least sporadically obtained.

A combination between a three-dimensional MHD model of the solar wind away from the sun and the potential magnetic field close to it [4.34] was used to compute a simulated Faraday rotation profile during the outbound pass of *Helios 1* after first perihelion in 1975. The observed Faraday rotation profile was found to agree closely in one broad longitude band, but to disagree strongly in another well-defined longitude band. These results thus give direct evidence for major shortcomings in the potential field models and, even though we think less probably so, in the solar wind model.

4.3.4 The Fine Structure of the Sector Boundary

On the gross scale considered in the previous section the sector boundary is characterized by a relatively sharp boundary at which the field direction usually reverses in time periods ranging from a few minutes (Fig. 4.8) to several days. On a smaller time resolution, down to a few seconds, the field generally shows a complex structure.

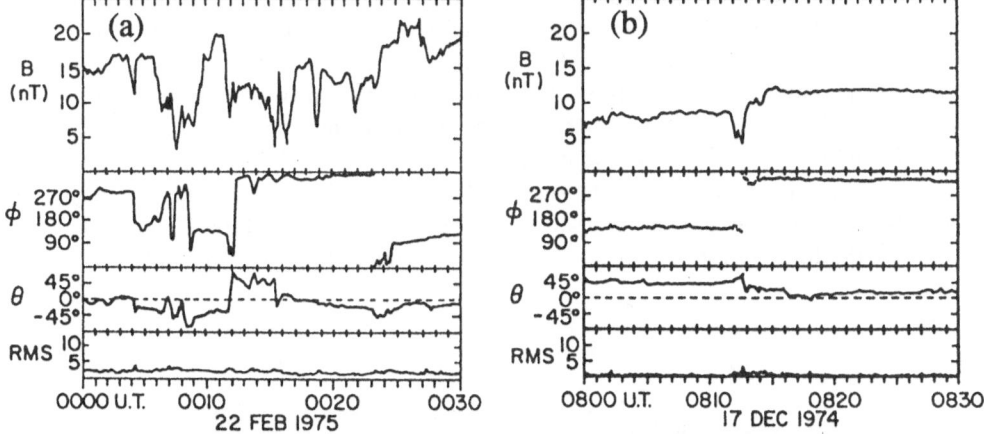

Fig. 4.8a,b. High-resolution (8 s averages) display of fine-scale current sheet traversals during two half-hour intervals in February 1975 and December 1974. In (a) there were two traversals for which 120 was satisfied. In (b) a single traversal is shown [4.9]

It was soon recognized [4.8] that sector transitions often consist of a number of field polarity reversals over a period of hours, with the individual changes occurring within that period usually being relatively abrupt and often accompanied by field magnitude depressions. In these cases any attempt to identify the boundary plane orientation by the minimum variance analysis is complicated by the complex fine structure of the transitions. So the real question has to do with the way a sector boundary is defined. In other words, is each large-angle directional discontinuity occurring in a sector transition region an individual full or partial crossing of the current sheet, or should the boundary be defined as the total region of transition from the initial to the final, opposite, polarity status?

In an unbiased approach [4.9] both individual case studies and a statistical survey of the data from the primary *Helios 1* mission were made. In these studies 105 cases of partial polarity transition were identified for which the rotation angle of the field in the plane of maximum variance (i.e. the plane perpendicular to the minimum variance direction) was greater than 120°. These events tend to cluster around the sector boundaries and are quite uniformly distributed in the angular interval 120° to 180°. The observations suggest that sector boundaries are often complex structures in which the transition in magnetic field direction between both sides occurs partly in large discontinuous changes in direction and partly in smooth variations.

Also, in most cases these transitions show a hierarchy of complex structures on the scale of several hours or even days down to the resolution of the magnetometers used, i.e. spanning over five orders of magnitude in scale. In the majority of cases fine-scale fluctuations in the boundary orientation within a single transition were observed: although, in general, the orientations are more or less random, there are cases when the series of consecutive crossings in a given transition suggest either wavelike motions of the sheet or a rippled or corrugated surface structure (Fig. 4.9). The apparent "wavelength" of the corrugations is of the order of 0.05–0.1 AU. The individual crossings are mainly more like tangential discontinuities (in 80% of the cases) than rotational discontinuities. In this latter case, there is evidence of some continuity of field lines across the sector boundary, with a remarkable minimum of the field strength, which might be indicative of field line reconnection. The most probable value of the thickness τ of the heliospheric current sheet [4.9] is about 3×10^4 km, i.e. more than an order of magnitude larger than the thickness of the "ordinary" tangential discontinuities. Only in a few cases was τ greater than 10^5 km; and in one case

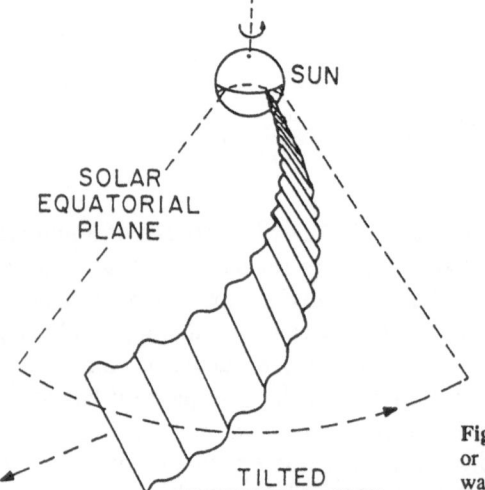

SUN

SOLAR EQUATORIAL PLANE

TILTED CURRENT SHEET SEGMENT

Fig. 4.9. Sketch illustrating small-scale ripples or corrugations superimposed on a large-scale warped current sheet. A tilted spiral segment of the heliospheric sheet is shown for the case of ideal sinusoidal ripples [4.9]

it exceeded 10^6 km. Finally, the average inclination of the sheet normal relative to the ecliptic plane was 58°, while the azimuths were not consistent with the expected perpendicularity to the mean spiral orientation: they were distributed in a range lying between the radial direction and the normal to the spiral, suggesting a tendency of the sheet normal to rotate toward the radial direction.

A similar study was made [4.55] on 14 current-sheet crossings by *Helios 2* in early 1976, selected if they occurred in less than 1 hour and if they were associated with clear rotations ($\geq$ 130°) of the IMF ecliptic component. For each event minimum variance direction and minimum component direction were compared: coincidence of the two directions would mean that the field rotation occurs in the same plane of the field lines on the two sides of the sheet, as expected for tangential discontinuities. Actually the angle between the two directions was always a few degrees and its maximum value was 19°. The thickness τ was again on the order of a few 10^4 km. No indication was found, within the limits of the small number of events, of temporal or distance variations of the physical characteristics of the boundaries. Although no evidence of oscillations, as reported above [4.55], is obtained from *Helios 2* data, there is still some degree of coherence between the inclinations to the ecliptic for some consecutive groups of crossing. On the average the same mean tilt angle of 58° was estimated.

The finding of almost no radial dependence of the orientation of the sector boundaries in conditions of nearly identical levels of solar activity [4.55] supports the view that their gross structure is already configured relatively close to the sun, rather than in the expanding solar wind region.

4.4 Summary and Outlook

In summary, the observations collected by the *Helios* missions in the inner heliosphere have shown that the solar magnetoplasma (solar wind and magnetic field) beyond 0.29 AU expands into the interplanetary medium in substantial agreement with the predictions of Parker's model. The radial structure of the field parameters in the inner heliosphere matches well with that observed beyond 1 AU, so that a satisfactory physical description of the gross structure of the heliosphere beyond 0.3 AU, close to the ecliptic, is available.

Helios observations have been instrumental in proving the close relationship between the interplanetary field structure and the coronal hole pattern evolution on the sun; it is now well established that the roots of the observed IMF are in the high-velocity solar wind source regions associated with the coronal holes: the early observations of magnetic sectors and the dominant polarity effect have been correctly understood.

Very low frequency field fluctuations have been extensively studied and the existence of structures such as broad discontinuities taking place in days to minutes, as well as of wavelike structures propagating outward from the sun, has been proved. But how and when these waves are generated is still an open question.

The magnetic energy density in the far interplanetary medium is a small fraction of the particle energy density, so that the field characteristics and evolution are strictly related to the solar wind properties. Then, most of the open questions addressed to understanding the solar wind have analogous ones for the understanding of the magnetic field. One specific remark on the magnetic field measurements is that their time resolution is usually higher than that of the solar wind, so that exploiting the full potential of field measurements requires a parallel effort to increase the particle resolution. In our discussions this has in fact not been a limitation, since our attention was concentrated on the gross field structure.

New and really significant knowledge of the interplanetary magnetic field is to be obtained along two lines. The first is that begun using *Helios*, i.e. that of getting closer and closer to the sun to finally have direct access to the source region where the acceleration and the consequent expansion processes of the solar wind take place. No mission like the proposed *Solar Probe* is yet in sight; however, the *SOHO* mission will give pertinent and interesting new information, even though deprived of direct plasma and field observations.

The other line of advancement has to do with the real three-dimensional structure, i.e. that of the field and solar wind properties above and below the ecliptic, in particular at high latitudes and on the polar regions of the sun. With the exception of *Pioneer 10*, no other spacecraft has yet explored the region beyond a few degrees of ecliptic. The first step in filling the gap will be the launching of the *Ulysses* mission in 1990.

References

4.1   Akasofu, S.I., C.D. Fry, Heliospheric current sheet and its solar cycle variations, J. Geophys. Res., **91**, 13679–13688, 1986.
4.2   Alfvén, H., Electric currents in cosmic plasmas, Rev. Geophys. Spa. Phys., **15**, 271–284, 1977.
4.3   Bavassano, B., R. Bruno, Large-scale solar wind fluctuations in the inner heliosphere at low solar activity, J. Geophys. Res., **94**, 168–176, 1989.
4.4   Bavassano, B., M. Dobrowolny, G. Fanfoni, F. Mariani, N.F. Ness: Statistical properties of MHD fluctuations associated with high speed streams from Helios-2 observations, Solar Phys., **78**, 373–384, 1982.
4.5   Behannon, K.W., Heliocentric distance dependence of the interplanetary magnetic field, Rev. Geophys. Spa. Phys., **16**, 125–145, 1978.
4.6   Behannon, K.W., L.F. Burlaga, J.T. Hoeksema, L.W. Klein, Spatial variation and evolution of heliospheric sector structure, J. Geophys. Res., **94**, 1245–1260, 1989.
4.7   Behannon, K.W., L.F. Burlaga, A.J. Hundhausen, A comparison of coronal and interplanetary current sheet inclination, J. Geophys. Res., **88**, 7837–7842, 1983.
4.8   Behannon, K.W., F.M. Neubauer, Investigation of sector boundary fine structure between 0.3 and 1 AU, in *Solar Wind Four*, Burghausen, Sept. 1978, Report MPAE-W 100-81-31, 179–184, 1981.
4.9   Behannon, K.W., F.M. Neubauer, H. Barnstorf, Fine-scale characteristics of interplanetary sector boundaries, J. Geophys. Res., **86**, 3273–3287, 1981.
4.10   Bruno, R., L.F. Burlaga, A.J. Hundhausen, Quadrupole distortion of the heliospheric current sheet in 1976 and 1977, J. Geophys. Res., **87**, 10339–10346, 1982.
4.11   Bruno, R., L.F. Burlaga, A.J. Hundhausen, K-coronameter observations and potential field model comparison in 1976, J. Geophys. Res., **89**, 5381–5385, 1984.

4.12 Burlaga, L., E. Barouch, Interplanetary stream magnetism: kinematic effects, Astrophys. J., **203**, 257–267, 1976.

4.13 Burlaga, L.F., A.J. Hundhausen, Xue-Pu Zhao, The coronal and interplanetary current sheet in early 1976, J. Geophys. Res., **86**, 8893–8898, 1981.

4.14 Goldstein, B.E., J.R. Jokipii, Effects of stream associated fluctuations upon the radial variation of average solar wind parameters, J. Geophys. Res., **82**, 1095–1105, 1977.

4.15 Hansen, S.F., C. Sawyer, R.T. Hansen, K-corona and magnetic sector boundaries, Geophys. Res. Letters, **1**, 13–15, 1974.

4.16 Hoeksema, J.T., J.M. Wilcox, P.H. Scherrer, The structure of the heliospheric current sheet in the early portion of sunspot cycle 21, J. Geophys. Res., **87**, 10331–10338, 1982.

4.17 Hoeksema, J.T., J.M. Wilcox, P.H. Scherrer, The structure of the heliospheric current sheet, 1978–1982, J. Geophys. Res., **88**, 9910–9918, 1983.

4.18 Jokipii, J.R., Fluctuations and the radial variation of the interplanetary magnetic field, Geophys. Res. Letters, **2**, 473–475, 1975.

4.19 Klein, L.W., L.F. Burlaga, N.F. Ness, Radial and latitudinal variations of the interplanetary magnetic field, J. Geophys. Res., **92**, 9885–9892, 1987.

4.20 Levine, R.H., M.D. Altschuler, Representations of coronal magnetic fields including currents, Solar Phys., **36**, 345–350, 1974.

4.21 Levine, R.H., M. Schulz, E.N. Frazier, Simulation of the magnetic structure of the inner heliosphere by means of a non-spherical source surface, Solar Phys., **77**, 363–392, 1982.

4.22 Levy, E.H., The interplanetary magnetic field structure, Nature, **261**, 394–395, 1976.

4.23 Mariani, F., N.F. Ness, L.F. Burlaga, B. Bavassano, U. Villante, The large-scale structure of the interplanetary magnetic field between 1 and 0.3 AU during the primary mission of Helios 1, J. Geophys. Res., **83**, 5161–5166, 1978.

4.24 Mariani, F., N.F. Ness, L.F. Burlaga, S. Cantarano, Variations of the interplanetary magnetic field intensity between 1 and 0.3 AU, Space Research, **XVI**, 675–680, 1976.

4.25 Mariani, F., U. Villante, R. Bruno, B. Bavassano, N.F. Ness, An extended investigation of Helios 1 and 2 observations, the interplanetary magnetic field between 0.3 and 1 AU, Solar Phys., **63**, 411–421, 1979.

4.26 Musmann, G., F.M. Neubauer, F.O. Gliem, R.P. Kugel, Shock identification-computer on board of the spacecraft Helios 1 and 2, IEEE Trans. Geosci. Electr., **GE-17**, 92–95, 1979.

4.27 Musmann, G., F.M. Neubauer, E. Lammers, Radial variation of the interplanetary magnetic field between 0.3 and 1.0 AU, J. Geophys. Res. **42**, 591–598, 1977.

4.28 Musmann, G., F.M. Neubauer, A. Maier, E. Lammers, Das Förstersonden-Magnetfeld Experiment (E2), Raumfahrtforschung, **19**, 232–237, 1975.

4.29 Neubauer, F.M., Recent results on the sector structure of the interplanetary magnetic field, in *Pleins Feux sur la Physique Solaire*, Proc. 2nd European Solar Phys. Conference, Toulouse 8–10 March 1978.

4.30 Nerney, S.F., S.T. Suess, Correction to the azimuthal component of the interplanetary magnetic field due to meridional flow in the solar wind, Astrophys. J., **200**, 503–509, 1975.

4.31 Nöthen-Klima, P., Macroscopic spatial variations of the interplanetary magnetic field from measurements by the space probes Helios 1 and 2, Diploma thesis (F.M. Neubauer, Academic Supervisor), University of Cologne, 1988.

4.32 Parker, E.N., Dynamics of the interplanetary gas and magnetic fields, Astrophys. J., **128**, 664–676, 1958.

4.33 Pneuman, G.W., M. Schulz, E.N. Frazier, On the reality of potential magnetic fields in the solar corona, Solar Phys., **59**, 313–330, 1978.

4.34 Riesebieter, W., F.M. Neubauer, A comparison of 3D solar wind predictions with observations, in *Pleins Feux sur la Physique Solaire*, Proc. 2nd European Solar Phys. Conference, Toulouse 8–10 March 1978.

4.35 Riesebieter, W., F.M. Neubauer, Direct solution of Laplace's equation for coronal magnetic fields using line of sight boundary conditions, Solar Phys., **63**, 127–133, 1979.

4.36 Roberts, D.A., M.L. Goldstein, L.W. Klein, W.H. Mattheus, Origin and evolution of fluctuations in the solar wind: Helios observations and Helios-Voyager comparisons, J. Geophys. Res., **92**, 12023–12035, 1987.

4.37 Rosenberg, R.L., P.J. Coleman, Heliographic latitude dependence of the dominant polarity of the interplanetary magnetic field, J. Geophys. Res., **74**, 5611–5622, 1969.

4.38 Saito, T., T. Oki, S.I. Akasofu, C. Olmsted, The sunspot cycle variations of the neutral line on the source surface, J. Geophys. Res., **94**, 5453–5455, 1989.

4.39   Scearce, C., S. Cantarano, N. Ness, F. Mariani, R. Terenzi, L. Burlaga, The Rome-GSFC magnetic field experiment for Helios A and B (E3), Raumfahrtforschung, 19, 237–240, 1975.

4.40   Scearce, C., S. Cantarano, N.F. Ness, F. Mariani, R. Terenzi, L.F. Burlaga, Rome-GSFC magnetic field experiment for Helios A and B, NASA/GSFC Tech. Report X-692-75-112, 1975.

4.41   Schatten, K.H., Current sheet model for the solar corona, Cosmic Electrodyn., 2, 232–245, 1971.

4.42   Schatten, K.H., J.M. Wilcox, N.F. Ness, A model of interplanetary and coronal magnetic fields, Solar Phys., 6, 442–455, 1969.

4.43   Schulz, M., Interplanetary sector structure and the heliomagnetic equator, Astrophys. Spa. Sci., 24, 371–383, 1973.

4.44   Schulz, M., E.N. Frazier, D.J. Boucher Jr., Coronal magnetic field model with non-spherical source surface, Solar Phys., 60, 83–104, 1978.

4.45   Sheeley Jr., N.R., J.W. Harvey, Coronal holes, solar wind streams and geomagnetic disturbances during 1978 and 1979, Solar Phys., 70, 237–249, 1981.

4.46   Slavin, J.A., E.J. Smith, B.T. Thomas, Large scale temporal and radial gradients in the IMF, Helios 1, 2, ISEE-3 and Pioneer 10, 11, Geophys. Res. Letters, 11, 279–282, 1984.

4.47   Smith, E.J., A. Barnes, Spatial dependences in the distant solar wind: Pioneer 10 and 11, in Solar Wind Five, Woodstock, Nov. 1982, NASA-CP, 2280, 521–535, 1983.

4.48   Smith, E.J., B.T. Thomas, Latitudinal extent of the heliospheric current sheet and modulation of galactic cosmic rays, J. Geophys. Res., 91, 2933–2942, 1986.

4.49   Smith, E.J., B.T. Tsurutani, L.R. Rosenberg: Observations of the interplanetary sector structure up to heliographic latitudes of 16, Pioneer 11, J. Geophys. Res., 83, 717–724, 1978.

4.50   Suess, S.T., Comment on "A comparison of coronal and interplanetary current sheet inclination" by K.W. Behannon, L.F. Burlaga, A.J. Hundhausen, J. Geophys. Res., 89, 11059–11060, 1985.

4.51   Suess, S.T., E. Hildner, Deformation of heliospheric current sheet, J. Geophys. Res., 90, 9461–9468, 1985.

4.52   Suess, S.T., B.T. Thomas, S.F. Nerney, Theoretical interpretation of the observed interplanetary magnetic field radial variation in the outer solar system, J. Geophys. Res., 90, 4378–4382, 1985.

4.53   Thomas, B.T., E.J. Smith, The Parker spiral configuration of the interplanetary magnetic field between 1 and 8.5 AU, J. Geophys. Res., 85, 6861–6867, 1980.

4.54   Thomas, B.T., E.J. Smith, The structure and dynamics of the heliospheric current sheet, J. Geophys. Res., 86, 11105–11110, 1981.

4.55   Villante, U., R. Bruno, Structure of current sheets in the sector boundaries, Helios 2 observations during early 1976, J. Geophys. Res., 87, 607–612, 1982.

4.56   Villante, U., R. Bruno, F. Mariani, L. Burlaga, N.F. Ness, The shape and location of the sector boundary surface in the inner solar system, J. Geophys. Res., 84, 6641–6648, 1979.

4.57   Villante, U., F. Mariani, The latitudinal dependence of the IMF polarity during 1975–81, Lett. Nuovo Cim., 36, 313–315, 1983.

4.58   Villante, U., F. Mariani, R. Cirone, Helios 1 + Helios 2, a summary of IMF observations performed in the inner solar system during 1975–1981, Nuovo Cim., 5C, 497–506, 1982.

4.59   Villante, U., F. Mariani, P. Francia, The IMF sector pattern through the solar minimum, two spacecraft observations during 1974–1978, J. Geophys. Res., 87, 249–253, 1982.

4.60   Villante, U., M. Vellante, The radial evolution of the IMF fluctuations, a comparison with theoretical models, Solar Phys., 81, 367–374, 1982.

4.61   Wilcox, J.M., A.J. Hundhausen, Comparison of heliospheric current sheet structure obtained from potential magnetic field computations and from observed polarization coronal brightness, J. Geophys. Res., 88, 8095–8096, 1983.

4.62   Wilcox, J.M., N.F. Ness, Quasi-stationary corotating structure in the interplanetary medium, J. Geophys. Res., 70, 5793–5805, 1965.

5. Interplanetary Dust

Christoph Leinert and Eberhard Grün

5.1 Introduction

The existence of a cloud of dust particles that pervades interplanetary space has long been known from the presence of its optical manifestation, the zodiacal light. This dim glow, best seen about one hour after sunset or before sunrise, when its brighter inner parts are above the horizon, was one of the first phenomena recognized as originating in interplanetary space. Given its conspicious appearance at low geographic latitudes (Fig. 5.1) one may wonder that there were no

Fig. 5.1. Cone of zodiacal light as seen one hour after sunset on the evening of 13 May 1983 from the top of Mauna Kea, Hawaii. Venus, at the top of the cone, and the setting crescent moon, 19° above the sun, approximately delineate the ecliptic

Physics and Chemistry in Space - Space and Solar Physics, Vol. 20
Physics of the Inner Heliosphere I   Editors: R. Schwenn · E. Marsch
© Springer-Verlag Berlin Heidelberg 1990

records of it until the end of the 17th century, when Cassini (1683) announced its discovery and offered the correct explanation: it is sunlight, scattered off myriads of small particles orbiting the sun. Towards the sun it continues with increasing brightness; the innermost parts can be seen during solar eclipses or from space as the F component (showing Fraunhofer lines) of the solar corona. Photometric measurements have shown that it covers the whole sky, but it is brightest in a band along the ecliptic. Around the antisolar point it brightens again; this *Gegenschein* can be seen by the naked eye under exceptional conditions.

From the conical shape of the zodiacal light both on the evening and morning sides of the sun a flattened, lenticular shape for the dust cloud was inferred, as visualized in Fig. 5.2. Details of the three-dimensional distributions are still poorly known; ellipsoidal density contours with axial ratio $1:7$ are a simple and not grossly inadequate description. Along the ecliptic the interplanetary dust cloud was found to extend well beyond Jupiter and to increase in density towards the sun roughly in inverse proportion to heliocentric distance.

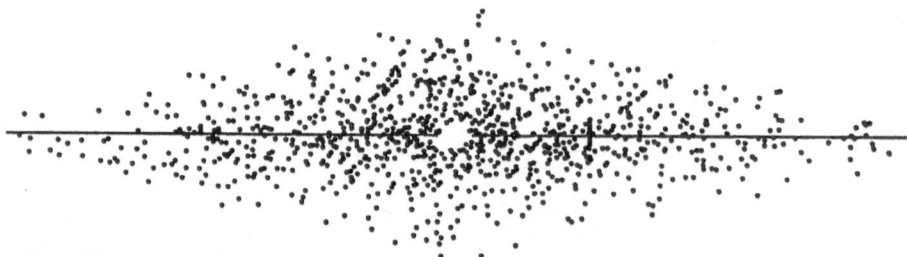

Fig. 5.2. Schematic cross section through the interplanetary dust cloud perpendicular to the ecliptic, shown as a solid line. The position of the earth is marked by a vertical bar. The sun would be at the center of the dust-free zone, which is shown enhanced by factor of five, for clarity

The cloud is extremely tenuous, with an optical depth along a tangent to the earth's orbit of only $\tau = 10^{-7}$ to 10^{-6}. At 1 AU the concentration of the optically important particles with radii 1–$1000\,\mu$m is a few particles per km^3. The corresponding mass density of $10^{-19}\,\mathrm{kg\,m^{-3}}$ is comparable to that of the solar wind; the orbital velocities of ≈ 10–$42\,\mathrm{km\,s^{-1}}$, and hence the energy density, are of course much smaller. Estimates for the mass of the whole cloud are 10^{16}–10^{17} kg, equivalent to the mass of one large comet, an almost negligible quantity within the whole solar system. The dust cloud may have been much more massive in earlier times. But it is not a direct offspring of the accreting disk which presumably surrounded the newborn sun and represented the first step in forming the present planetary system. Rather we will see that it may be considered a grandchild deriving from and being fed by first-generation bodies.

The lifetime of interplanetary dust particles is limited. The braking force associated with the absorption and scattering of solar radiation by an orbiting particle, called the Poynting–Robertson effect, gradually reduces the size of the orbit until after 10^4–10^5 years the particle passes close enough to the sun to evaporate. In addition, the momentum of solar wind ions impacting on the particle adds a similar drag force that is about one third of the Poynting–Robertson drag. Collisions between dust particles create smaller debris which is more strongly subject to the action of the drag forces than the original larger particles. Submicron-sized pieces will even be immediately blown out of the planetary system by solar radiation pressure. These are called β-meteoroids after the usual designation of the ratio of radiation pressure to gravitational attraction. They are believed to be mainly produced in the vicinity of the sun, where the dust concentration is highest; they compete with evaporation in being the final step of dust grain destruction and constitute the main mass loss mechanism in the interplanetary dust cloud.

The total mass loss is of the order of $10^4 \, \mathrm{kg\,s^{-1}}$. If there were no new supply for these losses, interplanetary dust and zodiacal light would disappear within the typical lifetime for dust particles, being a spectacle just set up in time for the human race to admire. The commonly preferred view, however, is that interplanetary dust should be a persistent feature of the solar system, particularly since an adequate source for its replenishment is known to exist: the comets. Unfortunately, although most researchers in the field consider comets to be the main source of interplanetary dust, the seemingly obvious transfer of cometary material to interplanetary space has not yet been worked out quantitatively.

In the context of the solar system, interplanetary dust is young. Nevertheless it should contain some primordial material, conserved in a comet nucleus since the beginnings of the solar system and recently set free. But the vast majority of particles, and even many of those newcomers, will have suffered significant alteration by cosmic rays, solar wind induced chemical reaction, sputtering, and collisions. In contrast to interstellar dust, interplanetary particles do not grow from molecules but are mostly ground down from pebbles and boulders. Therefore they are not expected to contain unaltered information on the origin of solar system.

The word "dust" suggests a size range with radii of 1–100 μm. In interplanetary space this is but a small section of the total distribution, which continuously covers the interval from submicron-sized particles to kilometer sized comets and asteroids, the smaller ones, however, always being much more numerous. Particles with radii less than 100 μm are also called micrometeoroids, particles in the size range 100 μm (10^{-8} kg) to centimeters, meteoroids. In the term "dust cloud" however, we include, for simplicity, both dust and meteoroids. Where no distinction is meaningful we use the two words synonymously. The dividing line of 100 μm, although somewhat arbitrary, can be given a physical meaning. For dust the lifetime is primarily limited by the Poynting–Robertson effect, and collisions in the cloud have the net effect of producing new dust. For the larger meteoroids, collisions dominate the lifetime and, in summary, constitute the major loss process. Even large particles are rare; they will not be discussed. The probability is that boulders of 1 m radius or larger will hit the surface of the earth at a rate of less than 1 per month [5.75].

To avoid confusion in further parts of our text we insert here a common definition: "meteoroids" are the particles in interplanetary space, "meteors" the phenomena occurring when they hit the earth's atmosphere, and "meteorites" those parts of a meteoroid which have survived the passage through the atmosphere and can be found on the ground.

The physics of interplanetary dust in a sense is very simple, because crude approximations are the rule and demanding theoretical treatments the rare exception. In another sense, it is also quite complicated, because peculiar parameters such as shape, composition, and structure have to be considered, difficult-to-characterize interactions, such as collisions, exist, and important quantities or boundary conditions are poorly known. This enforces the above-mentioned use of approximations.

The prospects for knowing the structure of the interplanetary dust cloud and understanding it as the result of an interplay between producing, transforming, and destroying processes have improved considerably over the last two decades. This was mainly due to observational results. For our discussion, which tries to approach this kind of understanding, it is therefore crucial to know the strong points and the limitations of different measurement techniques.

5.2 Methods of Observation

The more important methods, illustrated with a few experimental results, are presented below. None of them is able to provide all parameters of interest for describing the status of a dust particle. Fortunately they are different enough to complement each other.

5.2.1 Collection of Individual Particles

A collection is only possible at low particle velocities relative to the collecting surface, typically less than a few $km\,s^{-1}$. Occasionally this condition may be met (see Fig. 5.3) when a particle hits the detector of a space-borne instrument or a collecting surface [5.157]. The most promising approach is to search for the particles in the earth's atmosphere where they have been decelerated gradually

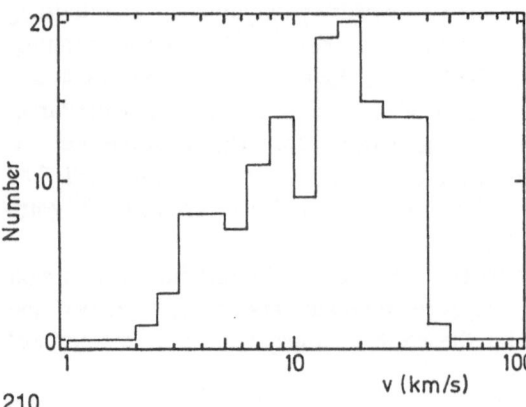

Fig. 5.3. Distribution of impact speed of interplanetary particles as measured by the micrometeoroid detectors on board *Helios 1* [5.70]

during their entry from interplanetary space. At stratospheric altitudes (≥ 20 km) nearly all of the particles with radii less than $5\,\mu$m are terrestrial aerosols or produced by rocket exhaust, but many in the size range 10–$100\,\mu$m have been shown to be of interplanetary descent on the basis of chemical composition, content of helium implanted by the solar wind, and the existence of "solar flare track" crystal defects [5.14]. Even larger particles are rare and quickly settle to the ground.

A particularly successful apparatus, where the particles were caught in a coating of silicon oil on quartz rods, was flown by Brownlee in balloons and high-flying aircraft. Other methods are summarized in his review in the book *Cosmic Dust* (1978). Among present developments the quasicollection by capture cells [5.128] deserves attention. It is used on the shuttle-launched first Long Duration Exposure Facility (LDEF), which hopefully will be brought back to earth in 1989. Such an experiment has also been flown on Salyut 7 [5.12].

5.2.2 Impact Ionization Detectors

Impact ionization detectors register the electrons and ions produced during high-velocity impacts of dust grains on a target surface. Let us denote the mass of a dust particle by m_{pr} and its impact velocity by v_{pr}. The collected charge Q may be expected to depend mainly on the kinetic energy $m_{pr}v_{pr}^2$; in fact laboratory calibration yielded $Q \sim m_{pr}v_{pr}^\alpha$, with $\alpha = 2.5$–3.5, depending on detector geometry. Empirically, the velocity of the projectile can be determined from the rise time of the charge pulses within a factor of 1.5–2; the scatter in m_{pr} is then within a factor of 6–10 [5.67]. With known orientation of the detector at the time of the impact, the direction from which the particle must have come can be constrained. This gives coarse orbital information. If coupled with a mass spectrometer (Fig. 5.4) the detector also yields information on chemical composition [5.35, 100]. Impact ionization detectors like the one on *Helios* are sensitive to particles in the mass range 10^{-18} kg to 10^{-12} kg for impact velocities of 20 km s^{-1}. The upper limit is selected electronically because only a few larger particles are expected to hit a target area of typically 0.01 to 0.1 m^2 during the usual duration of a mission of a few years.

5.2.3 Penetration Detectors

Penetration detectors are considered to provide a reliable and particularly simple *in situ* detection of dust particles [5.127]. To penetrate a foil, a particle must have a minimum kinetic energy, seen in the direction perpendicular to the surface. Particles of lower specific density have a lower ability to penetrate, because for given mass the impact affects a larger area. Specifically, at an impact velocity of 20 km s and for a density of 0.5 g/cm^3 the minimum mass necessary to penetrate $25\,\mu$m of stainless steel is 8×10^{-12} kg, while for $230\,\mu$m of aluminium it is 2×10^{-10} kg. These examples refer to experiments on the space probe *Pioneer 10*

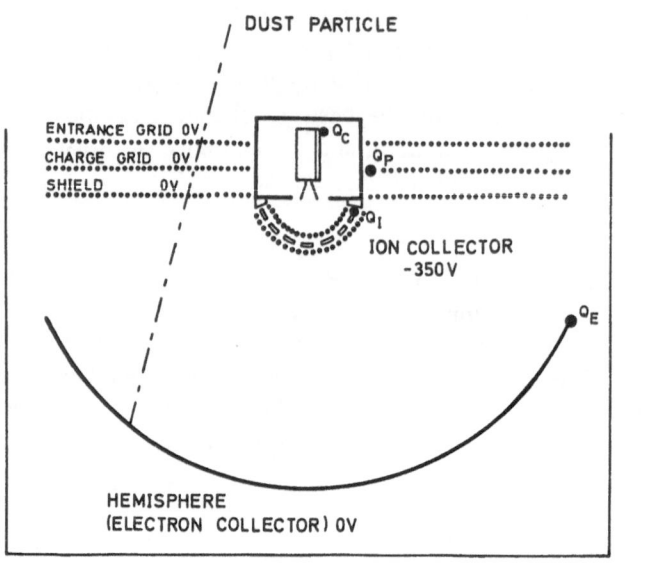

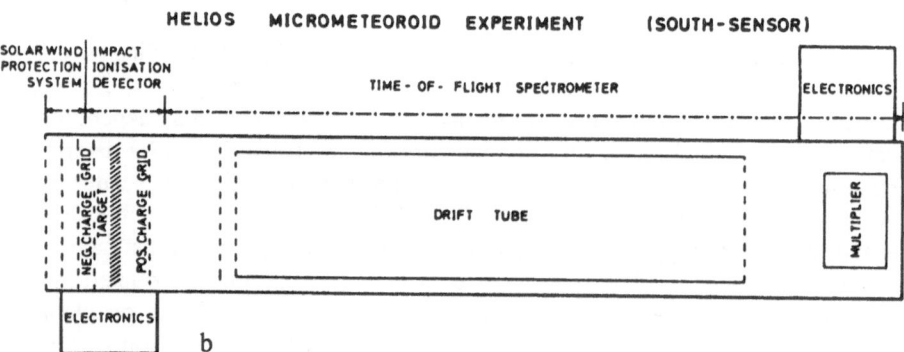

Fig. 5.4a,b. Schematic drawings of impact ionization detectors. (a) Simple detector to determine mass, velocity, and pre-impact particle charge. Q_p, Q_E, Q_I, Q_c indicate measurements of induced particle charge, electron charge, ion charge and a redundant measurement of the ion charge by a channeltron. Flown on the *HEOS-2* satellite (without Q_p, Q_c), scheduled for the *Ulysses* and *Galileo* space probe missions. (b) Analyzer for mass, velocity, and rough chemical analysis flown on the *Helios* solar probes. A similar analyzer with improved resolution of the mass spectrum was part of the *Giotto* cometary mission

[5.93], where the penetration simply leads to loss of pressure in a pressure cell, and to satellite *Pegasus* [5.142]. An empirical formula for penetration depth as function of particle and foil properties was given by Pailer and Grün [5.146] and further discussed by Carey et al. [5.23]. Because in interplanetary space large particles are rare, most of the detected particles have masses near the threshold mass.

5.2.4 Lunar Microcraters

The moon, like other planetary bodies without atmosphere, is being hit continuously by interplanetary dust particles. In the case of the moon, these velocities may go up to $72\,\mathrm{km\,s^{-1}}$, the sum of the heliocentric velocity of the moon and the velocity of escape from the planetary system. Hence the surface of lunar samples is covered by numerous high-velocity impact craters with sizes from below $1\,\mu m$ to several millimeters (Fig. 5.5). Laboratory simulation (Fig. 5.6) led to relations between crater diameter and projectile mass, which depend on velocity and specific particle density, both for the diameter of the central pit and for the about four times larger surrounding spallation zone [5.88]. A typical value for the ratio of pit diameter to particle diameter is 5. Except for quite oblique incidence the impact craters will be round, but the ratio crater depth to diameter increases with increasing specific density of the particle. Attempts to recover the chemical composition of impacting meteoroids from residues in or near the crater pits could not compete with the analysis of whole particles collected in the stratosphere.

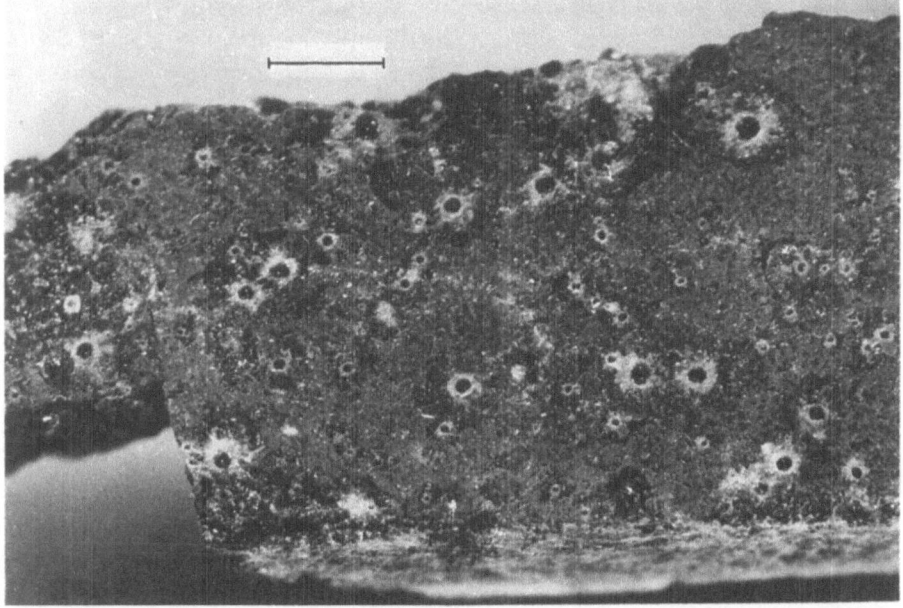

Fig. 5.5. Cratered surface of lunar sample 60015.35. The scale bar has a length of 1 mm

5.2.5 Meteors

When interplanetary particles shoot into the earth's upper atmosphere they heat and may melt or evaporate, at least partly, at the same time ionizing and exciting the surrounding air. This is particularly true for the larger particles, for

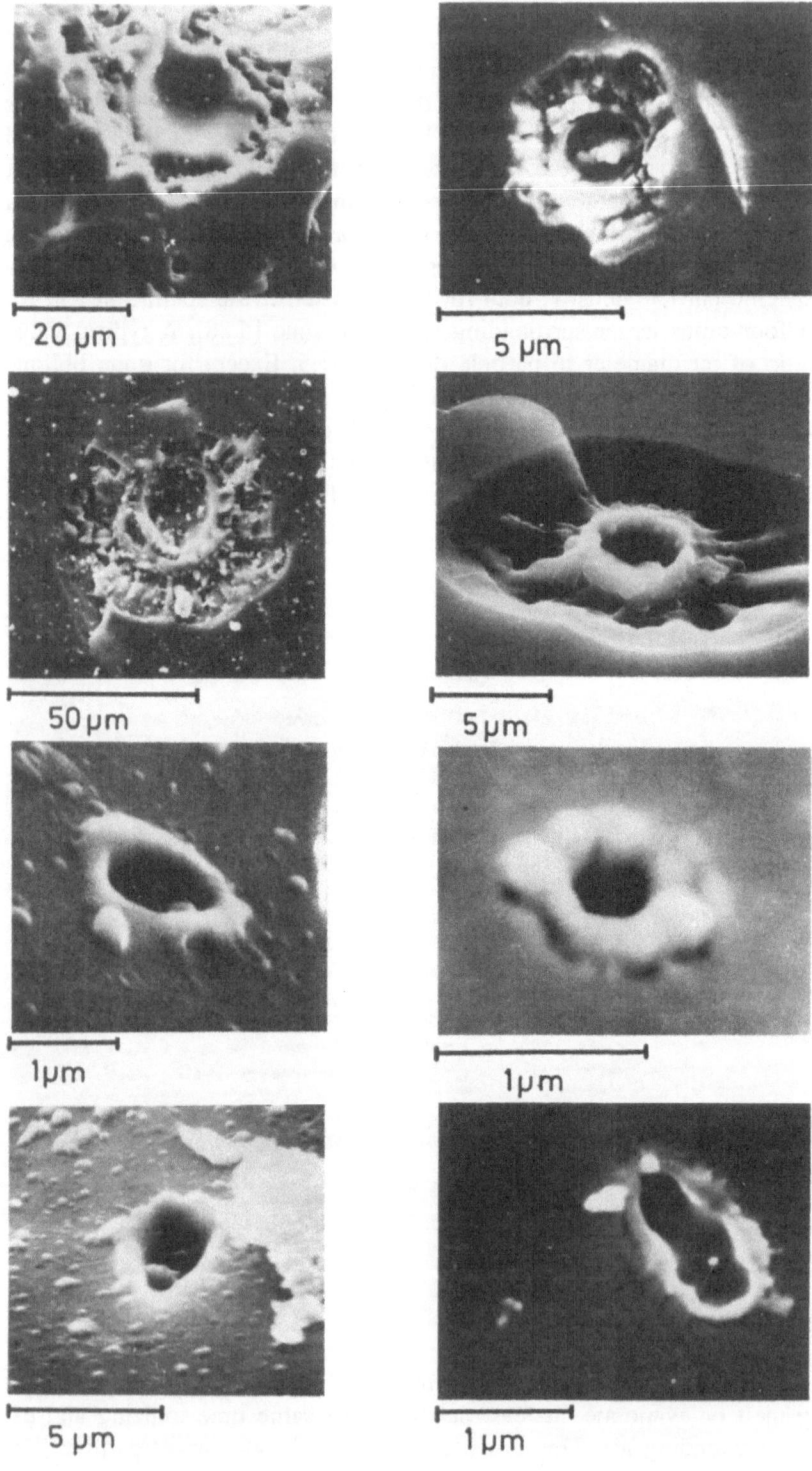

20 µm

5 µm

50 µm

5 µm

1µm

1µm

5 µm

1 µm

214

Fig. 5.6. (see opposite page for legend)

which deceleration starts to be effective deeper in the atmosphere and therefore is a more abrupt process. The associated luminous phenomenon is the optically visible "meteor". However, the sensitivity of detection is greatest for the developing trail of ionization which can be detected by radar for particles larger than 10^{-8} kg. Simultaneous observations of such a "radio meteor" from several stations yield the orbit and mass of the incident particle. For particles larger than 10^{-7} kg the luminous vapour can be photographed through a telescope. For such "photographic meteors" the deceleration during their flight through the atmosphere allows, in addition, the measurement of specific density, while for radio meteors it has to be deduced from the height at which the meteor starts to radiate [5.16, 90, 171]. The mass determination of meteor particles depends on the accuracy with which the ionization efficiency or the luminous efficiency of the entry process is known, and therefore should be considered with caution. Meteoroids in the size range of several mm and larger produce bright enough trails to be seen by the naked eye as shooting stars.

5.2.6 Zodiacal Light Observations

Two kinds of radiation are emitted by interplanetary dust. The first is scattered light from the incident solar radiation which constitutes the visible and near infrared part. The second is thermal radiation of the dust particles according to the temperature which they acquire in equilibrium with the solar radiation. This dominates the spectrum longward of 5 μm. We start by discussing the scattering.

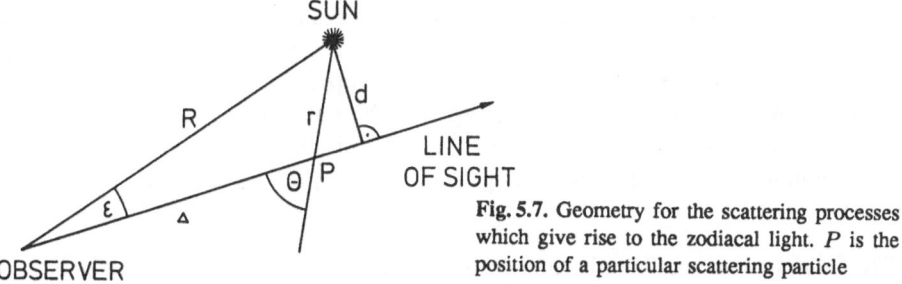

Fig. 5.7. Geometry for the scattering processes which give rise to the zodiacal light. P is the position of a particular scattering particle

Scattered Light. The intensity measured at an angle ε from the solar direction is made up from numerous individual scattering processes to which all dust particles along the line of sight contribute (Fig. 5.7), although not all 10^{22} covered by a typical 10^{-3} sr field of view at the same instant (because a typical photon flux is only 10^6 cm^{-2} s^{-1}). For each wavelength the observed intensity depends on solar irradiance, spatial density $n(r)$ of the dust particles, and average cross section $\sigma(\theta)$ for scattering by an angle θ. This average extends over all individual

---

Fig. 5.6. Comparison between lunar (*left*) and simulated impact craters (*right*). High-velocity impacts in brittle materials like lunar rocks or glass produce a central pit and an extended spallation zone around it. For submicron-sized craters (*lower rows*) no spallation zone develops [5.143]

particles in a volume around the position determined by the values of r and θ. For an extended brightness like the zodiacal light the distance to the individual scattering particles does not enter the calculation, since the attenuation $\sim 1/\Delta^2$ is offset by the larger number of particles ($\sim \Delta^2$) contained in the field of view. With azimuthal symmetry in the plane of the ecliptic, the intensity observed at distance R and angle ε from the sun is given by the brightness integral

$$I(\varepsilon, R) = F_0 R_0^2 \int_0^\infty \frac{n(r)\sigma(\theta, r)}{r^2} d\Delta \quad [\mathrm{W\,cm^{-2}\,sr^{-1}\,\mu m^{-1}}] \tag{5.1}$$

or, changing to integration over scattering angle θ, by [5.108]

$$I(\varepsilon, R) = \frac{F_0 R_0^2}{R \sin \varepsilon} \int_\varepsilon^\pi n(r)\sigma(\theta, r) d\theta \quad [\mathrm{W\,cm^{-2}\,sr^{-1}\,\mu m^{-1}}] . \tag{5.2}$$

Here, F_0 is the solar irradiance ($\mathrm{W\,cm^{-2}\,\mu m^{-1}}$) at $R_0 = 1$ AU. To evaluate (5.2) the units of the different quantities have to be chosen in a consistent way, e.g. σ in $[\mathrm{cm^{-2}\,sr^{-1}}]$, n in $[\mathrm{cm^{-3}}]$, and R, R_0 in [cm].

For simplicity, the wavelength dependence of I, F_0, and σ has not been shown explicitly in the formulae. Since $r = R \sin \varepsilon / \sin \theta$, relation (5.2) has a tempting property: *if* $\sigma(\theta)$ is independent of r and *if* n varies as power law $r^{-\nu}$, then the integral over θ does not depend on R, and for all viewing directions the intensity of zodiacal light will vary with the position of the observer as follows:

$$I(\varepsilon, R) \sim \mathrm{const.}(\varepsilon) \cdot R^{-\nu-1} . \tag{5.3}$$

With the help of this relation the exponent of the radial dependence of the dust distribution would follow immediately from observations at different heliocentric distances. This emphasizes the importance of space probe observations like the ones with *Pioneer 10* and *Helios*. Chemical composition, size, and shape of the particles, which determine the scattering properties, cannot be determined directly from zodiacal light observations but have to be discussed on the basis of models for brightness distribution, wavelength dependence, and polarization of zodiacal light.

Thermal Emission. The thermal emission of interplanetary dust, which is reradiation of absorbed solar energy, has a more direct relation to particle properties because the complication of angular dependence is missing. The brightness integral now has the form

$$I(\varepsilon, R, \lambda) = \int_0^\infty n(r)E(\lambda)d\Delta \quad [\mathrm{W\,cm^{-2}\,sr^{-1}\,\mu m^{-1}}] \tag{5.4}$$

and is dominated by the average emissivity $E(\lambda)$ of the particles, which depends on their size distribution, temperature, and optical properties:

$$E(\lambda) = \langle \pi s^2 B(T, \lambda) Q_{abs}(s, \lambda) \rangle [\mathrm{W\,sr^{-1}\,\mu m^{-1}}] . \tag{5.5}$$

Here $B(T, \lambda)$ is the Planck function and $Q_{abs}(s, \lambda)$ the absorption efficiency, which is largely determined by chemical composition; "$\langle\rangle$" denotes the average over the particle size distribution. The particle temperature T is determined from the condition of equilibrium between absorption and emission, where $F_\odot(R, \lambda)$ is the solar irradiance at heliocentric distance R (AU),

$$\int_0^\infty F_\odot(R, \lambda)\pi s^2 Q_{abs}(s, \lambda)d\lambda = \int_0^\infty 4\pi s^2(\pi B(T, \lambda))Q_{abs}(s, \lambda)d\lambda . \quad (5.6)$$

The extra factor π in front of the Planck function is the effective solid angle into which each surface element emits. The equilibrium temperature is size dependent [5.153]. Based on the fine observations of infrared satellite *IRAS*, average dust temperatures at 1 AU of 238 K, 257 K and 275 K were derived [5.34, 47, 83]. This side of zodiacal light observations still has much potential.

F-corona. In principle, coronal observations present nothing new, the F-corona also being composed of scattered light in the visual region and thermal emission in the infrared. However, the occurrence of very small scattering angles, of dust temperatures well in excess of 1000 K, and the existence of evaporated dust-free zones promise valuable additional information. Therefore this small angular interval (0°–5°) should not be passed by in the study of interplanetary dust [5.173]. These observations are difficult because they imply measurements close to the solar limb and because of the superimposed highly variable K corona, which is due to light scattering by electrons. The review by Koutchmy and Lamy [5.102] shows that for these reasons progress since the report of Blackwell et al. [5.13] has been moderate. Clearly, good new space observations are needed; these might be obtained on dedicated satellite missions such as the *SOHO* project.

Since zodiacal light observations contain information on many relevant properties of interplanetary dust, we summarize here some of the basic results obtained at 1 AU (Fig. 5.8), rather than presenting them piecewise in the different contexts. However, the spectrum of zodiacal light, which is close to the solar spectrum from 200 nm to the near infrared, will be presented in Sect. 5.8 (Fig. 5.27). Typical measurement techniques are discussed by Leinert [5.108] and Weinberg and Sparrow [5.174]. Concerning the discrepancies in infrared brightness of nearly a factor of two in Fig. 5.8, we prefer the values of satellite *IRAS* over the rocket and balloon experiments of Murdock and Price [5.139] and Salama et al. ([5.154], not shown), because this data base is much more extended, particularly in temporal coverage.

In any case, zodiacal light observations give an average over a large ensemble of particles along the line of sight. Although they do not yield detailed information on interplanetary dust they provide different kinds of gross information [5.174]; e.g., brightness is proportional to number, the color is blue for particles smaller than the wavelength, dielectric particles give negative polarization at small and large scattering angles; to produce a Gegenschein, particles must be dielectric or have a porous surface, polarization in the Gegenschein is only possible if particles are elongated and aligned, etc. A broad range of examples was

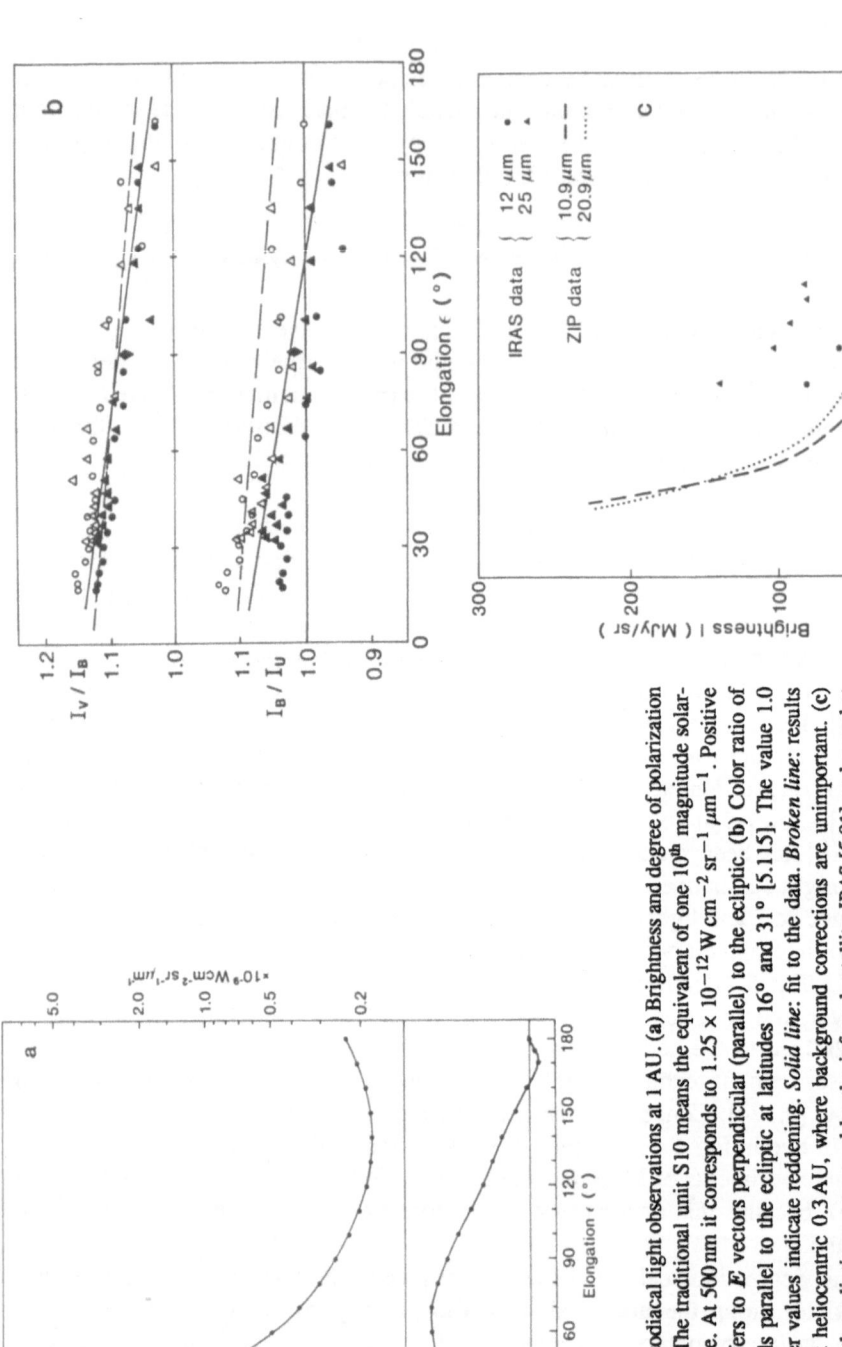

Fig. 5.8a–c. Summary of zodiacal light observations at 1 AU. (a) Brightness and degree of polarization along the ecliptic [5.53]. The traditional unit S10 means the equivalent of one 10th magnitude solar-type star per square degree. At 500 nm it corresponds to $1.25 \times 10^{-12}\,\text{W cm}^{-2}\,\text{sr}^{-1}\,\mu\text{m}^{-1}$. Positive (negative) polarization refers to E vectors perpendicular (parallel) to the ecliptic. (b) Color ratio of the zodiacal light in bands parallel to the ecliptic at latitudes 16° and 31° [5.115]. The value 1.0 refers to solar color, larger values indicate reddening. *Solid line:* fit to the data. *Broken line:* results of same measurements at heliocentric 0.3 AU, where background corrections are unimportant. (c) Infrared brightness along the ecliptic as measured by the infrared satellite *IRAS* [5.81] and a rocket experiment [5.139]. 1 MJy/sr corresponds to 2.52, 2.08, 0.69, and $0.48 \times 10^{-12}\,\text{W cm}^{-2}\,\text{sr}^{-1}\,\mu\text{m}^{-1}$ for the wavelengths of 10.9, 12, 20.9, and 25 µm (from [5.47])

calculated by Giese [5.60]; other examples, with emphasis on small particles, by Giese et al. [5.62]. On the other hand, if specific models of interplanetary dust are proposed, zodiacal light observations often present a sharp criterion against which such models can and should be tested. They are more judge than detective.

5.2.7 Comparison

Different methods of observation refer to different size ranges of interplanetary dust particles with not too much overlap (Fig. 5.9). This makes a comparison debatable because of the possibility of size effects. The use of larger impact ionization detectors in space probes *Galileo* and *Ulysses*, which have a chance also to catch and measure particles with radii $> 10\,\mu$m (mass $> 10^{-11}$ kg) will hopefully soon strengthen the comparison between direct detection and zodiacal light measurements. One other point is that so far only crude orbital information is available for dust particles. To improve this situation, penetration and impact detectors with time-of-flight measurement [5.9] or triangulating optical detection sensors should be revived with increased sensitivity and reliability, respectively.

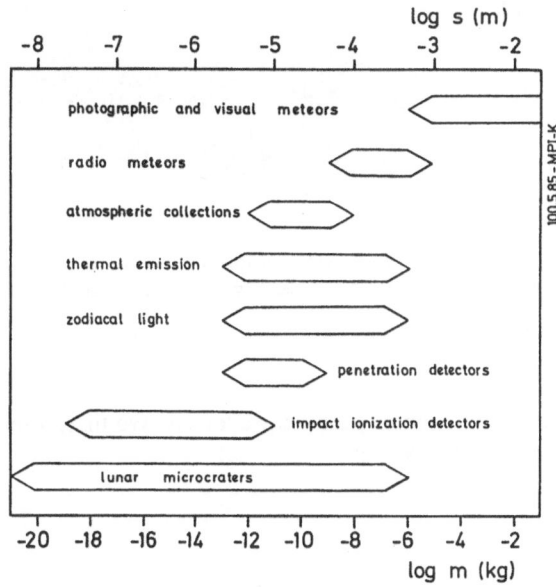

Fig. 5.9. Comparison of the size ranges covered by different observational methods

5.3 Particle Properties

Properties are best known now for the particles collected in the stratosphere, i.e. the size range of diameters $3\,\mu$m to about $100\,\mu$m. This happens to be the range which contributes a large fraction to the zodiacal light as well as to the total mass flux of interplanetary dust. These particles, at hand for detailed laboratory

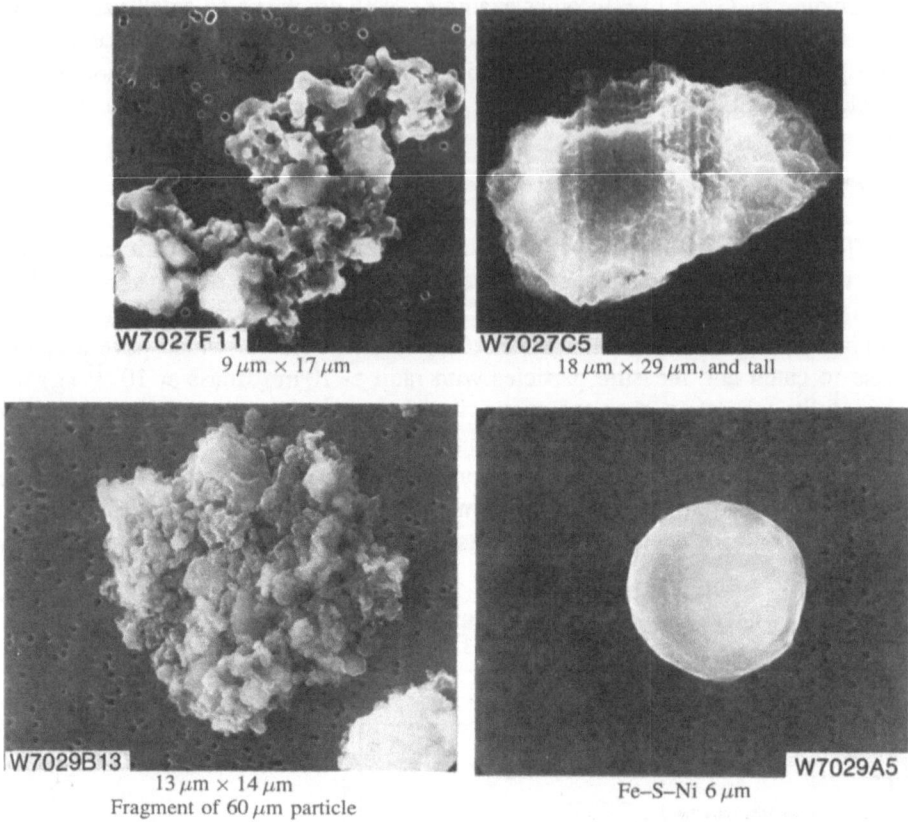

W7027F11
$9\,\mu m \times 17\,\mu m$

W7027C5
$18\,\mu m \times 29\,\mu m$, and tall

W7029B13
$13\,\mu m \times 14\,\mu m$
Fragment of $60\,\mu m$ particle

W7029A5
Fe–S–Ni $6\,\mu m$

Fig. 5.10. Interplanetary dust particles collected above 20 km height in NASA's cosmic dust program. From over 200 catalogued specimens we show three chondritic particles in various shapes and degrees of compactness and one Fe–S–Ni sphere

studies, are the cornerstone of our knowledge of particle properties. We therefore start with a description of their general properties.

5.3.1 Elemental Composition

Brownlee [5.17] grouped the collected particles in three categories:

1. Chondritic particles. These represented the majority (60%) of the collected particles (Fig. 5.10). They often appear to be aggregates of smaller ($0.1\,\mu m$) grains and show, within a factor of two, "cosmic", i.e. chondritic, abundance for the more important elements (Fig. 5.11). Specifically, the high carbon content of several percent makes them dark and similar to the most undifferentiated and therefore most primitive meteorites, the carbonaceous chondrites of classes C1 and C2, although the texture of the dust does not

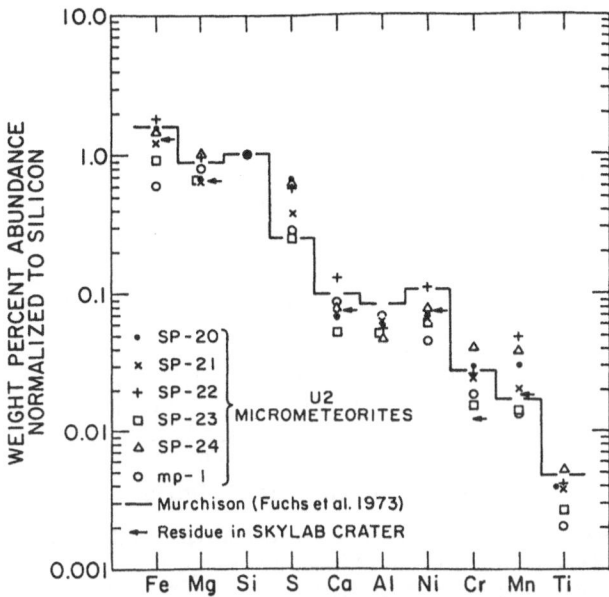

Fig. 5.11. Relative abundance of important elements in micrometeoroids: comparison of six chondritic aggregate micrometeorites and residue in a crater found on *Skylab* with the Murchison carbonaceous chondrite meteorite [5.17]

show the fibrous structures common in those meteorites. This has been taken as argument that they formed in a different environment to that of the chondritic meteorites.

2. A large fraction (30%) of the collected particles were of composition iron–sulfur–nickel, where the nickel is a few percent addition to an iron–sulfide mineral. These particles are among the smallest collected. Some have the form of spheres, which may be the result of melting during entry into the atmosphere (Fig. 5.10), but also platelets or single crystals occur.

3. About 10% of the collected particles were classified as mafic silicates consisting mainly of the minerals olivine or pyroxene.

Mass spectroscopic measurements by the *Helios* micrometeoroid experiment, although difficult to calibrate because of uncertainties in the ionization efficiencies of different elements, indicated two broad types of particle composition, compatible with groups 1 and 2 discussed above (Fig. 5.12).

The collections did show a few fluffy, very low density aggregate particles. It may be hypothesized that more of them exist in interplanetary space but cannot be collected because they are generally too fragile to survive atmospheric entry.

Surprisingly, then we know comparatively much about properties as complicated as elemental composition and mineral content of interplanetary dust. The analysis of the spectra of photographic meteors [5.129] has also found a carbonaceous composition for those larger particles. This then appears to be the dominating solid material in the interplanetary dust cloud. Taken together with

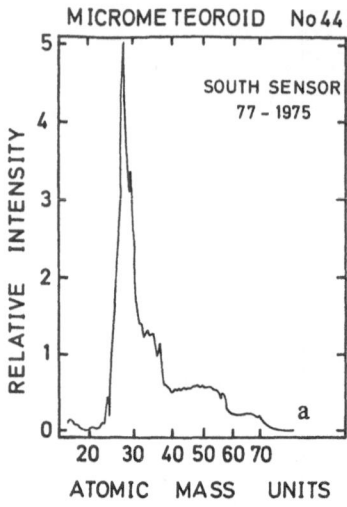 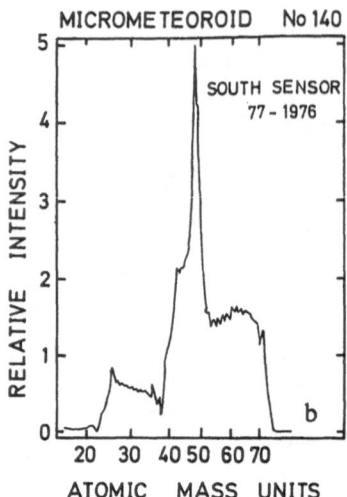

Fig. 5.12a,b. Mass spectra measured by the *Helios* micrometeoroid impact analyzer. The left spectrum indicates chondritic composition (dominant masses are 16 to 35 amu), the right spectrum is compatible with an iron-rich composition (dominant masses are 35 to 74 amu)

the findings of the comet Halley missions (Sect. 5.9.3), it supports the notion that comets are the main source, but this is no proof: many asteroids show surface reflectivities and hence compositions similar to carbonaceous chondrites [5.179].

5.3.2 Shape

The shape of the collected dust particles is quite complicated, but at least they are roughly equidimensional. This was already concluded from the morphology of lunar microcraters, which are round with a few exceptions. Elongated particles with axial ratios in excess of 4 would create elongated craters [5.6]. The equidimensional shape therefore is expected to hold also for smaller and larger particles than the collected ones. It gives some justification to the attempts to estimate physical effects on the dust particles using calculations for spheres, for which the formalism of the Mie theory allows exact results.

5.3.3 Optical Properties

The *albedo* of the interplanetary particles must be low on the average if the collected particles form a representative subgroup, because of the high carbon content of their chondritic material. The albedo could be estimated more quantitatively by reflection measurements on the actual collected particles. Historically, a comparison between the brightness of the Fraunhofer corona, then taken to be diffracted light only, and the zodiacal light at larger elongations gave the first indication that the albedo of interplanetary dust might be only a few percent

[5.51]. Nowadays, a direct comparison of the reflected and thermally emitted components of zodiacal light [5.47, 82] give an average albedo of $A = 0.08$–0.09 for interplanetary dust at 1 AU. This is in line with the composition of the collected dust particles; a typical reflectivity of chondritic meteorites is 10%. The *scattering function* of interplanetary dust is shown in Fig. 5.28, Sect. 5.8, where its derivation from zodiacal light measurements and its implications are discussed. However, referring to these results, one has to consider that the optical properties, in particular albedo, may vary with heliocentric distance (Sect. 5.6.3).

5.3.4 Density

The traditional view is that the density of meteoroids decreases with increasing particle size. Verniani [5.171] found the average density of faint radio meteor particles with mass 10^{-7} kg to be 0.8 g cm^{-3} [5.94]. The collected dust particles fit into this scheme, most of them having estimated densities of 1–3 g cm^{-3}. Le Sergeant d'Hendecourt and Lamy [5.119] derived densities from depth-to-diameter ratios of lunar microcraters corresponding to particle masses 10^{-18} kg to 10^{-7} kg. Their results are compatible with an average density of 2.5 g cm^{-3} but increasing systematically from larger to smaller particles. All these findings are consistent with a cometary origin: the larger particles, after evaporation of adhering ices, would be expected to contain voids and to be of loose structure; breakup and sublimation would lead to smaller, less porous elements.

But this compelling picture has to be modified a little. Photographed meteorite falls allowed a calibration of meteor mass determination, indicating that about 60% of the photographic and visual meteor particles (10^{-6} kg to 0.1 kg) have densities $\varrho \geq 1$ g cm^{-3}, which greatly reduces the trend of density with size. But still there remain meteor particles, e.g. the Draconids, with very low densities: $\varrho = 0.02$ g cm^{-3} [5.26]. Also, Smith et al. [5.163] and Nagel et al. [5.140] found that 30% to 40% of lunar microcraters are very shallow, which points to low-density projectiles with $\varrho \leq 1$ g cm^{-3}. In addition, the *Helios* micrometeoroid experiment obtained information on particle densities by comparing the impact rate of two sensors, one of which was covered with a thin plastic film. For given mass and impact velocity only particles with a minimum density are able to penetrate. On this basis it was estimated from the different rates observed with the two sensors [5.71] that about 10% to 20% of the meteoroids with masses 10^{-18} kg to 10^{-12} kg have densities below 1 g cm^{-3}. Summarizing all this observational evidence, we conclude that at most every third or fourth meteoroid is of low density while for the majority the average density appears to increase from about 1.5 g cm^{-3} for photographic meteor particles to about 3 g cm^{-3} for micron-sized dust.

5.4 Physical Processes

Qualitatively most physical processes acting on interplanetary dust particles are self-evident. It is a different question to assess quantitatively their importance and to assure oneself that the calculated effects do work as predicted.

5.4.1 Gravity

The dominant force for all meteoroids, except perhaps the smallest submicron-sized particles (see below), is the solar gravitational attraction

$$F_{grav}(r) = \gamma \frac{M_\odot m}{r^2} \tag{5.7}$$

where $\gamma = 6.67 \times 10^{-11} \, kg^{-1} \, m^3 \, s^{-2}$ is the gravitational constant and $M_\odot = 1.99 \times 10^{30}$ kg the solar mass. Smaller gravitational interactions occur with the planets. Resonant perturbations are of less importance than for large bodies (e.g. asteroids) because usually the drift due to the Poynting–Robertson effect soon drives particles out of resonant regions. However, close approaches to planets perturb particles of all sizes in the same way. According to Öpik [5.145] significant orbit changes occur inside a distance D from the planet if

$$D = a_p (m_p/2M_\odot)^{1/3} \, , \tag{5.8}$$

with a_p being the mean distance of the planet from the sun and m_p the mass of the planet. Venus, Earth, and Mars have such spheres of influence with radii of about 0.01 AU, whereas for the major planets they increase to about 0.5 AU. The concept of spheres of influence is a simplifying guideline. Resonant, therefore effective, interactions may also occur for particles well outside this range [5.21, 95]. Orbit perturbations by planets have been observed for meteor streams [5.89].

5.4.2 Radiation Pressure

Scattering or absorption of solar radiation by an interplanetary dust particle inevitably leads to momentum transfer and hence to a radiation pressure force directed radially outward [5.22]. For spherical particles it has the value

$$F_{rad} = \int_0^\infty Q_{pr}(s, \lambda) \pi s^2 \frac{F_\odot}{c \, d\lambda} \, , \tag{5.9}$$

s being the radius of the particle, $F_\odot (W \, cm^{-2} \, sr^{-1} \, \mu m^{-1})$ the solar irradiance, and Q_{pr} (≈ 1 for radii $\geq 1 \, \mu m$) an efficiency factor which can be calculated from Mie theory. Radiation pressure is characterized by the ratio of its strength to gravitational attraction, which for spherical particles has the value [5.40]

$$\beta = \frac{F_{rad}}{F_{grav}} = 5.7 \times 10^{-7} \frac{\langle Q_{pr} \rangle}{\varrho s(m)} \, , \tag{5.10}$$

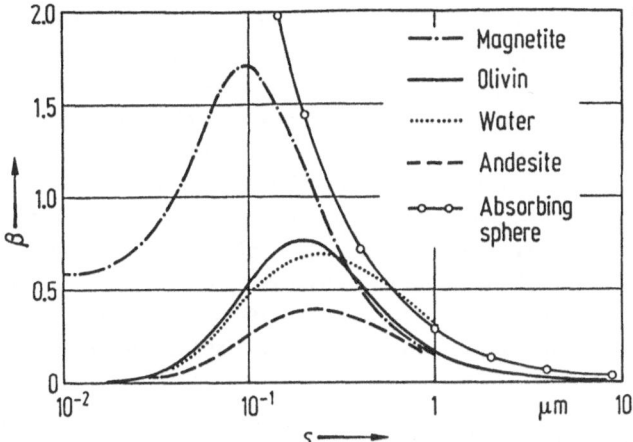

Fig. 5.13. Ratio β of radiation pressure to gravity as function of grain radius s. The curve labeled "water" is also valid for water ice. Assumed densities ($g\,cm^{-3}$) for the five materials are, from top to bottom, 5.2, 3.3, 1.0, 3.3, 2.0 g cm^{-3} [5.160]. The curve "absorbing sphere" corresponds to the approximation of (5.10); the others have been calculated using Mie theory

where the efficiency factor has to be averaged over all wavelengths, weighted by the solar spectrum. The result is that for a sphere of 1 μm radius gravitation is largely balanced by solar radiation. Radiation pressure may even surpass gravity ($\beta > 1$), but not to the extent suggested by (5.10), since very small particles are poor scatterers and absorbers (see Fig. 5.13). However, a particle in an eccentric orbit may be driven out of the solar system even if radiation pressure neutralizes only a fraction of the gravitational pull, because for an eccentricity $e > 1-2\,\beta$ the kinetic energy for particles set free in perihelion exceeds the magnitude of the

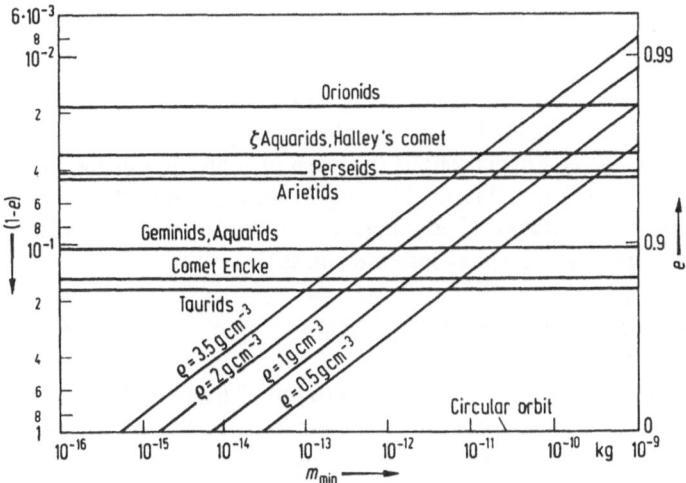

Fig. 5.14. Minimum mass necessary for a perfectly absorbing particle to remain in bound orbit under the action of radiation pressure when released at perihelion. e is the eccentricity of the parent object's orbit, ϱ the density of the particle [5.40]. The eccentricities of several meteor streams are indicated

potential energy. (Note that there is essentially no effect of radiation pressure as long as a particle is part of a larger body.) This applies in particular to particles of cometary origin, since comets release most material close to perihelion at high eccentricities. Figure 5.14 shows the minimum mass a dust particle must have for a given eccentricity to remain in bound orbit under these conditions. For comet Halley with $e = 0.967$ this mass is 1.5×10^{-10} kg (20 μm radius). This effect limits the direct supply of small dust particles by comets. On the other hand, considering the particles responsible for the zodiacal light, which are mostly larger than 10 μm and in orbits of moderate eccentricity, radiation pressure is a small, unimportant correction to gravity. Really? This statement would be true without reservation only if radiation pressure were directed strictly radially.

5.4.3 The Poynting–Robertson Effect

Seen from the moving dust particle, the incident solar radiation is displaced by an angle v_{tan}/c from the radial direction, where v_{tan} is the tangential component of the orbital velocity. Seen from outside, the particle motion causes scattering and thermal emission to have a forward component. The result is the same: a braking force, on the average over one orbit directed opposite to the direction of motion, equal in first order to v/c times the radiation pressure force for a particle in a circular orbit. The induced changes in the orbital elements are [5.178]

$$
\begin{aligned}
\frac{da}{dt} &= -\alpha \frac{2 + 3e^2}{a(1 - e^2)^{3/2}} \\
\frac{de}{dt} &= -\frac{5\alpha e}{2a^2(1 - e^2)^{1/2}} \ , \\
\frac{di}{dt} &= 0
\end{aligned}
\tag{5.11}
$$

where the constant has the value $\alpha = 3.55 \times 10^{-10} \langle Q_{\text{pr}} \rangle / s(m) \cdot \text{AU}^2/\text{yr}$, which means that the Poynting–Robertson effect, like radiation pressure, is stronger for smaller particles. The result is a synchronous decrease of a and e, during which the expression $e^{4/5}/a(1 - e^2)$ remains unchanged. In other words, the particle orbits get smaller and more circular. For a particle already in circular orbit at heliocentric distance r the time to spiral into the sun is

$$
\tau_{\text{pr}} = 700 \cdot s(\mu\text{m}) \cdot \varrho(\text{g/cm}^3) \cdot r^2(\text{AU}) / \langle Q_{\text{pr}} \rangle \text{ years} \ ,
\tag{5.12}
$$

or several 10^4 years for a typical dust particle (10 μm, 3 g cm^{-3}, 2 AU). The inward drift, measured by the change in semimajor axis, is twice as fast for eccentricity $e = 0.5$ than for a particle in circular orbit.

The impact of solar wind ions also exerts an outward pressure on orbiting dust particles. It is more than three orders of magnitude weaker than radiation pressure. But the aberration angle $v_{\text{tan}}/v_{\text{sw}}$, where v_{sw} is the velocity of the solar wind, is much larger, making the resulting "ion impact" drag comparable to the Poynting–Robertson effect. Because of some corotation of solar wind plasma, the addition to the Poynting–Robertson drag force is smaller for particles in prograde than in retrograde orbits. Mukai and Yamamoto [5.137] give values of about 30% and 60%, respectively, for magnetite grains. The lifetime given in (5.12) will be decreased by this amount. In addition, charged dust particles will feel a force due to passing solar wind ions. This "Coulomb pressure" is negligible under interplanetary conditions (Table 5.1).

Table 5.1. Comparison of various forces acting on a dust particle under typical interplanetary conditions (at 1 AU, solar wind with $400\,\mathrm{km\,s^{-1}}$ and 6×10^6 protons/m^3, $B = 5nT$, particles charged to 10 V, $\varrho = 3.5\,\mathrm{g\,cm^{-3}}$, $Q_{\mathrm{pr}} = 1$). Subscripts refer to gravity, radiation pressure, Poynting–Robertson effect, ion impact pressure, ion impact drag, Coulomb pressure and Lorentz force [5.53]. Forces are given in Newtons

$s/\mu\mathrm{m}$	F_{g} in [N]	F_{rp} in [N]	F_{PR} in [N]	F_{ip} in [N]	F_{id} in [N]	F_{Cp} in [N]	F_{L} in [N]
0.1	-8.7×10^{-20}	1.4×10^{-19}	1.4×10^{-23}	5.0×10^{-23}	3.8×10^{-24}	10^{-26}	1.6×10^{-19}
1	-8.7×10^{-17}	1.4×10^{-17}	1.4×10^{-21}	5.0×10^{-21}	3.8×10^{-22}	10^{-24}	1.6×10^{-18}
10	-8.7×10^{-14}	1.4×10^{-15}	1.4×10^{-19}	5.0×10^{-19}	3.8×10^{-20}	10^{-22}	1.6×10^{-17}
100	-8.7×10^{-12}	1.4×10^{-13}	1.4×10^{-17}	5.0×10^{-17}	3.8×10^{-18}	10^{-20}	1.6×10^{-16}

5.4.4 The Lorentz Force

The motion of interplanetary dust particles is influenced by interplanetary magnetic fields because the particles are charged. The main competing mechanisms which charge a dust particle are photoemission of electrons by absorption of solar ultraviolet radiation and capture of impinging solar wind electrons. Protons may be neglected because of their slower thermal motion. Since solar wind density nearly follows an inverse square law decrease with heliocentric distance, these potentials are independent of distance and reflect only the state of the solar ultraviolet emission and the local solar wind conditions. The prediction is that under average conditions particles will be charged to 6–10 V positive with little dependence on particle material [5.104, 177]. Measurements of space probe charging in interplanetary space confirm these predictions.

With the *Helios* micrometeoroid experiment it was tried for the first time to detect charges carried by individual dust grains. For four out of the seven biggest particles ($m > 10^{-9}$ g) the measurements indicated significant positive charges ($q > 1.8 \times 10^{-13}$ C). The charge on a spherical particle is given by $q = 4\pi\varepsilon_0 U s$, with the permittivity $\varepsilon_0 = 8.859 \times 10^{-12}$ C/Vm, the surface potential U, and the grain radius s. Even if one considers the uncertainty of the mass determination (by a factor of 10) the charges observed by *Helios* imply either particles of very low densities ($\varrho < 10^{-2}\,\mathrm{g\,cm^{-3}}$) or surface potentials around $100\,\mathrm{V}$. This question needs further study by future space experiments.

Uncertainty in particle charge is not the only reason effects on particle trajectories resulting from the interplanetary magnetic field are not well understood. A major source of uncertainty lies in the complicated structure and variability of this field. Strictly speaking, the electromagnetic force exerted on the particle,

$$F_{\mathrm{L}} = q V_{\mathrm{rel}} \times B \;, \tag{5.13}$$

where V_{rel} is the relative velocity between particle and the field $B = \mu_0 H$, is a Lorentz force only in the frame moving with the solar wind. In the rest frame of the particle it is an electric force of identical strength. The overwhelming part of the relative velocity is due to solar wind motion. The Lorentz force increases only proportionally to particle radius, much slower than gravity, and should become negligible for particles with $10\,\mu\mathrm{m}$ radius. This can also be seen from the fact that the gyration radius for such a particle in the interplanetary magnetic field would be $10^6\,\mathrm{AU}$, its gyration frequency would correspond to $70\,000$ years. However, for particles $\leq 10^{-7}\,\mathrm{m}$ the Lorentz force is the dominating force. It has a radial component which drives such particles outside $10\,\mathrm{AU}$ solar distance within 100 days [5.156], unless this job has already been performed by radiation pressure. Such particles should not exist in interplanetary space in appreciable quantities.

Particles in low inclination orbits experience alternating signs of the Lorentz force because of the interplanetary sector structure where the polarity of the magnetic field alternates between neighboring sectors. Because of stochastic variations in sector length this leads to a random walk in inclination of the particle orbit. The resulting dispersion, $\sqrt{\sum \Delta i^2}$, predicted by Parker [5.147], and more correctly calculated by Morfill and Grün [5.131], is $0.3°$ for a particle of $1\,\mu\mathrm{m}$ radius near $1\,\mathrm{AU}$ after 3×10^3 years. It decreases strongly with radius and heliocentric distance, being proportional to $s^{-3}r^{-2}$. More systematic treatments by Consomagno [5.28], Barge, Pellat, and Millet [5.7], and Wallis and Hassan [5.172] gave no improvement, predictions differing by factors of ten and more. The reason for this lies in different assumptions for the power spectrum of the disturbing force for periods in excess of 10 days, where observations are not very reliable because of the long time intervals involved [5.132].

Dust particles in orbits with higher inclination spend a fraction of their time in the unipolar magnetic fields at heliographic latitudes $> 15°$. At the times of sunspot maximum the polarity of this field changes. In the beginning of solar cycle no. 21, from 1969 to 1980 the magnetic polarity at the solar north pole was positive; under these conditions interplanetary dust particles experienced a focusing effect, a continuous decrease in inclination, by $5°$ during one solar cycle for a particle with $1\,\mu\mathrm{m}$ radius [5.131]. The effect will change sign with every field reversal accompanying the solar cycles. It is doubtful that the eleven-year changes have enough influence on the average inclination of dust particle orbits of $i \approx 30°$ to be noticed individually. But considering, as above, the random walk in inclination over many polarity reversals one finds a strong effect of dispersion in inclinations, at least for particles $1\,\mu\mathrm{m}$ in size (Table 5.2).

Table 5.2. Spread in orbital inclination due to Lorentz scattering during the Poynting–Robertson life time of a particle. r = starting distance, r_0 = 1 AU [5.132]

s	$T_{PR}(y)$	i (degrees)	$\langle i^2 \rangle^{1/2}$ (degrees) Sector Structure	$\langle i^2 \rangle^{1/2}$ (degrees) Solar Cycle
$1\,\mu$m	3×10^3	0	$0.25 \times \sqrt{r/r_0}$	
		30		22.5 $\times \sqrt{r/r_0}$
		60		11.4
$10\,\mu$m	3×10^4	0	$8.0 \times 10^{-3} \times \sqrt{r/r_0}$	
		30		0.71 $\times \sqrt{r/r_0}$
		60		0.36
$100\,\mu$m	3×10^5	0	$2.5 \times 10^{-4} \times \sqrt{r/r_0}$	
		30		2.3×10^{-2} $\times \sqrt{r/r_0}$
		60		1.1×10^{-2}

The effects of the Lorentz force on inclination are important since in contrast to radiation forces they do change orbital planes.

5.4.5 Evaporation

Figure 5.15 shows temperatures expected for interplanetary particles. Evaporation must play a role close to the sun. Dust-free zones due to evaporation were predicted with an extent, given in solar radii, of $24R_\odot$ for iron, $10R_\odot$ for olivine, $4R_\odot$ for obsidian, $1.5R_\odot$ for graphite or quartz [5.103, 153]. Mukai and Yamamoto [5.136] calculated that a particle of suitable size and material would spend a prolonged time just outside the evaporation zone, oscillating in and out under the interplay of evaporation, radiation pressure, and Poynting–Robertson drag. It remains doubtful that this effect will lead to a measurable enhancement of spatial density. Most particles in this region will probably evaporate completely. Nevertheless, with the temperatures of melting materials in the range 1000–1500 K [5.153] the edges of dust free zones should be observable in the near infrared. Indeed, structures of enhanced infrared emission have been observed (Fig. 5.16) and have been interpreted as the edges of dust free zones [5.124, 148], but so far the observations are not coherent enough to be sure about it [5.102, 125].

Pure ice is predicted to evaporate at heliocentric distance 3 AU [5.77], where its equilibrium temperature is 170 K. If contaminated with dark material the limit may be farther out than 4 AU. Ice particles should not be expected in the inner heliosphere. However, they may be the main constituent of interplanetary dust in the outer solar system.

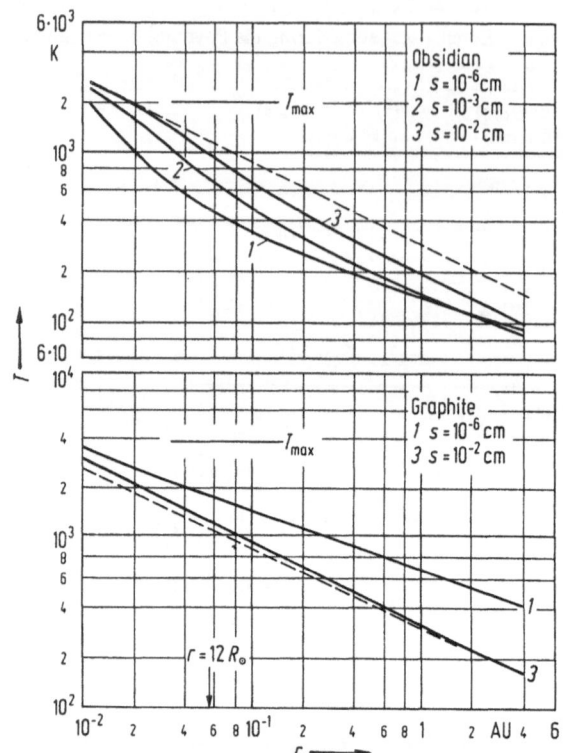

Fig. 5.15. Calculated grain temperature T as function of heliocentric distance, r, for different particle sizes and materials. T_{max} is the melting temperature. The black-body dependence (*dashed line*) is

$$T = 280°(1\ \text{AU}/r)^{1/2}\ \text{K}$$
[5.153]

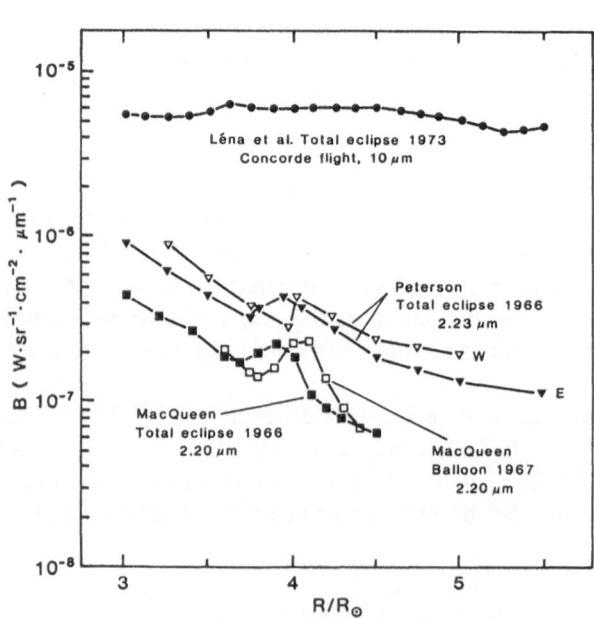

Fig. 5.16. Infrared structures observed in the solar corona [5.124, 148] (taken from [5.102])

5.4.6 Sputtering

Sputtering, the ejection of atoms from the surface of a dust particle by impinging fast solar wind ions, is unimportant for stony materials. The erosion of lunar material due to sputtering was found to be only 0.03 Å per year [5.6]. However, on ice particles sputtering has a 100 times higher efficiency [5.107]. Therefore the lifetime of ice particles in the outer solar system is much shorter than that of mineral grains.

5.4.7 Collisions

In collisions between interplanetary meteoroids the impact speed is usually high enough to result in fragmentation of one or both particles. We approximate the average impact speed $\langle v(r) \rangle$ by

$$\langle v(r) \rangle = v_0 \left(\frac{r}{r_0} \right)^{-0.5} , \tag{5.14}$$

with $v_0 = 20\,\mathrm{km\,s}$ at $r_0 = 1\,\mathrm{AU}$ and dependence on heliocentric distance r as given by Kepler's law. In such a collision the smaller particle always gets destroyed. A catastrophic collision, i.e. fragmentation of both particles, only occurs if the mass ratio is not too large, i.e. if

$$m_{\mathrm{projectile}} \geq \frac{1}{\Gamma(v)} M_{\mathrm{target}} , \tag{5.15}$$

where the maximum allowable mass ratio depends on velocity and was found experimentally for basalt to be $\Gamma(v) = 500\,v^2$ (km s) [5.57]. This may or may not also be typical for interplanetary particles. If the projectile mass is smaller, the larger particle is eroded by the impact cratering process, a kind of sputtering on a macroscopic scale. Dohnanyi [5.37] showed that for the meteoroids of concern to us ($m < 10^{-3}\,\mathrm{kg}$) erosive collisions are much less important than catastrophic collisions, so that we may limit our discussions to the latter case.

The rate $C(m, r)$ of catastrophic collisions of a meteoroid of mass m at heliocentric distance r can now be calculated by adding the probabilities that the meteoroid will encounter during the following second a projectile particle large enough to disrupt the meteoroid. This integral has to be taken over all projectile masses m_p larger than the minimum mass given in (5.15):

$$C(m, r) = \int_{m/\Gamma(v(r))}^{\infty} \sigma n(m_p, r) \langle v(r) \rangle \, dm_p . \tag{5.16}$$

Here $n(m_p, r)$ is the number density of particles of mass m at heliocentric distance r.

The cross section is taken as the total area inside the circle where the particles touch, $\sigma = \pi(s + s_p)^2$, where s and s_p are the radii of the meteoroid and projectile particles, respectively. The collisional lifetime for this particle then is defined as the reciprocal of the average collision rate,

$$\tau_c(m, r) = 1/C(m, r) \,. \tag{5.17}$$

Further the added mass of those particles with masses between m_1 and m_2 which are destroyed by catastrophic collisions and therefore lost to this mass interval, amounts per second to

$$\dot{M}(m_1, m_2) = -\int_{m_1}^{m_2} mn(m, r)C(m, r)dm \,. \tag{5.18}$$

Equation (5.18) is valid for a unit volume at heliocentric distance r. However, it does not take care of fragments which might still be large enough to qualify as particles of the mass range in question. This "production" of particles by collision is usually handled separately.

Dohnanyi [5.36] showed that an interesting conclusion can be drawn from the above relations, assuming only conservation of mass. If the mass distribution $n(m)$ is steep, i.e. if for any mass of a particle there are many small projectiles which are able to disrupt it, then much will be lost from any given size interval. If the distribution is flat there is a surplus of large targets to produce fragments, with the result that the number of smaller particles is increased. Balance between production and destruction – excluding the largest particles, which cannot be replaced – is achieved for $n(m) \sim m^{-11/6}$, a mass distribution which was observed for the smaller asteroids [5.36]. A particle population with an exponent in the mass distribution of $\alpha > 11/6$ will be found to decay under the influence of catastrophic collisions alone; an extra source of particles is required to maintain such a steep distribution. This is, as we will see, the case for meteoroids of mass $m \geq 10^{-8}$ kg. For a population index $\alpha < 11/6$ we have to expect a build up of particles with small masses, always provided that the basalt breakup is valid for meteoroids. This second case may be applicable to the smaller interplanetary meteoroids.

For quantitative calculations of collisional effects, the size of the biggest fragment and the size distribution of fragments have to be known. Laboratory experiments [5.57, 58] give

$$M_{\text{biggest}} \approx 2M_{\text{target}}/v^2 \tag{5.19}$$

for projectiles with masses above the catastrophic fragmentation limit [5.38] and an exponent η for the power law of fragment mass distribution of

$$\eta \approx -1.8 \,. \tag{5.20}$$

The majority of created fragments therefore are quite small, but the mass remains concentrated in the larger ones.

Figure 5.17 shows a photograph of the remainders of such an artificial collision, typically performed with targets of 0.01–1 kg, projectiles of 10^{-5} to 10^{-4} kg and impact speeds of several km s^{-1}. The extrapolation to interplanetary conditions refers to size range, velocities, and, particularly, material. Quantitative treatments of collisions therefore should be taken with some caution.

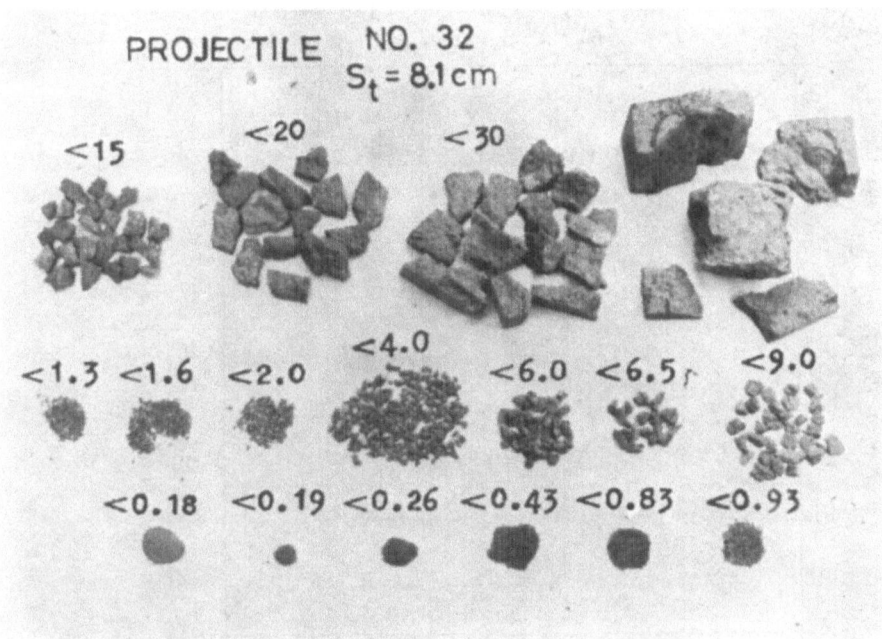

PROJECTILE NO. 32
$S_t = 8.1$ cm

Fig. 5.17. Fragments recovered from the impact of a 0.37 g polycarbonate projectile on a 1.4 kg basalt target (cube of 8.1 cm), sorted according to size [5.57]. Sizes are given in mm

5.5 Mass and Size Distribution

Specific information on the size distribution of micrometeoroids originates from various sources: microcraters on lunar samples, *in situ* detectors, and meteor observations. Among these, lunar microcrater counts are unrivalled in statistical accuracy and mass range covered. To facilitate physical discussion, and to allow a comparison with other methods, the results of crater counts are usually presented as fluxes. To do this transformation one either has to know the time interval during which the respective rocks were exposed to interplanetary dust impacts, or one has to calibrate the relative count number at a specific point. Among the two flux curves shown in Fig. 5.18, the one derived by Fechtig et al. [5.52] from several lunar samples uses the solar flare track method to determine the exposure ages; the other one by Grün et al. [5.72], which is derived from a different set of lunar microcrater counts, uses *in situ* satellite measurements for the absolute calibration of flux values. The flux curves shown are cumulative fluxes, which represent the flux [m^{-2}s^{-1}] of particles greater than a given mass: $F(m) = \int_m^\infty f(m')dm'$. This presentation has been selected because it allows a direct comparison with penetration detectors, like the well established measurement on the *Pegasus* satellite [5.141]. For our discussion we prefer the higher fluxes of the later results because of their absolute calibration by *in situ* measurements.

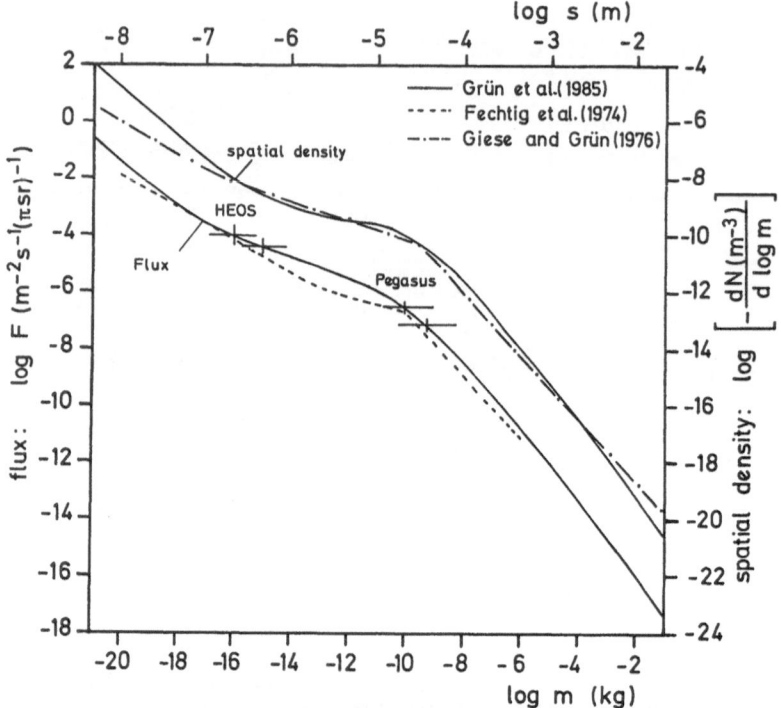

Fig. 5.18. Flux and size distribution of interplanetary dust. *Lower curves:* Cumulative particle flux on a spinning flat plate at 1 AU in the ecliptic. *Upper curves:* Size distribution and spatial density of interplanetary dust. The unit is particles per m^3 per decade in mass (factor 2.15 in radius)

But we included the flux curve of Fechtig et al. since it was a major breakthrough in its time, and since it has been used for a number of applications which we refer to in the following sections. Information on the larger masses ($> 10^{-8}$ kg) comes from meteor observations [5.176]. Thus the size range fully covers the particles responsible for the zodiacal light.

5.5.1 Particle Populations

The flux curves show two remarkable features. First, there is a knee in the curves at masses of 10^{-10} kg (radii of about 30 μm) to 10^{-8} kg. This corresponds to that part of the size distribution, where both the mass and the cross section per logarithmic mass (or size) interval are at maximum (Fig. 5.19). Therefore it identifies the typical size of an interplanetary dust grain. We will see later that the maxima occur in that part of the mass distribution where the transition occurs from collision-dominated lifetimes of the bigger particles to the Poynting–Robertson-dominated lifetimes of the smaller ones, and where the lifetime against these combined effects is at maximum. Therefore increased loss of small particles due to the Poynting–Robertson effect may play a role in creating the bend in the flux curve. Another possible cause for the decreasing slope of the mass

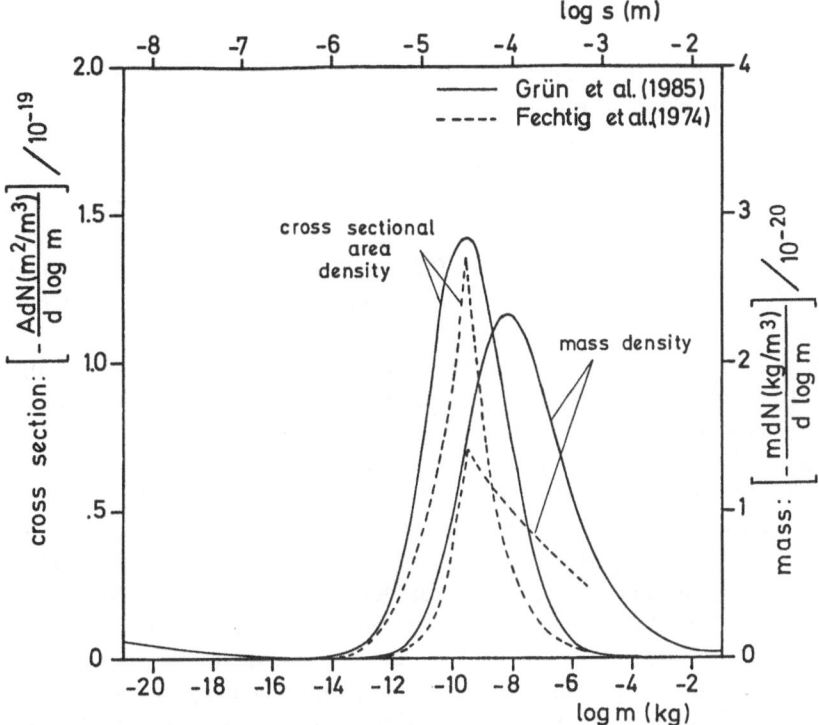

Fig. 5.19. Distribution of cross section and mass density with particle size. The results refer to 1 AU in the ecliptic and are shown for both flux curves of Fig. 5.18

distribution at smaller masses could be the limitation of input of material from comets on bound orbits because of the increased action of radiation pressure on smaller particles (see Figs. 5.13,14). At present this is no more than a conjecture. But the changing slope is an important feature of the flux curve and in future certainly deserves more effort to explain and understand it.

The second feature of the mass distribution which requires explanation is the steepening of the flux curve for masses less than 10^{-16} kg (radii $\sim 0.2\ \mu$m). This upturn shows the existence of another component of interplanetary dust: it is thought to be due to β meteoroids, collisional debris driven by radiation pressure out of the solar system on hyperbolic orbits [5.180]. They were detected with the penetration impact detector on space probes *Pioneer 8* and *9* [5.11] as particles coming from the general solar direction, after – ironically – first having been discussed away as spurious measurements due to solar-induced disturbances [5.10]. Further proof of their existence came from the impact ionization detectors on the *Helios* space probes, which showed an excess of small particles from the solar direction. One would like to know more about them, but there is no doubt that they exist.

5.5.2 Spatial Density

The measured flux is related to the spatial density of the particles by their velocity relative to the detector. The translation of fluxes to spatial densities (Fig. 5.18) was done with an average impact velocity of 20 km s; the transformation from mass to radius with a specific density of $2.5\,\mathrm{g\,cm}^{-3}$. This may give a slight underestimate of the radii at the high mass end of the curves. To characterize the size distributions we note that the differential curve labeled "Giese and Grün" varies with s as $\sim s^{-2.7}$ for radii $< 0.2\,\mu\mathrm{m}$, $\sim s^{-2.0}$ up to radii of $30\,\mu\mathrm{m}$, and $\sim s^{-4.3}$ for larger particles, where the regime around the break at $30\,\mu\mathrm{m}$ is optically most important. It is interesting to see that early attempts to determine the size distribution by interpreting the Fraunhofer corona as diffracted light [5.51, 91] led to a not unrealistic exponent of -2.0. As a useful consequence of the size distributions, the distributions of cross-sectional area and mass are shown in Fig. 5.19. The resulting values for total cross-sectional area and total mass density at 1 AU then are $5 \times 10^{-19}\,\mathrm{m^2/m^3}$ and $1 \times 10^{-19}\,\mathrm{kg/m^3}$ for the Grün et al. [5.73] flux curve and about a factor of two less for the curve adopted by Fechtig et al. [5.52]. Hanner [5.76] showed that in the latter case a geometric albedo of $A = 0.24$ for the interplanetary dust particles would be necessary to match the observed zodiacal light brightness. The spatial densities derived by Grün et al., on the other hand, led to $A = 0.13$. This is close to the values suggested by studies of. collected micrometeoroids ($A < 0.1$), by analogy to carbonaceous meteorites ($A \approx 0.10$), or by infrared emission of zodiacal light ($A \approx 0.07$–0.09 [5.47, 81, 82]), and gives indirect support for the correct normalization of their flux curve.

Of course, quantitative knowledge of the spatial density of dust at present cannot be more accurate than the calibration of the *Pegasus* experiment, to which the new flux curve is fitted. This leaves an uncertainty of about a factor of two. We have to wait and see whether and how missions like the Long-Duration Exposure Facility (*LDEF*) will improve this situation [5.27].

5.5.3 β Meteoroids and the Discussion About Small Particles in Interplanetary Space

The uncertainty is larger in the flux of β meteoroids. This results from the difficulty of the measurement. But this also fits into the history of the search for small particles in the zodiacal light, which followed many unexpected turns.

First, the very high degree of polarization initially found in the zodiacal light had to be explained by electron scattering, requiring as many as 300 electrons/cm^3 at 1 AU. After 1960, it became clear that the early observations were in error and that the polarization of zodiacal light was lower; also the first space probes measured the electron density to be only 5–10 cm^{-3}. Then it was realized that scattering by spherical particles alone was sufficient to explain brightness and polarization of zodiacal light [5.59]. For a while, submicron dielectric particles appeared to be a good candidate for interplanetary dust. However, Giese et al.

[5.62] demonstrated that this would lead to a color of zodiacal light bluer than the solar spectrum, while the observations showed the opposite trend. Indeed, the similarity between zodiacal light and solar spectra from $0.2\,\mu m$ to $2.4\,\mu m$ (Fig. 5.27) is independent evidence that the scattering particles must be larger than visible wavelengths. Otherwise the reduced scattering efficiency for wavelengths larger than the particle radius would lower the spectrum at longer wavelengths.

Maybe submicron particle models only became fashionable because, around the same time, the first space experiments indicated a very high concentration of such small particles around the earth. Soon after, the measurement techniques of those early experiments proved to be unreliable, and Shapiro et al. [5.155] presented theoretical arguments against such concentrations of small particles. Once more, numerous small particles in interplanetary space were postulated when Lillie [5.122] derived an immense upturn of the spectrum of zodiacal light at wavelengths shorter than 200 nm. But subsequent experiments with more careful subtraction of airglow and galactic components of UV radiation showed that the spectrum of zodiacal light tends to be even redder than that of the sun down to 180 nm, and gave restrictive upper limits for still shorter wavelengths (see Fig. 5.27). The question appeared settled when the lunar crater counts discussed in the previous sections led to a fairly definite size distribution with comparatively few submicron particles.

However, the discussion about small particles was revived again. The abundance of very small microcraters on selected lunar samples [5.134] led Le Sergeant d'Hendecourt and Lamy [5.119] to postulate a very high flux of submicron-sized meteoroids which would contribute as much as one third to one half of the zodiacal light [5.105]. The Rayleigh scattering of these small particles would have been difficult to reconcile with the reddish color of zodiacal light (Figs. 5.8b and 5.27). The difficulty was resolved when Zook et al [5.183] showed that most of the very small lunar microcraters are probably due to secondary ejecta produced from nearby larger impacts. However, the flux of β meteoroids is still of crucial importance for the discussion of mass balance and collisional evolution of the interplanetary dust cloud. Since microcraters on lunar material are not well suited for studying this question, future *in situ* experiments should be designed specifically to answer it.

5.5.4 Mass Influx

As in Fig. 5.18, the determination of meteoroid fluxes with exposure ages determined by the solar flare track method generally leads to smaller values than from *in situ* experiments [5.79]. Unless one of the involved calibrations is in error, this means that the meteoroid flux may have been less on the average over the last 10^4 to 10^5 years than at present. Such a discussion is only possible because the size and flux distribution at 1 AU today is one of the best known quantities of interplanetary dust. It allows the calculation of a daily dust influx on the earth of 40 tons. This is close to earlier determinations [5.90, 141, 176, 182].

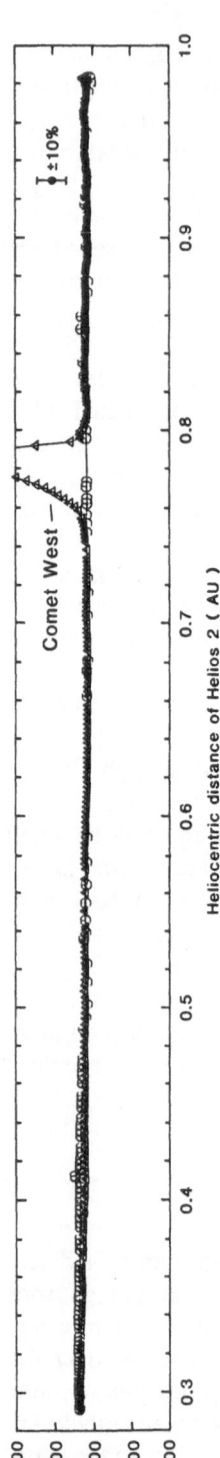

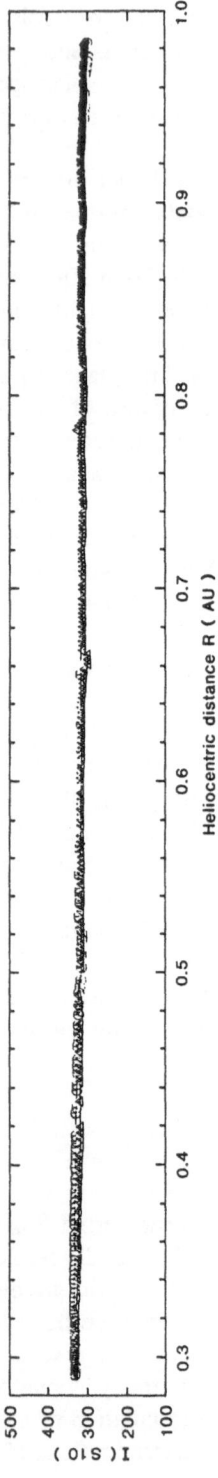

Fig. 5.20. Repeatability of observed zodiacal light brightness over one complete orbit of *Helios* 2. The data sets (taken at $\beta = 31°$, $\epsilon = 34°$) have been corrected to 1 AU with the radial gradient and plane of symmetry given in the text. Triangles refer to the inbound half of the orbit ellipse; circles (mostly hidden by the triangles) to the outbound part. Azimuthal variations would show as splitting of the two halves of the orbit

5.6 Spatial Distribution

5.6.1 Symmetries

Azimuthal symmetry of the interplanetary dust distribution, i.e. independence of heliocentric ecliptic longitude, or rotational symmetry around an axis through the sun perpendicular to the ecliptic, is assumed when interpreting zodiacal light data. The main argument in favor of this symmetry are appealing simplicity and smooth repeatability of space observations of zodiacal light from satellite *D2B* and particulary from space probes *Helios 1* and *2* along the orbit (Fig. 5.20). Anyway, symmetry is expected to be better for micron sized than for millimeter-sized particles because of the much stronger action of the dissipating forces, which tend to smear out local enhancements and reduce the importance of gravitational resonances. The existence of seasonal dependence of meteor flux by ±10–20% therefore need not contradict an azimuthal symmetry of zodiacal light. Stohl [5.165] suggested that a large part of this seasonal dependence of meteors may be due to one very wide meteor stream.

The plane of symmetry of interplanetary dust is expected to delineate the plane of symmetry of the dominant perturbing forces (see Table 5.3). For gravitational perturbation it would be expected to coincide with the invariable plane of the solar system determined mainly by the orbital angular momentum of the large planets, $\Omega = 108°$, $i = 1.6°$ [5.19], and for disturbances by interplanetary magnetic field and solar wind with the solar equator, $\Omega = 70°$, $i = 8.3°$. Determinations so far rest on remote sensing zodiacal light observations. The situation is most clear inside 1 AU, where the *Helios* space probes found $i = 3.0° \pm 0.3°$, $\Omega = 87° \pm 5°$. Being based on a method of vanishing differences, balancing brightnesses east and west of the sun, this result should be reliable (Fig. 5.21). Essentially the same result has been found by Misconi [5.130] from earthbound observations at small elongations.

Table 5.3. Inclinations and nodes of various planes (epoch 1990)

	Inclination to the ecliptic (i)	Ecliptic longitude of the ascending node (Ω)
Solar equator	7.3°	70°
Invariable plane	1.6°	107°
Jupiter's orbit	1.3°	100°
Mars' orbit	1.8°	49°
Venus' orbit	3.4°	77°

For the region outside 1 AU the situation is more controversial. While *Helios* data should still be applicable out to about 1.5 AU, Dumont and Levasseur-Regourd [5.44] found a value of $i = 1.6°$, $\Omega = 107°$ from the position of maximum brightness, supported by *D2B* measurements of seasonal variation of zodiacal light brightness at the north ecliptic pole, which also was interpreted by a symmetry to the invariable plane. This has been taken as an indication that the

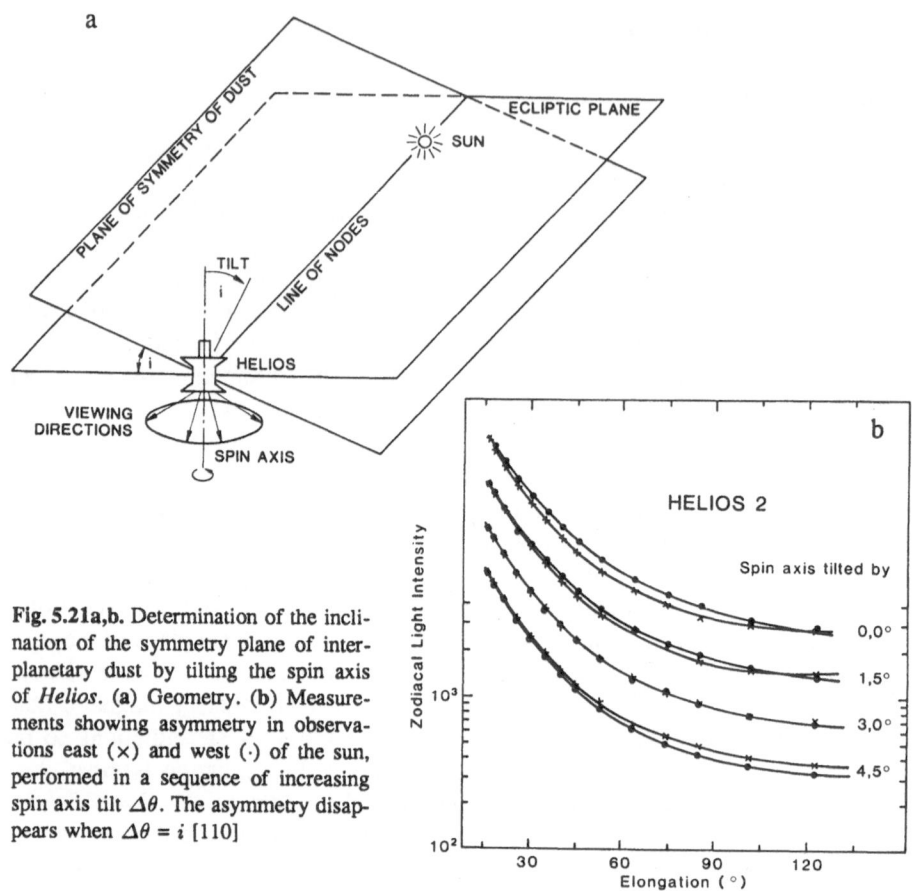

a

PLANE OF SYMMETRY OF DUST

ECLIPTIC PLANE

SUN

TILT
i

LINE OF NODES

HELIOS

VIEWING
DIRECTIONS

SPIN AXIS

b

HELIOS 2

Spin axis tilted by

Zodiacal Light Intensity

10^3

10^2

0,0°

1,5°

3,0°

4,5°

30   60   90   120
Elongation (°)

Fig. 5.21a,b. Determination of the inclination of the symmetry plane of interplanetary dust by tilting the spin axis of *Helios*. (a) Geometry. (b) Measurements showing asymmetry in observations east (×) and west (·) of the sun, performed in a sequence of increasing spin axis tilt $\Delta\theta$. The asymmetry disappears when $\Delta\theta = i$ [110]

"plane" of symmetry is warped, following the influence of the nearest planets [5.130]. *IRAS* observations so far give no unique answer. The seasonal variations of pole brightness at 1 AU give $i = 2°–2.5°$, $\Omega = 70 \pm 5°$, while from the position of brightness maxima the value for outside the earth's orbit is $i = 1.3°$, $\Omega = 45°$ [5.34, 82]. Critical rediscussions of the *IRAS* data [5.46, 151] yielded values of $i = 1.5°–1.7°$, $\Omega = 80–90°$. The plane of symmetry therefore may be more or may be less warped; it does not in general coincide with the invariable plane of the solar system.

This seemingly exaggerated discussion about small differences is necessary if one wants to identify the physical processes involved. On the other hand, if one compares the average plane of the asteroids, $i = 0.7°$, $\Omega = 88°$, or short-period comets, $i = 2.7°$, $\Omega = 80°$ [5.20], there is the distinct possibility that the symmetry of dust to a significant extent derives from the distribution of its sources. So far no serious attempt has been made to understand the symmetry plane or planes of the interplanetary dust cloud.

240

5.6.2 Out-of-ecliptic Distribution

For the following discussion, the difference between plane of symmetry and plane of the ecliptic is small enough to be neglected because the half width of the distribution perpendicular to the ecliptic is much larger. The out-of-ecliptic distribution of interplanetary dust is best inferred from the brightness distribution of zodiacal light. The zodiacal light brightness is highest in the ecliptic, decreasing towards the ecliptic poles. For a given angular distance from the sun, a minimum of zodiacal light brightness is reached on the great circle passing through the sun and the ecliptic poles. This line is called the "solar" or "helioecliptic" meridian. On circles around the sun, the brightness decrease from the ecliptic to this meridian is by a factor of 3 or 4 for elongations between 15° and 90°. Reliable space observations exist for specific cuts through the brightness distribution [5.111, 116]. However, the summary of Dumont's careful ground-based observations is mostly used in this analysis because they cover the whole hemisphere [5.49, 121]. Various functional forms have been used to describe the decrease of dust concentration outside the plane of symmetry, in order to fit the observed decrease of zodiacal light intensity towards the ecliptic poles. Usually, for simplicity, height dependence and radial dependence are separated and the distribution of dust is assumed to be of the form

$$n(r, z) = r^{-\nu} f(\beta_\odot) , \qquad\qquad (5.21)$$

where z is the height above the ecliptic and $\beta_\odot = \arcsin(z/r)$ is the heliocentric ecliptic latitude. With such a formula the brightness distribution of the zodiacal light is predicted to be independent of heliocentric distance, as found in good approximation in the inner solar system by the *Helios* space probes [5.115]. The defunct second spacecraft of the International Solar Polar Mission (the remainder is now called the *Ulysses* mission), which should have carried a zodiacal light photometer, would have allowed the three-dimensional distribution of the zodiacal cloud to be probed [5.61]. With this more direct method missing, the first thing now being investigated is whether the simple form of (5.21) is suitable for describing the zodiacal light [5.64, 65]. In the meantime, simple representations, such as ellipsoids with axial ratio 1 : 7 [5.43] or an exponential decrease $f(\beta_\odot) = \exp(-2.1|z/r|)$ [5.115], may be taken as an approximation as well as more custom tailored dependences [5.65, 123]. Typical isodensity contours are shown in Fig. 5.22. It is doubtful whether more detailed fits and models should be made at present, given the uncertainties in the heliocentric dependence of important physical parameters such as albedo or particle size.

In any case the spatial density of interplanetary dust 0.25 AU above the earth's orbit is still half of the value in the ecliptic [5.65]. The concentration to the plane of symmetry is not particularly strong. To evaluate the implications of this finding it is helpful to consider the out-of-ecliptic distribution together with the corresponding inclination distribution of particle orbits. This will be done in Sect. 5.7.2.

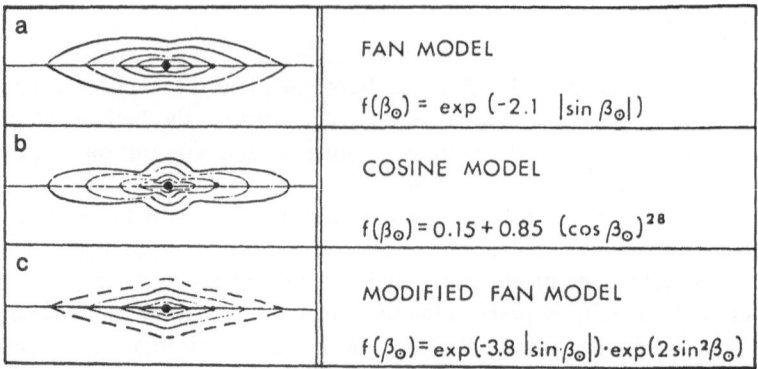

a		FAN MODEL
		$f(\beta_\odot) = \exp\left(-2.1 \ |\sin \beta_\odot|\right)$
b		COSINE MODEL
		$f(\beta_\odot) = 0.15 + 0.85 \ (\cos \beta_\odot)^{28}$
c		MODIFIED FAN MODEL
		$f(\beta_\odot) = \exp(-3.8 \ |\sin \beta_\odot|) \cdot \exp(2 \sin^2 \beta_\odot)$

Fig. 5.22. Isodensity contours for different models of out-of-ecliptic distribution. Density steps are in factors of two [5.65]

5.6.3 Radial Dependence

In contrast to other parameters, fairly definite predictions exist for this quantity. The natural distribution against which to compare the measurements is the power law $n(r) \sim r^{-1}$, which would result as equilibrium under the action of the Poynting–Robertson effect if the particles were injected outside the region considered. Strictly speaking, this is true only for circular orbits or if the source is at such a large heliocentric distance that the particle orbits during the inward drift have become circularized to a sufficient degree. The argument goes as follows. In equilibrium the number of dust particles $\dot{S}$ created at the source per second is equal to the flux of particles through any sphere inside the source. Particularly at heliocentric distance r we have

$$\dot{S} = 4\pi r^2 n(r) V_{\text{drift}} , \tag{5.22}$$

where the radial drift velocity due to the Poynting–Robertson effect is proportional to $1/r$ (5.11). Hence $n(r) \sim r^{-1}$. This is the minimum slope. A smaller increase towards the sun could be achieved only by destroying particles. On the other hand, if the source region is extended over the planetary system, the total production $\dot{S}$ outside r gets the larger the closer we come to the sun, and $n(r)$ gets correspondingly steeper (see Sect. 5.9).

Observationally, the radial dependence of interplanetary dust can best be studied by zodiacal light observations from deep space probes. In the inner solar system the *Helios* solar probes found the intensity to increase with decreasing heliocentric distance as

$$I(\varepsilon, R) \sim \text{const.}(\varepsilon) R^{-2.3 \pm 0.05} \tag{5.23}$$

(Fig. 5.23). According to (5.3) the brightness increase is steeper by R^{-1} than the spatial distribution of dust, and therefore this translates into $n(r) \sim r^{-1.3}$, if there is no change in average scattering function of particles with heliocentric

242

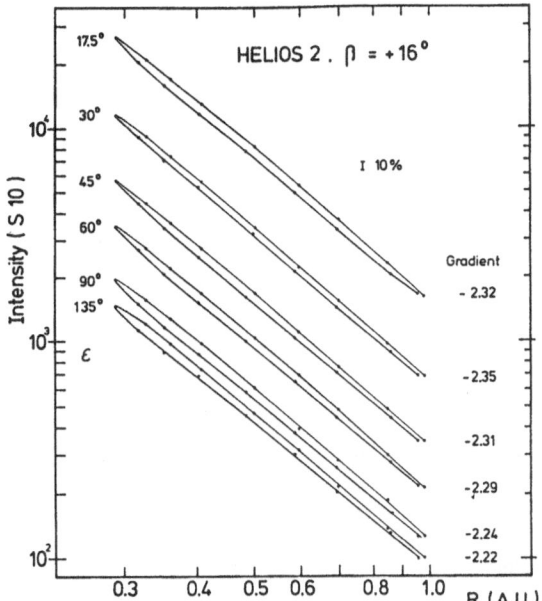

Fig. 5.23. Increase of zodiacal light intensity with heliocentric distance as observed by *Helios* [5.113]. For all elongations it follows the same power law $R^{-\nu}$, with $\nu = 2.3$. The difference between inbound (*upper leg*) and outbound (*lower leg*) parts of the orbit is due to the inclination of the plane of symmetry with respect to the ecliptic

distance. This relation is expected to hold between 0.1 AU and 1.5 AU, the region which significantly contributes to the brightness measured by *Helios*. A similar analysis from the outgoing space probe *Pioneer 10* indicates a slightly steeper decline from 1 AU to 3 AU, approximately $n(r) \sim r^{-1.5}$ [5.30, 169], with the final analysis yet to come. With the assumption of a slowly varying scattering function in (5.2), earthbound observations of brightness increase towards the sun give nearly the above result for the inner solar system. The exact value for the exponent of the power law of 1.2 was chosen by Dumont and Sanchez [5.48] for esthetic reasons. Similar results were obtained by Dumont and Levasseur-Regourd [5.45] from a new inversion technique (see Sect. 5.8).

The particle impact experiments add complexity to this picture. Inside 1 AU the impact ionization detectors on board *Helios 1* and *2*, which face the ecliptic, found the particle flux to increase with decreasing heliocentric distance until 0.5 AU, followed by a decrease, while impact ionization detectors looking at high ecliptic latitudes on the very same spacecraft found a power-law increase of flux, compatible with zodiacal light observations. This emphasizes the difficulty of deducing spatial distributions on the basis of a few dozen to a few hundred individually registered particles.

Outside 1 AU the situation gets worse. We neglect *Pioneer 8/9* measurements between 0.9 and 1.1 AU, because the short range in heliocentric distance makes it difficult to assess the results in terms of a radial gradient. The results of the penetration detector on *Pioneer 10/11* are considered reliable, and they indicate that the concentration of 5 μm (10^{-12} kg) particles remains roughly constant between 1.5 and 18 AU [5.92, Fig. 5.24]. Zodiacal light observations do not contribute much to this part of the discussion, because zodiacal light brightness becomes

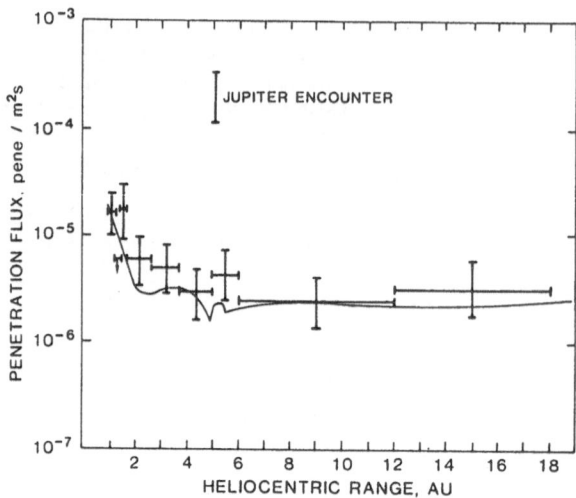

Fig. 5.24. Penetration flux of particles with mass $> 8 \times 10^{-13}$ kg, measured by *Pioneer 10* as function of heliocentric distance [5.92]. The solid line is a prediction for constant spatial density and particles in randomly inclined high-eccentricity orbits

negligible beyond the asteroid belt (2.6 AU [5.78]). The conclusion reached from the observations of radio meteors is again different. It is a limitation of these observations that they only refer to particles crossing the earth's orbit. Without the unobservable orbits the information on the *a–e* distribution is incomplete, the missing parts of the *a–e* diagram therefore have to be filled in by extrapolation. When this was done by Southworth and Sekanina [5.164] they found that the space density of mm-sized meteoroids peaks in the asteroid belt, decreasing on both sides, but increasing again inside 0.7 AU. Cook [5.30] strongly argues that this maximum outside 1 AU cannot be due to extrapolation to unobservable orbits, and therefore must be real.

We then have the uncomfortable situation that three measurement techniques, representative of different size regimes with little overlap, give three different answers. If they all are right, the size distribution at 4 AU must be very different from the one described in Sect. 5.5. Of course, in such a comparison, we have to remember that zodiacal light experiments are basically measuring scattering cross sections, related to particle numbers by size and albedo. It was therefore possible to force agreement by a suitable choice of a very low albedo varying with heliocentric distance [5.30]. Measurements of infrared thermal emission, which indicate a normal average particle albedo of $A \approx 0.09$ now make this type of fitting unattractive. We leave a resolution of the discrepancies at larger heliocentric distances to the dust detectors on the upcoming *Galileo* and *Ulysses* space probes. But we repeat that a constant dust concentration independent of heliocentric distance cannot be in equilibrium under the action of the Poynting–Robertson effect. It shows rather that a destructive mechanism such as collisions or evaporation of ice grains [5.181] must be dominating. For the comparatively small sizes around 8 μm radius sampled by the *Pioneer 10/11* penetration experiment this is a somewhat puzzling conclusion, not in line with the picture we are trying to set up. We come back to this point in Sect. 5.9.

But even close to 1 AU there is a reason to discuss the spatial distribution. Strictly speaking, the density law $\sim r^{-1.3}$ refers to the scattering cross section per unit volume, $n\sigma$. Which part of this increase is due to the spatial distribution and which part, if any, to scattering properties remains to be discussed. Several authors prefer to adopt for the spatial distribution the theoretically basic $1/r$ dependence. This is equivalent to requiring an increase of particle albedo towards the sun [5.123]. This would help to understand the fact that *Helios 1* and 2 found the polarization of zodiacal light to decrease by 30% when going from 1 AU to perihelion at 0.3 AU. Since decrease in polarization and increase in reflectivity are correlated in lunar samples [5.41], the same effect could be at work in interplanetary space. Inversion of *IRAS* infrared measurements [5.47, 86] gives such a gradient in albedo, although its numerical value is somewhat high. At first glance, this trend is a little surprising, because for lunar material irradiation with protons resulted in a darker surface, and the same might happen to interplanetary dust particles exposed to the solar wind. However, Mukai et al. [5.138] showed that sublimation and sputtering should make the interplanetary grains brighter, while they drift to smaller heliocentric distances under the action of the Poynting–Robertson effect. Changing albedo seems a plausible effect; the question remains how large this effect is.

On the other hand, the brightness distribution and color of zodiacal light were found by *Helios 1* and 2 not to vary with heliocentric distance. This implies the corresponding constancy of the particle scattering function, a fact well demonstrated by Lamy and Perrin [5.106], while their derived volume scattering functions are unfortunately not consistent with the observed increase of zodiacal light brightness at decreasing heliocentric distance. Actually, the constancy of scattering functions makes it probable that a large part of the extra brightness increase is a spatial density effect. In addition, we will see in Sect. 5.9 that collisions in the interplanetary dust cloud provide a significant input of dust all over the solar system, strongly peaked at small heliocentric distances. Quite naturally, then, a steeper equilibrium distribution than $1/r$ should result. For this reason, we keep in the following as a working model of spatial density the power law $n(r) \sim r^{-1.3}$ found from the *Helios* zodiacal light experiment. In any case, if the size distribution of interplanetary dust would strongly change with heliocentric distance, the concept of spatial distribution expressed by particle numbers would no longer provide an adequate description and would have to be replaced by a more general quantity such as, e.g., mass per unit volume.

5.7 Orbit Distribution

Information on meteoroid orbits is complex, the orbit distributions depending not only on size, but also on particle density. It is obtained from three independent sources: meteor observations, zodiacal light observations, and *in situ* measurements with satellites and deep space probes, among these mostly from *Pioneer 8/9* near the earth's orbit and from *Helios* between 0.3 AU and 1 AU.

5.7.1 Meteor Orbits

Meteors can be divided into a background of so-called sporadic meteors and superimposed meteor streams. Having in mind the relation between dust and comets we note that the sporadic meteors appear associated with the interplanetary dust cloud, while streams are related more closely to comets.

Sporadic Meteors constitute about 80% to 90% of all observed meteors. They fall into two major groups: the first has short-period orbits with low inclinations and high eccentricities similar to the earth-crossing Apollo asteroids and to short-period comets, while the orbits of the second with random inclinations resemble those of long-period comets. A breakup of these groups by observational methods, i.e. particle sizes, is shown in Table 5.4 [5.26, 90]. It can be seen that the percentage of low-inclination, short-period orbits does not change significantly with the mass of meteor particles. However if this group is divided further according to particle densities (as measured by the beginning height of the ionized trail [5.26]) then one finds that among the bright meteors the portion of asteroidal particles (high density, low beginning height) compared to cometary particles (low density, high beginning height) is higher than for faint meteors. As mentioned above, Stohl [5.165] suggests that the short-period cometary meteors may form a very wide stream (in time and radiant), which has some relation to the orbit of comet Encke. Radio meteors show no discrete levels of beginning heights, and therefore a separation into asteroidal and cometary particles is not justified.

In the second group the semimajor axes are much larger and consequently the eccentricities are close to 1.0. This is particularly the case for photographic meteors. For radio meteors only the random distribution of inclinations indicates a possible relation to long-period comets. Semimajor axes and eccentricities, however, are much reduced, perhaps because the orbits already evolved under the action of the Poynting–Robertson effect.

Meteor Streams constitute only a small fraction of the total meteor activity but they are easily recognized. First, their occurrence rate is strongly peaked, within days and even within hours, and then may exceed the sporadic meteor rate by large factors; second, they are annually recurring; and third within a stream all meteors have the same apparent radiant within about 1°. These observations prove that meteor stream particles move on almost identical orbits around the

Table 5.4. Classification of observed meteor orbits (from [5.26])

observational method	radar		super Schmidt cameras		small-camera meteors		Prairie Network fireballs	
typical mass range	10^{-8}–10^{-6} kg		10^{-6}–10^{-2} kg		10^{-4}–1 kg		10^{-1}–10^3 kg	
group	% obs.	characteristic orbit a e i	% obs.	characteristic orbit a e i	% obs.	characteristic orbit a e i	% obs.	characteristic orbit a e i
"asteroidal"	70	< 3 < 0.7 15°	< 1	2.4 0.64 15°	5	2.5 0.64 10°	32	2.4 0.68 6°
"short-period cometary"			69	2.3 0.66 2°	60	2.5 0.71 5°	46	2.3 0.65 5°
"long-period cometary"	30	∞ > 0.7 random	31	∞ 0.99 random	35	~ 20 0.97 random	22	~ 5 0.82 45°

sun but are more or less smeared out along the mean orbit. It should be noted that meteor streams consisting of smaller particles (radio meteors) show less contrast in flux to the sporadic background than those consisting of bigger particles (visual meteors). The orbital elements of streams vary over the same wide range as is the case for sporadic meteors [5.29].

Many meteor streams are related to comets, judged from the similarity of the orbits. This is strong evidence for a cometary origin for those particles. A famous example is the Leonid meteor shower associated with comet Tempel–Tuttle, which has an orbital period of about 33 years. The Leonid meteor stream also displays a 33-year periodicity of major activity (the last one was in 1966). This proves the association and indicates that this meteor stream is young and the ejected cometary material is not yet evenly smeared out along the orbit.

Because of perturbations by radiation pressure and Poynting–Robertson drag, the orbits of the parent comet and the associated meteor stream are not the same, even initially, and they will separate more in time. Within a meteor stream itself often a mass segregation is found which is the result of the radiation pressure at formation and radiation drag thereafter, as well as of collisions [5.37]. In addition, planetary perturbations and collisions with sporadic meteoroids eject particles out of the stream into the general meteor background. This accomplishes the transformation of cometary material to sporadic meteors, an important ingredient of our following discussion.

The recent identification of an earth-crossing asteroid, Phaeton (1983TB), as the probable parent body of the major Geminid meteor stream [5.54] reminds us that there may be no clear-cut distinction between comets and some of the asteroids.

5.7.2 Orbits Determined from Zodiacal Light Observations

Radial and latitudal dependence of the spatial meteoroid density derived from zodiacal light observations impose restrictions on the distribution of orbital elements $\tilde{f}(a, e, i)$ for these particles. Unfortunately, it is common to designate the spatial out-of-ecliptic distribution (5.21) also by the symbol "f"; these two functions should be clearly distinguished. The separation in spatial coordinates r and $\beta_\odot$ introduced in (5.21), can be shown to be equivalent to a separation of the orbit distribution,

$$\tilde{f}(a, e, i) = \tilde{f}_1(a, e)\tilde{f}_2(i) . \tag{5.24}$$

The distribution in semimajor axis is closely related to the questions of dust supply and therefore will be treated in Sect. 5.9. Here we focus on the distribution in inclination which is related to the out-of-ecliptic decrease of relative spatial density $f(\beta_\odot)$ with increasing $\beta_\odot$. The qualitative relation between the two functions is given by the inclination-dependent part of Haug's integral [5.80] [our Equation (5.30)]. $f_2(i)$ can then be derived by inversion of the resulting integral equation

$$f(\beta_\odot) =) \int_{i=\beta_\odot}^{\pi/2} \frac{\tilde{f}_2(i)di}{\sqrt{\sin^2 i - \sin^2 \beta_\odot}} . \tag{5.25}$$

This formula is derived for a random distribution of longitude of ascending node and argument of perihelion, a condition thought to be approximately fulfilled for interplanetary dust because of the remarkable azimuthal symmetry of zodiacal light.

Equation (5.25) is quite useful for discussing relations between inclination distribution of interplanetary dust particles and concentration of spatial density to the ecliptic, but it needs some comment. First, it looks different from the corresponding part of (5.30), but the expressions in the denominator can be converted one to the other. Second, it seemingly diverges for $\beta_\odot = 0$. However, one has to remember that for isotropic distribution of particle orbits $\tilde{f}_2(i) \sim \sin i$. This relation always holds for small values of i (if only $\tilde{f}_2(i)$ is smooth), thus resolving the problem of divergence. $\tilde{f}_2(i)$ describes the average orbit distribution in space. If one is interested in the deviation from isotropy, $\tilde{f}_2(i)/\sin i$ has to be considered instead. An observer in the ecliptic, because of his special position, happens to encounter interplanetary particles at a rate proportional to $\tilde{f}_2(i)/\sin i$.

Figure 5.25 shows both inclination distribution and deviation from isotropy, derived with the help of (5.25) from an out-of-ecliptic spatial density decrease $f(\beta_\odot) = \exp(-2.6|z/r|^{1.3})$. This form of $f(\beta_\odot)$ itself was a fit to a rocket photometry of the zodiacal light at elongations $\varepsilon = 15°$ to $30°$ [5.112]. Other models for $f(\beta_\odot)$ [5.65] are similar. For comparison, the curves for a simple approxima-

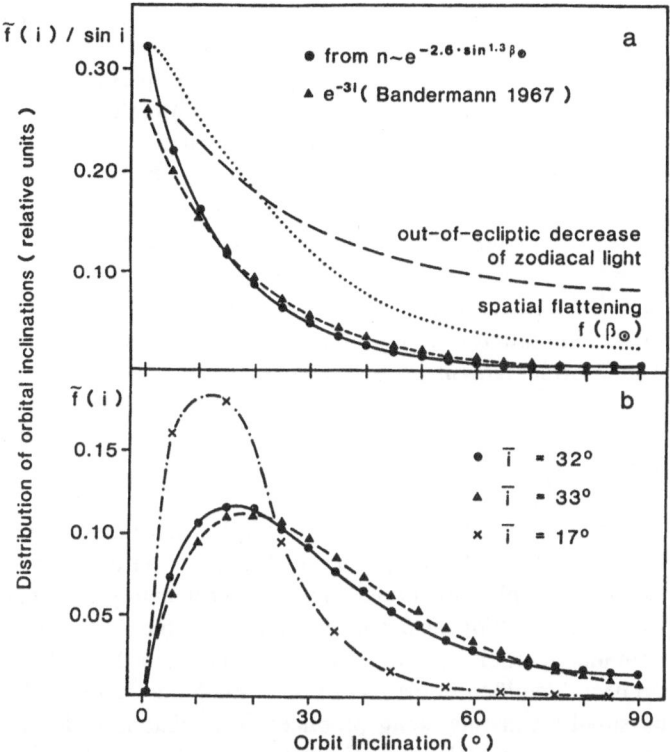

Fig. 5.25a,b. Distributions of orbital inclinations for interplanetary dust. (b) Inclination distributions $\tilde{f}(i)$ derived from zodiacal light observations [5.8, 112]. For comparison the distribution of radio meteor orbital inclinations is also included [5.164]. A random distribution would be $\sim \sin i$. (a) Deviation $\tilde{f}(i)/\sin i$ of inclination distribution from isotropy for the same two sets of zodiacal light observations. The primary observations (*dashed line* zodiacal light brightness along a circle around the sun, starting in the ecliptic) and the intermediate derived spatial distribution function $f(\beta_\odot)$ (*dotted line*) are also included for comparison. Note that $\tilde{f}(i)/\sin i$, as the least integrated quantity, has the steepest falloff rate

tion of zodiacal light brightness by Bandermann [5.8], $\tilde{f}_2(i) \sim (\sin i)\exp(-3i)$, are also shown. From Fig. 5.25 the average inclination of the particles causing the zodiacal light is $i \approx 32°$. This is far more than the values found for radio meteors (17°, [5.164]), bright photographic meteors (15°); fireballs (10°, [5.26]), short-period comets with aphelion less than 10 AU (12°, [5.150]), or asteroids (10°, [5.40]). (We discard the differing conclusions of Lumme and Bowell [5.123] because they neglect the difference between distributions in $\beta_\odot$ and inclination.) Average inclination appears to increase with decreasing size of the bodies, suggestive of an evolution from comets through meteor-forming bodies to dust. Burkhardt [5.21] and Gustafson et al. [5.74] studied how planetary perturbations change the inclination of dust orbits. Trulsen and Wikan [5.170] calculated that elastic collisions may transform orbits of high initial eccentricity into orbits of high average inclination. But their process may not be applicable to interplane-

249

tary space, where destructive collisions dominate [5.37, 120], and in addition it requires a buildup of a particular inclination distribution which does not fit the brightness distribution of zodiacal light [5.109]. Proposing long-period comets as a separate source for the high-inclination orbits faces the difficulty of keeping their ejected material on bound orbits and has the consequence that a large fraction of retrograde orbits would be introduced, which have not been observed.

In particular, measurements of Doppler shifts in the zodiacal light spectrum show no indication for particles in retrograde orbits [5.50] and have been used to set an upper limit of 5% for the scattering contribution by such particles in the ecliptic [5.96]. Doppler shift measurements in selected regions at high ecliptic latitude could sharpen this important criterion. The high average inclination of interplanetary dust particle orbits remains to be understood.

5.7.3 Orbits of Small Meteoroids

Orbital information on micrometeoroids ($m < 10^{-12}$ kg) has been obtained from *Pioneer 8/9*, *HEOS-2*, and the *Helios* spacecraft. Detectors on board these spaceprobes have observed a highly directional flux of interplanetary dust particles varying with particle sizes. *Pioneer 8* and *9* micrometeoroid experiments observed most of the smallest particles ($< 10^{-16}$ kg) to come from the solar direction. They have also been identified in the *Helios* data. Zook and Berg [5.180] interpreted them to be small fragments being produced in collisions inside the orbit of the observing spacecraft which leave the solar system because of the action of radiation pressure, and coined the term β meteoroids. Nowadays, an optical detection of β meteoroids appears feasible. Accurate Doppler shift measurements in the zodiacal light [5.50] show in the antisolar region the existence of receding material; this allows, in principle, the determination of the outward flux of β meteoroids.

Intermediate-sized particles (10^{-16} kg $< m < 10^{-13}$ kg) are observed by all experiments to arrive from the heliocentric spacecraft apex direction [5.85,73]. They move on low energy orbits (small a) with significant eccentricities ($e \approx 0.6$). These particles are believed to be either collisional products or particles originally emitted from comets on bound orbits of higher eccentricity, reduced to the present orbit by the Poynting–Robertson effect.

The biggest particles observed by *Helios* ($m > 10^{-13}$ kg), but also smaller ones of low density, have high-energy, high-eccentricity ($e \geq 0.7$) orbits [5.71]. Parts of their orbits are compatible with the criteria for hyperbolic orbits. They appear to form a separate class.

The study of dust particle orbits, which are of considerable importance for the discussion of origin and dynamics, is still in its infancy. Once the record is sufficiently complete, the by and large simple symmetry of zodiacal light should appear behind all the detailed complexity.

5.8 Compatibility of Dust Experiments with Zodiacal Light Observations

Above, we have repeatedly made use of specific zodiacal light observations to discuss certain aspects of the interplanetary dust cloud. The question still remains to be answered whether the knowledge of interplanetary dust accumulated from particle experiments is quantitatively compatible with zodiacal light observations. Features of the zodiacal light to be explained by a model of the interplanetary dust cloud are absolute value and brightness distributions of zodiacal light intensity, particularly along the ecliptic. In addition, polarization, color, spectrum, and thermal emission, have to be explained (Figs. 5.8 and 5.27). Heliocentric dependence and out-of-ecliptic distribution of interplanetary dust as well as size distribution and optical properties have to be known or assumed for this comparison.

Such a comprehensive model of the zodiacal light on the basis of known dust parameters was calculated by Röser and Staude [5.153]. They used the spatial distribution $n(r) \sim r^{-1.3}$ and the size distribution and particle concentration as derived by Giese and Grün [5.63] from the flux curve of Fechtig et al. [5.52]. The average scattering and absorbing properties of the dust particles are difficult to assess. Röser and Staude calculated them on the basis of Mie theory, i.e. for spherical particles. They used measured refractive indices of typical materials like obsidian, olivine, andesite, graphite, and magnetite in the wavelength range 0.15–100 μm to predict intensity, color, polarization, and thermal emission of zodiacal light over this whole range of wavelengths. Since they cut the size distribution at 100 μm, the small (6%) contribution of still-larger particles was neglected. They applied no fitting except changing the relative proportion of the five substances considered. This is therefore an absolute model in an acceptable sense. The assumption of spherical particles certainly is unrealistic, but on the average over a wide size distribution the approximation may be not unreasonable. The model succeeded astonishingly well (Fig. 5.26). In two places the assumption of spherical particles distorts the predictions. First, the polarization near $\varepsilon = 135°$ shows the high value typical of the haze bow, which does not appear

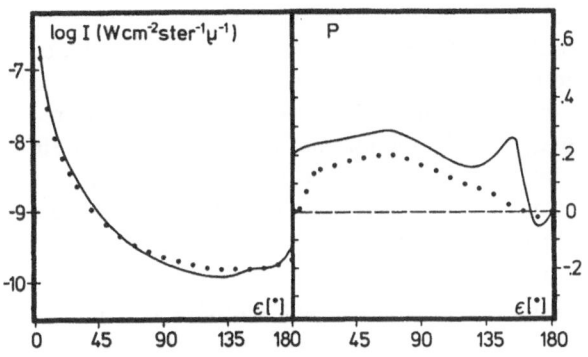

Fig. 5.26. Comparison of intensity and polarization of zodiacal light (*dotted lines*) with Röser and Staude's absolute model [5.153]. The adopted mixture is 10% obsidian, 20% Andesite, 20% Magnetite, and 50% Graphite

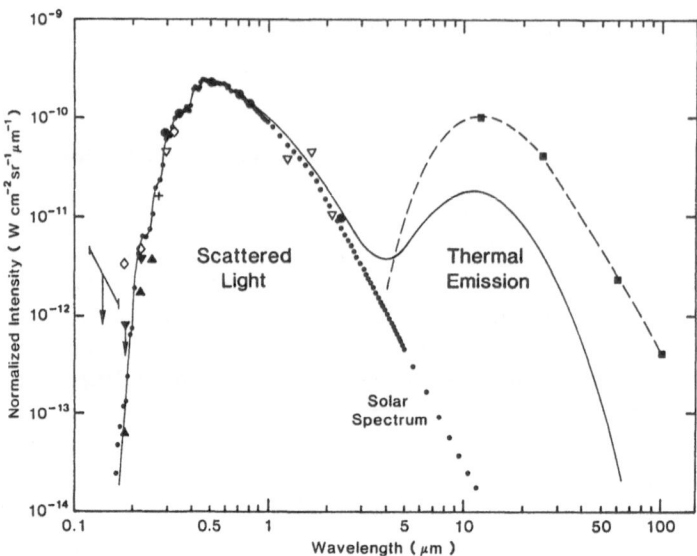

Fig. 5.27. Spectrum of the zodiacal light. The different observations have been normalized to an elongation of 90° in the ecliptic. A relative solar spectrum (*dotted line*) has been fitted to the measured values at 500 nm. The model of Röser and Staude is indicated by the thin solid line. Measurements are taken from Hauser et al. [5.81, ■], Frey et al. [5.55, ●], Nishimura [5.144, ▽], Maucherat-Joubert et al. [5.126, ◇], Pitz et al. [5.149, ▲], Morgan et al. [5.133, +], Cebula and Feldman [5.25, ▼], and the upper limit from Henry et al. [5.84]

in the zodiacal light; second, the albedo of the adopted mixture is comparatively high ($A = 0.35$, and even for graphite still $A = 0.25$). This is a consequence of the surface being smooth and therefore specular, and leads to an underestimate of infrared emission relative to scattering at visible wavelengths. One interesting detail of the model is that – contrary to earlier assumptions – in the F-corona as in zodiacal light most of the radiation is due to scattering, diffraction playing only a minor role. The model also succeeds in representing the spectrum and color of zodiacal light from 200 nm to the near infrared as observed in various space experiments (see Fig. 5.27). Dust experiments and zodiacal light experiments are compatible. The modeling of *IRAS* infrared observations [5.34] also succeeded as a fit, but did not derive from the first external inputs: it was based on a spatial distribution found from zodiacal light observations in the visual and on simplifying assumptions on size distribution. Satisfying models of infrared emission on the basis of measured size distribution and optical constants were obtained by Reach [5.151] and by Temi et al. [5.168]. However, to achieve a fuller understanding, it is strongly recommended that future modeling of zodiacal light should cover simultaneously both the visual regime and infrared thermal emission, as Röser and Staude's work basically did.

An obvious criticism of the above is that interplanetary particles as a rule are not spherical, so that the above results cannot be correct in detail, although we suppose that the main features are trustworthy. This has led to a second approach to check the compatibility with zodiacal light observations. Various techniques for inverting the brightness integral of the zodiacal light (5.1) were developed in order to solve for the volume scattering function $n(r)\sigma(r,\theta)$ and for the polarization $P_\sigma(r,\theta)$ of the light scattered by an angle θ [5.18, 42, 108, 158]. The comparison with dust experiments can then be made on these intermediate more physical quantities, and modeling of zodiacal light is avoided. In the simplest case, the volume scattering function is obtained from (5.1) by differentiating with respect to Δ and transforming to derivatives with respect to R or ε [5.42]:

$$n(R)\sigma(R, \theta = \varepsilon) = \frac{-R}{\text{const.}} \frac{\partial I}{\partial \varepsilon} \sin \varepsilon + \frac{R^2}{\text{const.}} \frac{\partial I}{\partial R} \cos \varepsilon$$

$$p_\sigma(R, \theta = \varepsilon) = I \frac{dp}{dI}\bigg|_{\text{along line of sight}} + p(R, \varepsilon)$$

(5.26)

Here, the quantities on the right are observed intensity and polarization of zodiacal light, while the quantities on the left give the deduced scattering properties of a unit volume. Heliocentric distance has been designated R to emphasize that measurement of the derivatives requires the physical presence of the observing instrument at this position.

Without measurements from space probes or assumptions on the spatial distributions of interplanetary dust it is only possible to determine the result for a scattering angle of 90° at 1 AU from the gradient of zodiacal light brightness,

$$n(1\,\text{AU})\sigma(1\,\text{AU}, \theta = 90°) = 7.8 \times 10^{-21} \text{m}^2 \text{ sr}^{-1}/\text{m}^3$$

(5.27)

[5.42]. This cross section for scattering by an angle of 90° should not be confused with the much larger total geometrical cross section given in Sect. 5.5.2. The numerical value is of diagnostic use. Cook's model [5.30] could only meet it with an incredibly low albedo of 0.007. Derivation of the full scattering function (Fig. 5.28) needs additional assumptions, e.g. a heliocentric dependence of $n\sigma(\theta)$ with power law $r^{-\nu}, \nu = 1.0, 1.2$, or 1.3, which means that the shape of the scattering function is assumed to be the same everywhere. Such scattering functions are shown by Dumont and Sanchez [5.48], Leinert et al. [5.112], Giese et al. [5.65]. Dumont and Levasseur-Regourd [5.45] used an ingenious method to avoid this assumption and to improve on the volume scattering function. This method, called "nodes of lesser uncertainty" can also be applied to infrared data, where it gave evidence of albedo variations with heliocentric distance [5.47]. A limitation of this method is that the effect of the necessary mathematical assumptions is not well known. Hong [5.87] derived scattering functions using an integral approach based on Henyey–Greenstein functions.

The above volume scattering function now has to be compared with dust properties. Weiss-Wrana [5.175] performed light scattering measurements on particles similar in size, texture, and appearance to the interplanetary dust specimens collected in the atmosphere. The average intensity and polarization of the scattering was in good agreement with the permitted range for the empirical scattering function (Fig. 5.28). This is important confirmation for the compatiblity between

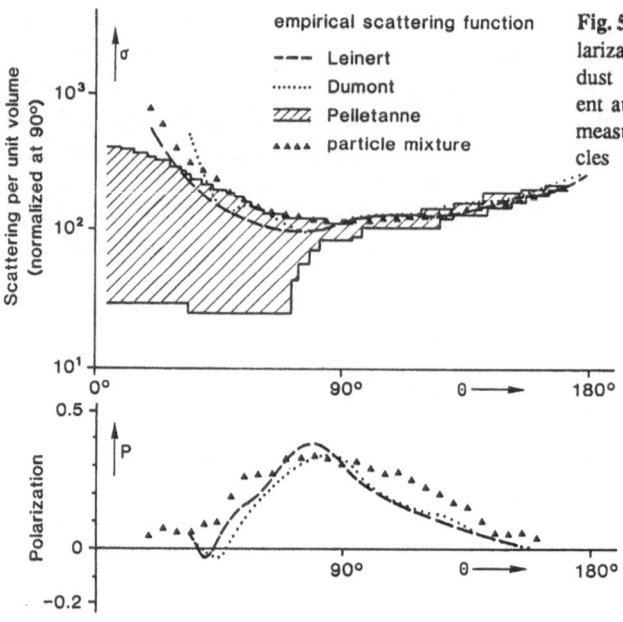

Fig. 5.28. Volume scattering and polarization functions of interplanetary dust at 1 AU, according to different authors, compared to laboratory measurements on a mixture of particles

dust experiments and zodiacal light measurements, free of arguments on spherical shape. A similar work for fragile particles composed of simulated interstellar core–mantle grains, but limited to unrepresentatively small equivalent radii of 1–2 μm, was presented by Greenberg and Gustafson [5.69]. It is common opinion that particles released from comets should have a similarly loose structure.

There is increasing activity both in theoretical investigations and laboratory experiments on light scattering from irregular bodies with rough surfaces [5.159]; see also conference book by Giese and Lamy [5.5]. These are expected to lead to a more definite interpretation of zodiacal light observations by establishing a better link between measured optical quantities and particle properties. This is important particularly for regions not accessible to space probes.

5.9 Sources and Sinks

Because of the limited lifetime of small particles in the solar system, the supply to the cloud of interplanetary dust must come from larger bodies. Two such groups are evident: asteroids and comets.

5.9.1 Asteroids as Sources of Dust

Collisions in the asteroid belt must occur. Dohnanyi [5.38] showed that the size distribution of kilometre-sized asteroids is as expected for a population of bodies in collisional equilibrium. The majority of the meteorites found on earth probably

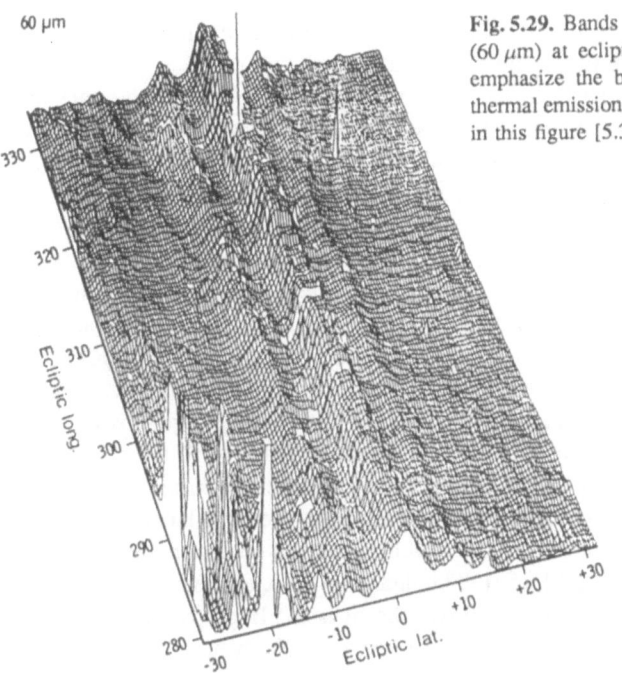

60 μm

330
320
Ecliptic long.
310
300
290
280

-30 -20 -10 0 +10 +20 +30
Ecliptic lat.

Fig. 5.29. Bands of increased infrared emission (60 μm) at ecliptic latitudes 0° and ±10°. To emphasize the bands the much larger general thermal emission distribution has been suppressed in this figure [5.32]

also derived from asteroids. The short cosmic ray exposure ages [5.99] show that they were recently set free in the collisional destruction of bodies at least 1 m in size. Along these lines Gillet [5.66] suggested the possibility of asteroidal origin for the zodiacal light. He calculated that collisions in the asteroid belt easily produce enough small particles. However, Dohnanyi [5.39], by extrapolating the size distribution of small asteroids, estimated that the total asteroidal contribution to interplanetary dust would be small. Afterwards asteroids were nearly forgotten as being a possible source of dust, until the *IRAS* sky survey showed weak narrow bands near 0° and ±10° ecliptic latitude, which are superimposed on the general thermal emission of interplanetary dust (Fig. 5.29). The position of the bands falls in the regions occupied by major asteroid families: Koronis, Eos, and Themis. Dermott et al. [5.33] showed that the bands could have indeed been produced by collisions between members of those families. If this explanation is correct, the entire asteroid belt should also make a substantial contribution to interplanetary dust. Further evidence in this direction was provided by Sykes [5.167] who discovered several additional dust bands. But in general it will be difficult to recognize dust originating from the asteroid belt as such.

5.9.2 Dust Production by Comets

Most probably comets are the main source of interplanetary dust. At least the splendid appearance of their dust tails (Fig. 5.30) shows that they are able to emit dust in large quantities. In addition, the meteor streams, thought to be the

255

remnants of disintegrated comets and sporadic meteors resulting from dissolved meteor streams, must produce dust in collisions.

From their orbital distribution short-period comets would be candidate sources of interplanetary dust. The supply necessary to keep the interplanetary dust cloud in steady state was calculated to be about 10 tons/s [5.38, 73, 176]. However, calculation of the amount of dust emitted by short-period comets gives only $250\,kg\,s^{-1}$ which remain in bound orbits [5.152]. Detailed studies on comets Encke and d'Arrest [5.161, 162] show that this may even be a considerable overestimate. Direct injection of dust from comets therefore does not appear to be the main supply for the interplanetary dust cloud, except if one suggests that one of the short-period comets was very much brighter a few thousand years ago, as Whipple did for comet Encke. But such speculations should be invoked only if other adequate explanations are missing. Of course, large new comets like that of 1577 could contribute [5.31], because they visibly shed a large amount of material. Unfortunately, because of their nearly parabolic orbits, essentially all of the produced dust will leave the solar system. Only the very largest pieces might remain in bound orbits. Integrated over all long-period comets no more than 400–$1500\,kg\,s^{-1}$ can be expected [5.56]. This leads us back to the idea propagated by Dohnanyi, and already mentioned above, that it is the large particles, meteoroids remaining from the gradual dissolution of comets, which by collisional grinding finally produce interplanetary dust. If an equilibrium exists, the time available to supply new material is quite different for different source mechanisms. For direct input of dust it is equal to the Poynting–Robertson lifetime (10^4–10^5 years). In the case of indirect input it is equal to the total time needed for dissolution of a comet, dispersion of the resulting meteor stream into the sporadic background, collisional breakup of meteoroids into dust particles, and Poynting–Robertson drift of this dust. The latter time is probably larger than the former by 1 to 2 orders of magnitude. This relieves the requirements for the steadiness of the supply and, because of the longer time available for the action of dissipating forces, leads to more thorough smoothing of the originally irregular input.

5.9.3 Cometary Dust

In 1986 three spaceprobes flew through the inner coma of comet Halley and made it possible to analyze chemically cometary dust particles directly. Mass spectra (Fig. 5.31) were obtained of the ions which were released upon impact of dust particles. These particle impact mass spectrometers [5.100] had a mass resolution > 100 and allowed the isotopic distinction of different elements. Figure 5.31a shows quite prominently the isotopic pattern of magnesium at masses 24 to 26 and silicon at masses 28 to 30. Somewhat surprising was the high abundance of light elements H, C, N, and O (Fig. 5.31b). Some particles solely consist of these elements. These light elements are enriched in dust particles from comet Halley relative to C1 chondrites by factors of 2 to 10 and are almost as abundant as in the solar photosphere, with the exception of hydrogen. This enrichment of the

Fig. 5.30. Coma and dust tail of comet Halley, photographed on 16 March 1986 by K. Birkle with the Schmidt telescope on Calar Alto, Spain

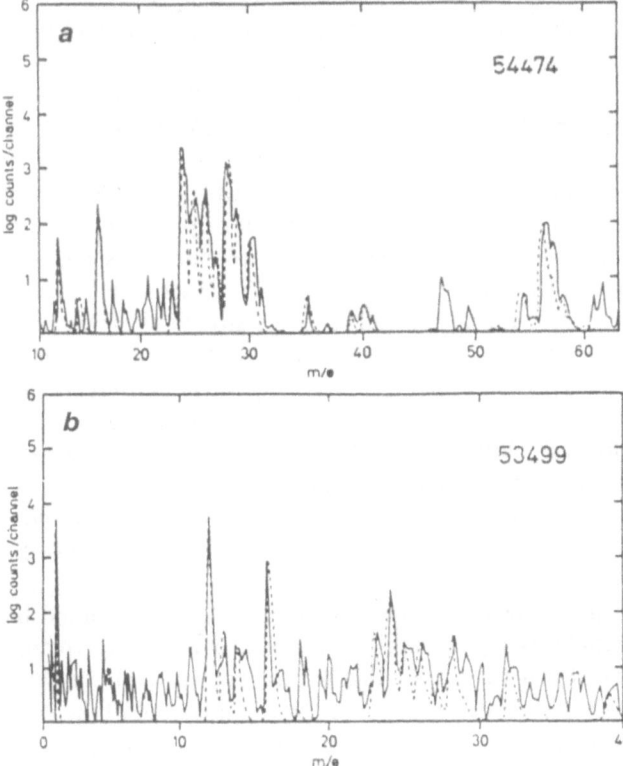

Fig. 5.31a,b. Chemical composition of submicron dust particles released by comet Halley. (a): Mass spectrum dominated by "silicates". (b): Mass spectrum dominated by light elements. The abscissa is in atomic mass units and is different for the two spectra [5.97]

257

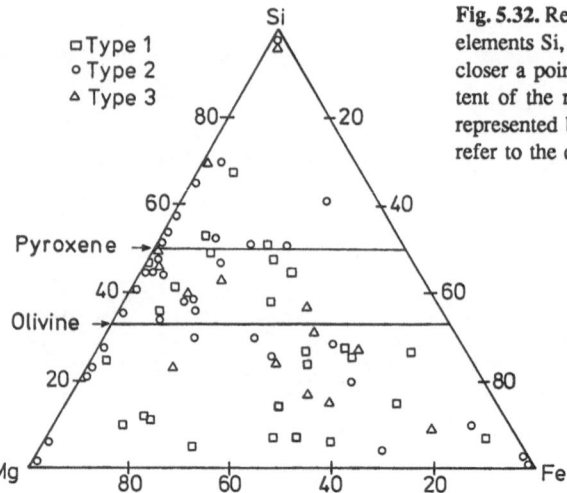

Fig. 5.32. Relative abundance of the rock-forming elements Si, Mg, Fe in cometary dust grains. The closer a point is to a corner, the higher the content of the respective element is in the particle represented by this point. The different symbols refer to the quality of the data [5.98]

"CHON" elements, which are present as refractory organics, identifies Halley's dust as the most unaltered early solar system material ever analyzed.

The bulk composition of the major rock-forming elements Mg, Si, and Fe is similar to that of chondritic meteorites [5.97, 98] within the accuracy of the measurement (a factor 2 in absolute abundances). The variation of the abundances of these three elements is large between different grains (Fig. 5.32). However, some general features of the composition of the rock forming elements are recognized: iron-rich pyroxenes and olivines have not been observed, whereas the group of grains close to Fe = 0, Mg = Si = 50 indicates the survival of primary enstatite, which has not readily reacted with iron oxides to form olivine. This again is evidence for a "low temperature", i.e. primitive nature of Halley's dust grains. In this respect cometary particles are related to individual structural elements of interplanetary grains collected in the stratosphere. The loss of the more volatile light elements in the collected particles marks the most significant difference between these two particle types. It may be speculated that for cometary grains loss of volatile elements due to evaporation and sputtering occurred, and may have been accompanied by a compaction of the remaining refractory material and a change in the optical properties of the grains [5.138]. However, an asteroidal origin for a large part of the dust grains collected in the atmosphere cannot be excluded.

In this context, the particle model proposed by Greenberg [5.68] should be seriously considered. Its emphasis is on establishing a link between interplanetary dust and the processes of star formation. Figure 5.33 shows what the basic building block of such a particle made up of interstellar grains could look like. Obviously their properties would be similar to the cometary particles just discussed.

A direct proof for the existence of cometary dust in interplanetary space has been given by infrared satellite *IRAS*, which imaged enhanced dust distribution

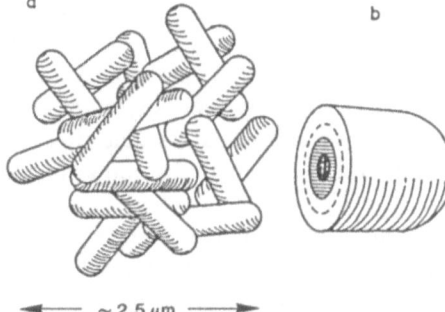

a   b

Fig. 5.33a,b. Hypothetical particle made up of interstellar grains [5.68]. The cross section shows, from circumference to center, two layers of ice, one layer of organic refractory material, and a silicate core

◄——— ~ 2.5 µm ———►

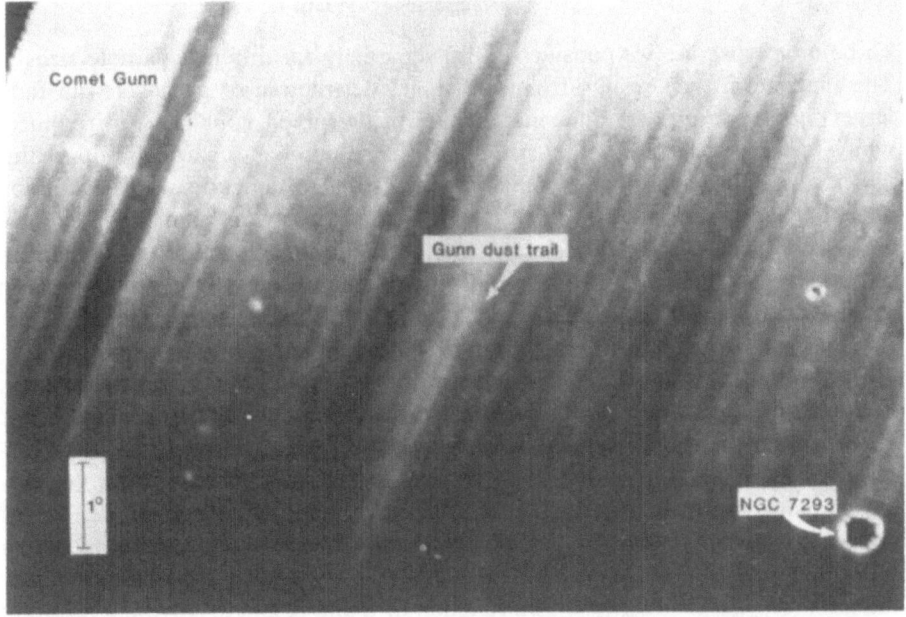

Comet Gunn

Gunn dust trail

1°

NGC 7293

Fig. 5.34. Dust enhancement in the orbital plane of comet Gunn, detected by *IRAS* infrared measurements [5.166]

in the orbital planes of several comets (Fig. 5.34). On some of these pictures the enhancement can be seen ahead of the comet, too, which is probably dust released in earlier orbital revolutions.

With all this evidence on the presence of cometary dust in interplanetary space one may wonder that only a third or a quarter of the small interplanetary bodies show the properties typically associated with cometary material: low density, fragile structure. This question is still open. Clarifying this point could well answer the additional question of how much comets and asteroids each contribute to interplanetary dust.

5.9.4 Input and Spatial Distribution – the Formal Relation

We want to discuss whether the observed spatial distribution can be understood as a consequence of dust supply by collisions between meteoroids, or indirect input, as we called it above. Calculations based on individual particle trajectories are still rare [5.21,74]. We therefore adopt a continuum description of the interplanetary dust cloud.

We start with a general consideration. If in a given volume of space at time t there are N particles, and shortly afterwards at time $t + \Delta t$ there are $N + dN$ particles, then the change in particle number can be written as

$$dN = N_{\text{incoming}} - N_{\text{outgoing}} + \Delta t \frac{\partial N}{\partial t_{\text{input}}} - \Delta t \frac{\partial N}{\partial t_{\text{loss}}} . \tag{5.28}$$

To be more specific, we consider (5.28) separately for different particle sizes s. The input term then results from collisional destruction of particles with radii larger than s during the time interval Δt in the given volume. Any fragment which happens to have the right size increases the number of particles with radius s. The losses are also due to collisions; they refer to the collisional destruction of particles of a size s during time interval Δt in the given volume. In this local description the Poynting–Robertson effect does not appear as a loss mechanism, but as a law which modifies the transport of particles in and out of the given volume.

Since the Poynting–Robertson effect is best described in terms of changing orbit ellipses, it is convenient to discuss (5.28) in terms of orbital parameters a, e, i. Three parameters are sufficient, since we assume azimuthal symmetry. We neglect Lorentz forces, and then have $di/dt = 0$. The equation is now essentially two dimensional. In the a–e plane the Poynting–Robertson drift carries the particles along smooth lines which all finally approach the sun. These lines do not cross each other: the flow of particles in a–e–i space is simply along tubes defined by these lines. This enables us to give an explicit expression for the term $(N_{\text{incoming}} - N_{\text{outgoing}})$ as a function of a and e, and to solve the resulting differential equation, at least for the assumption of steady state $(dN = 0)$. Figure 5.35 helps to visualize the flow geometry in a–e space and indicates the range of orbits contributing to observations at a fixed heliocentric distance, chosen to be 2.0 AU. We now give a quantitative formulation of these relations. Let N_t be the total number of particles in interplanetary space and $N_t \tilde{f}(a, e, i, s)$ the number of particles with orbital parameters in the range $(a, a + da)$, $(e, e + de)$, $(i, i + di)$ and sizes in the range $(s, s + ds)$. Similarly, N^i is taken to be the total number of particles added per second to interplanetary space, and $N^i \tilde{f}^i(a, e, i, s)$ the number of added particles per second in the range $(a, a + da)$, $(e, e + de)$, $(i, i + di)$, $(s, s + ds)$. Then it may be shown [5.117] that the steady-state distribution $N_t f$ is related to the input distribution per second $N^i \tilde{f}^i(a, e, i, s)$ by

$$N_t \tilde{f}(a, e, i, s) = \int_{\tilde{a}=a}^{a_{\max}} N^i \tilde{f}^i(\tilde{a}, \tilde{e}, i, s) G P d\tilde{a} , \tag{5.29}$$

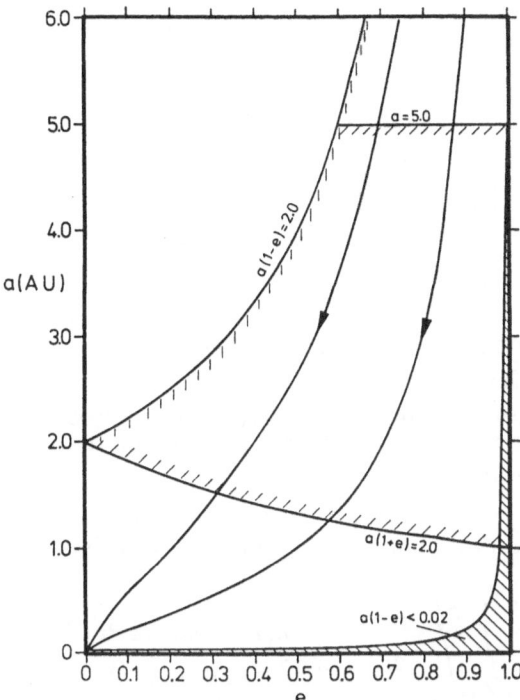

Fig. 5.35. Geometry of particle flow due to Poynting–Robertson effect in a–e space [5.117]. For a given heliocentric distance (here 2.0 AU) only particles with perihelion inside $[a(1-e) < 2.0]$ and aphelion outside this distance $[a(1+e) > 2.0]$ can be observed. Further restrictions of observable orbits are given by evaporation (perihelion < 0.02 AU) or the maximum size of the source region, assumed in this diagram to be 5 AU

which has a simple meaning. The integral is along the Poynting–Robertson drift path which passes through the point (a, e, i). Equation (5.29) says that all particles created along this drift path will eventually pass through this point. $\bar{f}(a, e, i, s)$ is therefore essentially the integral over all these input contributions $\tilde{f}^i(\tilde{a}, \tilde{e}, i, s)$, where $\tilde{a}, \tilde{e}$ are parameters along the drift path. G is a kind of geometry factor which has to do with the changing velocity of Poynting–Robertson drift. It depends on both the "coordinates" a, e and the two coordinates "$\tilde{a}, \tilde{e}$" at which a particle was put into interplanetary space. $P(a, e, \tilde{a}, \tilde{e})$ gives the probability that a particle released in orbit $\tilde{a}, \tilde{e}$, will reach orbit a, e. It is 1.0, i.e. no losses occur, under the action of the Poynting–Robertson effect alone. It will be less if particles are destroyed by collision during their drift from $\tilde{a}, \tilde{e}$ to a, e, and finally may approach zero. Figure 5.36 shows how strongly this quantity P depends on the ratio of the Poynting–Robertson and collisional lifetimes, where collisions were assumed to be the only loss mechanisms. For $\tau_{PR}/\tau_c = 0.1$ the particle number at a given point is influenced by the input over a wide range of heliocentric distances, for $\tau_{PR}/\tau_c = 10$ only by input in the immediate vicinity, as might be expected. With $\tau_{PR}/\tau_c = 0.1$ a particle at 0.8 AU has equal chances of being destroyed in a collisions or of evaporating after a drift close to the sun.

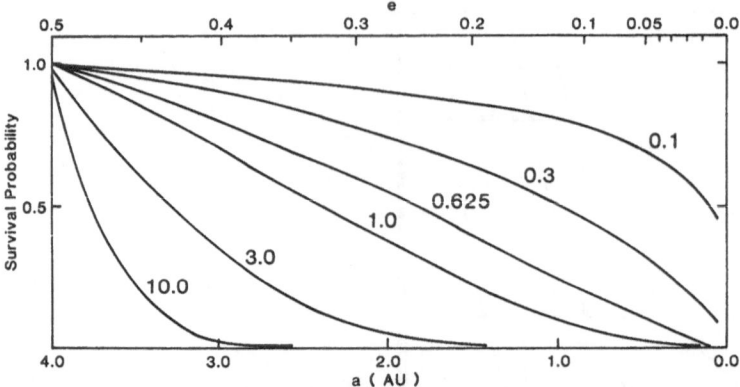

Fig. 5.36. Survival probability against collisions, $P = \exp[-1.6\tau_{PR}/\tau_c \log(e_{initial}/e_{final})]$, specifically shown for a particle starting its Poynting–Robertson drift at 4 AU with eccentricity 0.5. The curves correspond to different values of the ratio τ_{PR}/τ_c [5.117]

The transformation from orbital distribution f to spatial density n is given by Haug's integral [5.80]:

$$n(r, z, s) = \frac{N_t}{2\pi^3 r \cos\beta_\odot} \int_{a=r/(1+e)}^{r/(1-e)} \int_{e=0}^{1} \int_{i=\beta_\odot}^{\pi/2} \frac{\tilde{f}(a, e, i, s)da\,de\,di}{r^2(a/r)^{3/2}[2 - r/a - a/r(1 - e^2)]^{1/2}}$$

$$\times \frac{1}{[\sin^2 i - \tan^2 \beta_\odot \cos^2 i]^{1/2}}, \tag{5.30}$$

where $\sin\beta_\odot = z/r$. The limits of a, e, i integration select the orbits passing through point r, z (see Fig. 5.35). Transforming from a, e to other coordinates, (5.29–30) get simpler to handle on the computer but more difficult to visualize [5.15, 135].

If the interplanetary dust cloud is not in steady state, (5.29) is still valid for a given time if only the integral along the drift path takes the input for the suitable starting time of the particles now contributing the space density at position r, z. The qualitative conclusions of this section therefore should remain valid in a time-variable interplanetary dust cloud.

5.9.5 Input and Spatial Distribution – Consequences

With the formalism derived in the preceding section, a given spatial distribution of input can be tested to see whether it could produce the observed radial distribution $n(r) \sim r^{-1.3}$. Figure 5.37 shows that production of dust in a narrow region of heliocentric distance leads to unacceptable results. For nearly circular orbits, as might be the case for dust production in the asteroid belt, a radial distribution that varies as $1/r$ results. For quite elliptic orbits (not shown here) the slope of radial distribution changes by too much with heliocentric distance. Note that, contrary to intuition, in the steady state under the action of the Poynting–Robertson effect, the density in the source region is particularly low. If the asteroid belt were an

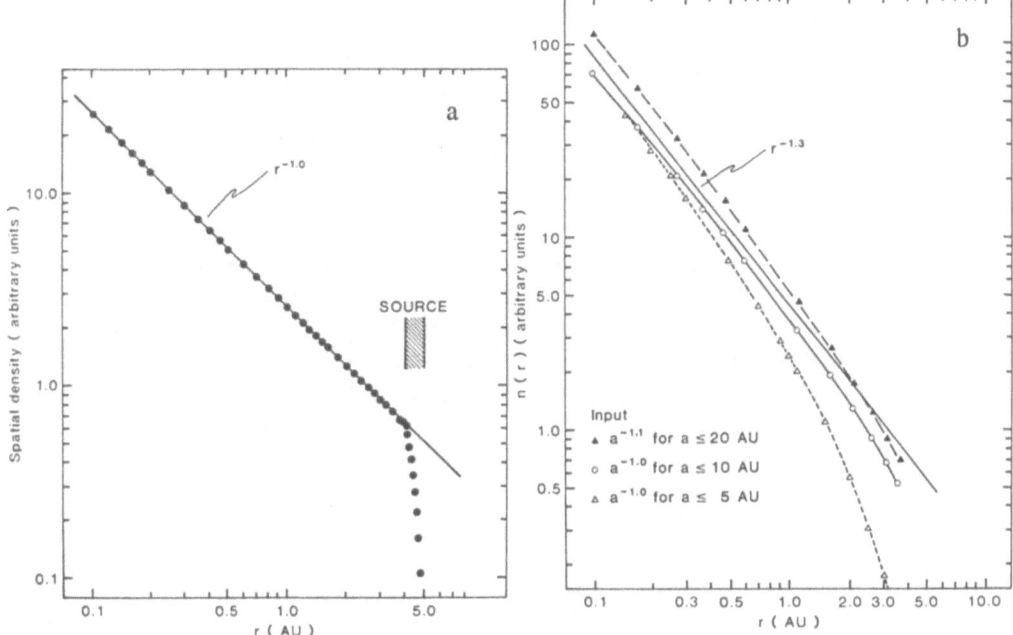

Fig. 5.37a,b. Equilibrium spatial distribution for interplanetary dust under the action of the Poynting–Robertson effect [5.117]. (a) Localized source, orbits nearly circular. (b) Extended input regions, average eccentricity $e = 0.5$

important source of interplanetary dust particles it would be just the wrong place to look for them. An analytical solution of (5.29) indicates the character of the required input. If the input distribution is infinitely extended and of the form $\tilde{f}(a, e) = a^{-x} g(e)$, with $x > 1$, then the resulting equilibrium spatial distribution is a power law with the same exponent, $n(r) \sim r^{-x}$. This helps to anticipate the results in more realistic cases. It also shows an interesting, self-stabilizing property of a spatial distribution with exponent $x = 1.3$. If the efficiency of interplanetary collisions increases like (velocity)$^{0.4}$ as is approximately the case, then the production rate of collisional debris has a dependence with heliocentric distance of $\sim r^{-3.3}$, which gives just the required input distribution $\tilde{f}(a) \sim a^{-1.3}$ to recreate the initial power law.

From a discussion of more realistic cases with limited input regions, it was found that an extended source region was necessary to produce the observed spatial distribution (see Fig. 5.37b). In this case about half of the input is within 1 AU and about a quarter still outside 4 AU, e.g. a distribution of semimajor axes $\sim a^{-1.1}$ from 0.1 AU to 20 AU. This type of extended source region for interplanetary dust quite naturally results from collisions. In particular, collisional destruction of radio meteoroids which are mostly hit by smaller particles provides approximately the necessary spatial distribution of input to lead to a steady-state density distribution $n(r) \sim r^{-1.3}$ [5.117]. Since the amount of material is also of the right order, this supports our concept of dust supply. This mass balance still has to be checked more closely.

5.9.6 Mass Balance on a Large Scale

Without collisional losses comparatively little dust is needed to maintain the present day density of interplanetary dust: about $250 \, \text{kg s}^{-1}$ outside 1 AU, and the same amount in the inner solar system. Collisional losses may increase this demand considerably, to several $10^3 \, \text{kg s}^{-1}$. The values given in Table 5.5 [5.117] are upper limits because collisions between dust particles also add to the input and only part of the fragments are really lost, i.e. converted to β meteoroids which quickly leave the solar system. On the other hand, our collisional model predicts that a large fraction of the collisional debris of larger meteor particles remains in the size range of dust, thus creating a sizeable input. Depending on the badly known spatial distribution outside 1 AU this collisional input globally is expected at least to balance the losses if not to surpass them by a considerable amount. We follow this question locally near 1 AU in order to avoid the uncertainties resulting from a lack of knowledge in the spatial distribution.

Table 5.5. Production rates $P(\text{kg s}^{-1})$ required to maintain the observed distribution of interplanetary dust

τ_{PR}/τ_c	$P(0.1 \, \text{AU} \leq a \leq 1 \, \text{AU}$	$P(a \geq 1 \, \text{AU})$	P (total)
0	260	210	470
0.3	440	350	790
1	980	780	1.8×10^3
3	2.8×10^3	2.2×10^3	5.0×10^3
10	9.2×10^3	7.3×10^3	1.7×10^4

5.9.7 Local Mass Balance

When we are talking of a balance between creation of dust per unit volume of space in collisions, which is proportional to (density)^2 of parent bodies, and losses by the Poynting–Robertson effect, which are proportional to density, it is immediately clear that this balance cannot easily exist over the whole sphere with radius 1 AU, because the production falls off more rapidly outside the ecliptic, up to ten times faster than the losses. Similar imbalances appear as a function of heliocentric distance or of particle size. We do not want to speculate on how this difficulty could be overcome but leave it for the moment as an open question and return to the ecliptic. A key piece of information for the discussion of mass balance is the lifetimes of particles of different size. Figure 5.38 shows the collisional lifetime at 1 AU, calculated from the observed size distribution of meteoroids, together with the Poynting–Robertson drift time (5.12). The lifetime of meteoroids with masses $\geq 10^{-8} \, \text{kg}$ is dominated by collisions. These particles, which represent more than half of the mass of the interplanetary dust cloud, will not change their orbits significantly by the Poynting–Robertson effect before they are involved in a collision.

The mass of particles destroyed by collisions per second in a volume of $1 \, \text{m}^3$ at 1 AU in the ecliptic is shown in Fig. 5.39 as function of size. How-

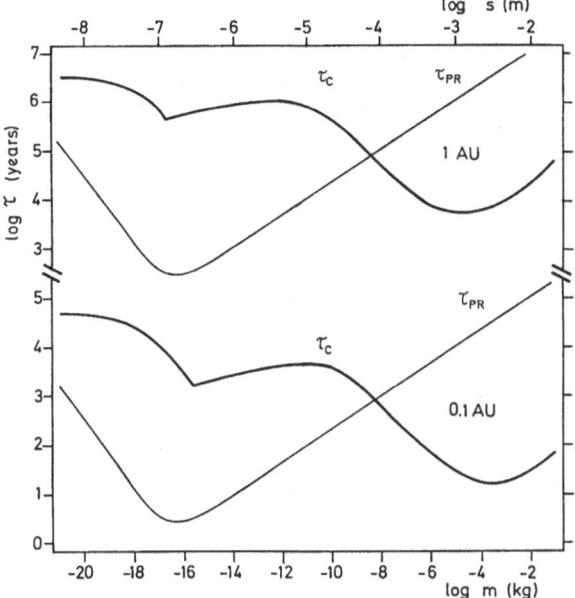

Fig. 5.38. Collisional lifetime τ_c compared with Poynting–Robertson lifetimes τ_{PR} at 1 AU and 0.1 AU heliocentric distance in the ecliptic. The kinks in τ_c near 10^{-16} kg are caused by the assumed cutoff of the meteoroid distribution at $m = 10^{-21}$ kg. The increase of τ_{PR} below 10^{-16} kg is due to the strong decrease of β for small dielectric particles (here: olivine). For absorbing particles it will be much less pronounced

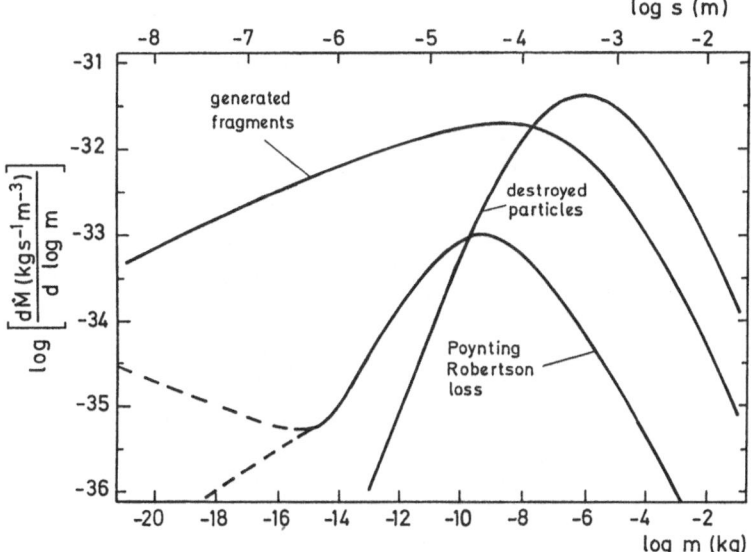

Fig. 5.39. Mass destroyed and fragments generated by collisions at 1 AU compared to the mass loss due to Poynting–Robertson effect

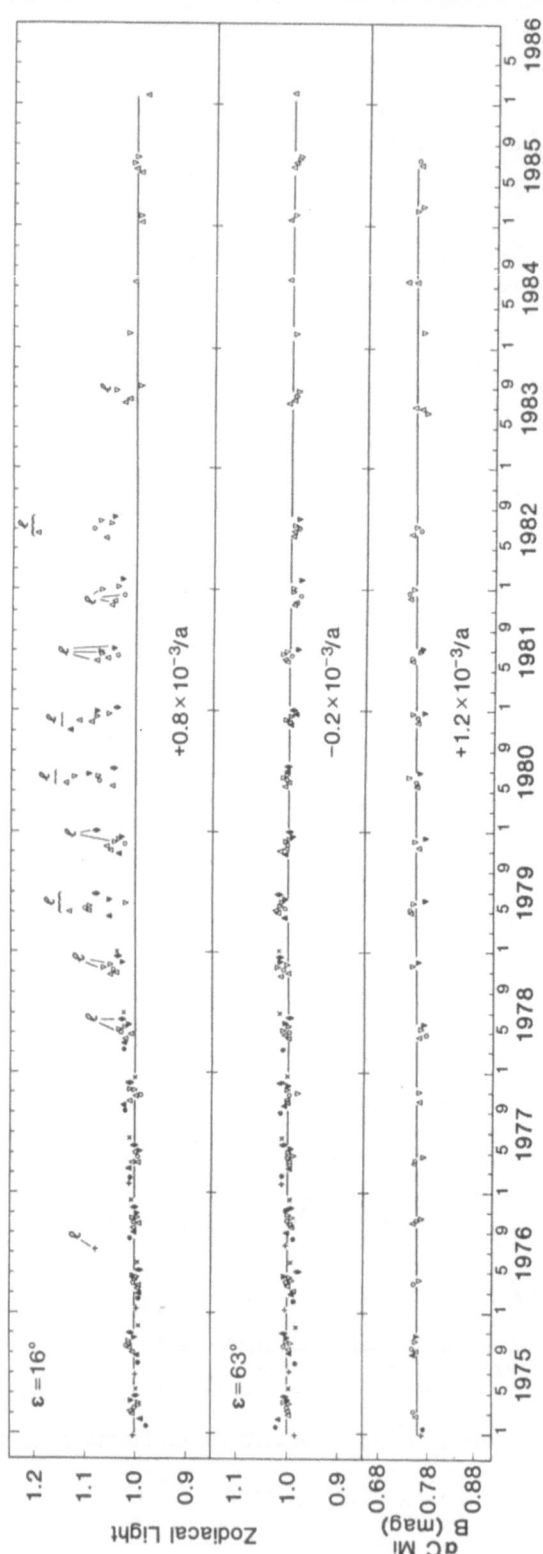

Fig. 5.40. Stability of the zodiacal light from December 1974 to February 1986 according to *Helios* [5.118]. To exclude instrumental drifts, measurements on a bright star are given for comparison. Each symbol refers to a specific orbital position of *Helios*. The only apparent brightness changes are those due to solar plasma cloud ejections (marked "*e*")

ever, this mass is not really lost but reappears in the form of small fragments, constituting a gain of particles in those size ranges. Comparing gains and losses, we see from Fig. 5.39 that the net effect of collisions is to produce dust particles ($m < 10^{-8}$ kg, radii $< 100\,\mu$m) at the expense of the larger meteoroids. Qualitatively this was to be expected (see Sect. 5.4.7). Quantitatively it means that 9×10^{-32} kg m^3 s^{-1} are added to interplanetary dust, while only 4×10^{-33} kg m^3 s^{-1} would be needed to replace the Poynting–Robertson losses in a spatial distribution with radial gradient $\sim r^{-1.3}$ [5.73]. This would mean that at 1 AU the spatial density of interplanetary dust would increase by 10% in 3000 years, at 0.1 AU, even, in only 30 years. This starts being an observable effect. The zodiacal light photometer on *Helios A* has monitored the zodiacal light brightness down to 0.1 AU over a period of more than 10 years. No systematic increase with time can be seen (Fig. 5.40), although an effect of 3% should be well detectable [5.118]. This stability is remarkable on its own, and we take it as limited evidence for a steady state of the interplanetary dust cloud.

On the other hand, we cannot neglect the fact that collision calculations appear to give a dust production rate 2 to 20 times greater (depending on height above the ecliptic) than needed to balance the Poynting–Robertson losses. The possibility of secular changes in the interplanetary dust cloud has to be considered

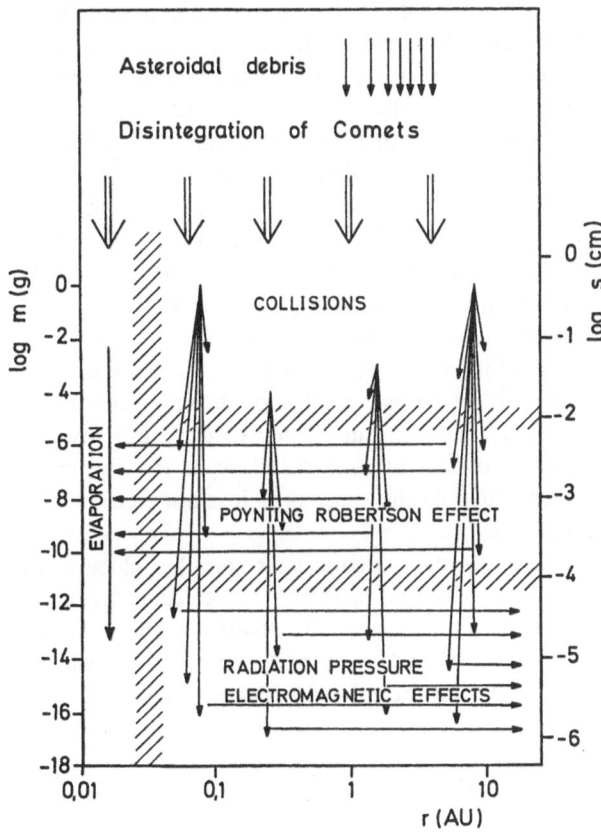

Fig. 5.41. Schematic sketch of dynamical effects changing meteoroid mass and heliocentric distance. The vapor of evaporated particles is ionized and carried away with the solar wind

seriously. But we have to remember that the calculations depend critically on the application of a collisional model derived from disruption of basalt rocks several cm in size. Interplanetary dust might react differently, producing more fine-grained debris, which would rapidly be lost as β-meteoroids. In view of the remaining uncertainties we note that the stability of the interplanetary dust cloud is not completely ruled out and may still give a resonable first-order description. In any case, measurements such as those in Figs. 5.20 and 5.40 are arguments for a temporally stable, well mixed spatial structure.

Figure 5.41 summarizes the picture resulting from our discussion. Most of the interplanetary dust is produced by collisions of larger meteoroids, which present a reservoir continually being replenished by disintegration of comets or asteroids. Part of it is lost by evaporation after being driven close to the sun by the Poynting–Robertson effect. The remainder is transformed by collisions to β meteoroids to be blown out of the solar system.

5.10 Attempt at a Judgement

In summary of this extended and sometimes critical review we want to emphasize that an increasing number of firm points appears in the area of interplanetary dust research. (1) The size distribution near the earth is one cornerstone, although quantitative results for β meteoroids unfortunately are still scarce. (2) Similarly the collection of particles in the stratosphere led to a solid picture of particle properties. (3) The exploration of comet Halley by space probes brings the relation between cometary and interplanetary dust from qualitative speculation to quantitative discussion. (4) The spatial distribution, derived mainly from deep space probe zodiacal light observations appears well founded for the inner solar system, although problems in detail remain and the information on z dependence has not yet been fully exploited. (5) In the outer solar system contradictions between different methods paralyze further understanding and await new clarifying experiments. (6) Dynamical studies indicate that collisions and the Poynting–Robertson effect dominate the evolution of individual particles but collisional processes among meteoroids are still not known well enough for a final discussion of mass balance. (7) Comets are an obvious source of dust, but just how they gave rise to the observed dust cloud is difficult to follow, as it is difficult to argue that asteroids cannot provide dust. At the same time it is not clear whether the cloud is in a state of equilibrium.

Nevertheless, we have this example of a cosmic dust complex close to our doorstep, ready for direct and indirect exploration. If not under such favourable conditions, where else should we be able to meet the challenge and achieve a thorough understanding of the origin, present state, and evolution of dust in a typical stellar system? We have begun but there is still a way to go.

Acknowledgements. This summary report was triggered by the results of the *Helios* dust experiments, and we therefore want to thank H. Elsässer and H. Fechtig, who initiated these experiments and provided generous support and important guidance throughout the project. No lesser thanks go to our colleagues who worked with us on the instruments: some during the construction and others during data evaluation; in particular we are indebted to P. Gammelin, J. Kissel, and E. Pitz. We also thank Ms. R. Wagner and Ms. R. Blythe for their careful typing of the many versions of this manuscript. H. Zook very accurately reviewed the text. We thank him for his constructive criticism.

References

Conference Proceedings and Review Books

5.1   *The Zodiacal Light and the Interplanetary Medium*, ed. by J.L. Weinberg, NASA SP-150, 1967.
5.2   *Interplanetary Dust and Zodiacal Light*, ed. by H. Elsässer, H. Fechtig, Lecture Notes in Physics, Vol. 48, Springer, Berlin, Heidelberg, New York, 1976.
5.3   *Cosmic Dust*, ed. by J.A.M. McDonnell, Wiley, Chichester, 1978.
5.4   *Solid Particles in the Solar System*, ed. by I. Halliday, B.A. McIntosh, D. Reidel, Dordrecht, Boston, London, 1980.
5.5   *Properties and Interactions of Interplanetary Dust*, ed. by R.H. Giese, P. Lamy, D. Reidel, Dordrecht, Boston, Lancaster, Tokyo, 1985.

Other Material

5.6   Ashworth, D.G., Lunar and planetary impact erosion, in Ref. [5.3], pp. 427–526.
5.7   Barge, P., R. Pellat, I. Millet, Diffusion of Keplerian motions by a stochastic force. II. Lorentz scattering of interplanetary dusts, Astron. Astrophys., **115**, 8–19, 1982.
5.8   Bandermann, L.W., Physical properties and dynamics of interplanetary dust, thesis, Univ. of Maryland, College Park, 1968.
5.9   Berg, O.E., F.F. Richardson, The Pioneer 8 cosmic dust experiment, Rev. Sci. Instrum., **40**, 1333–1337, 1969.
5.10 Berg, O.E., U. Gerloff, More than two years of micrometeorite data from two Pioneer satellites, Space Res., **XI**, 225–235, 1971.
5.11 Berg, O.E., E. Grün, Evidence of hyperbolic cosmic dust particles, Space Res., **XIII**, 1047–1055, 1973.
5.12 Bibring, I.-P., J. Borg, A. Katchanov, Y. Longevin, P. Salvetat, Y.A. Surkhov, B. Vassent, The COMET experiment: first results, Lunar and Planetary Science, **Conference, XIX**, 73–74, 1988.
5.13 Blackwell, D.E., D.W. Dewhirst, M.F. Ingham, The zodiacal light, Adv. Astron. Astrophys., **5**, 1–69, 1967.
5.14 Bradley, J.P., D.E. Brownlee, P. Fraundorf, Discovery of nuclear tracks in interplanetary dust, Science, **226**, 1432–1434, 1984.
5.15 Briggs, R.E., Steady-state space distribution of meteoric particles under the operation of the Poynting-Robertson effect, Astron. J., **67**, 710–723, 1962.
5.16 Bronshten, V.A., *Physics of Meteoric Phenomena*, D. Reidel, Dordrecht, Boston, Lancaster, 1983.
5.17 Brownlee, D.E., Microparticle studies by sampling techniques, in Ref. [5.3], pp. 275–336.
5.18 Buitrago, J., R. Gómez, F. Sanchez, The integral equation approach to the study of interplanetary dust, Planet. Space Sci., **31**, 373–376, 1983.
5.19 Burkhardt, G., On the invariable plane of the solar system, Astron. Astrophys., **106**, 133–136, 1982.
5.20 Burkhardt, G., Kollisionsfreie Dynamik von interplanetaren Staubteilchen, thesis, Univ. Heidelberg, 1986.
5.21 Burkhardt, G., Dynamics of dust particles in the solar system, in Ref. [5.5], pp. 389–393.
5.22 Burns, J.A., P.L. Lamy, S. Soter, Radiation forces on small particles in the solar system, Icarus, **40**, 1–48, 1979.
5.23 Carey, W.C., J.A.M. McDonnell, D.G. Dixon, An empirical penetration equation for thin metallic films used in capture cell techniques, in Ref. [5.5], pp. 131–136.

5.24 Cassini, J.D., Découverte de la lumière celeste qui paroist dans le zodiaque, Mem. Acad. Roy. Sci., Tome **VIII** (1666–1699), 119–209, Comp. Libraires, Paris, 1730.

5.25 Cebula, R.P., P.D. Feldman, Ultraviolet spectroscopy of the zodiacal light, Astrophys. J., **263**, 987–992, 1982.

5.26 Ceplecha, Z., Meteoroid populations and orbits, in *Comets Asteroids Meteorites. Interrelations, Evolution and Origins*, ed. by A.H. Delsemme, 143–152, Univ. of Toledo, 1977.

5.27 Clark, G.L., W.H. Kinard, D.J. Carter, J.L. Lones (Eds.), *The Long Duration Exposure Facility (LDEF) Mission 1 Experiments*, NASA SP-437, 1984.

5.28 Consolmagno, G., Lorentz scattering of interplanetary dust, Icarus, **38**, 398–410, 1979.

5.29 Cook, A.F., A working list of meteor streams, in *Evolutionary and Physical Properties of Meteoroids*, ed. by C.L. Hemenway, P.M. Millman, A.F. Cook, NASA SP-319, 183–191, 1973.

5.30 Cook, A.F., Albedos and size distribution of meteoroids from 0.3 to 4.8 AU, Icarus, **33**, 349–360, 1978.

5.31 Delsemme, A.H., The production rate of dust by comets, in Ref. [5.2], pp. 314–318.

5.32 Dermott, S.F., P.A. Nicholson, J.A. Burns, J.R. Houck, Origin of the solar system dust bands discovered by IRAS, Nature, **312**, 505–509, 1984.

5.33 Dermott, S.F., P.D. Nicholson, J.A. Burns, J.R. Houck, An analysis of IRAS' solar system dust bands, in Ref. [5.5], pp. 395–409.

5.34 Deul, E.R., R.D. Wolstencroft, A physical model for thermal emission from the zodiacal dust cloud, Astron. Astrophys., **196**, 277–286, 1988.

5.35 Dietzel, H., G. Eichhorn, H. Fechtig, E. Grün, H.J. Hoffmann, J. Kissel, The HEOS 2 and Helios micrometeoroid experiments, J. Phys. (E) Scientific Instrum., **6**, 209–217, 1973.

5.36 Dohnanyi, J.S., Collisional model of asteroids and their debris, J. Geophys. Res., **74**, 2531–2554, 1969.

5.37 Dohnanyi, J.S., On the origin and distribution of meteoroids, J. Geophys. Res., **75**, 3468–3493, 1970.

5.38 Dohnanyi, J.S., Interplanetary objects in review, statistics of their masses and dynamics, Icarus, **17**, 1–48, 1972.

5.39 Dohnanyi, J.S., Sources of interplanetary dust, asteroids, in Ref. [5.2], pp. 187–205.

5.40 Dohnanyi, J.S., Particle dynamics, in Ref. [5.3], pp. 527–605.

5.41 Dollfus, A., J.E. Geake, C. Titulaer, Polarimetric properties of the lunar surface and its interpretation. Part 3, Apollo 11 and Apollo 12 lunar samples, Proc. Sec. Lunar Sci. Conf., Geochim. Cosmochim. Acta., Suppl., **2** (3), 2285–2300, 1971.

5.42 Dumont, R., Phase function and polarization curve of interplanetary scatterers from zodiacal light photopolarimetry, Planet. Space Sci., **21**, 2149–2155, 1973.

5.43 Dumont, R., Ground-based observations of the zodiacal light, in Ref. [5.2], pp. 85–100.

5.44 Dumont, R., A.C. Levasseur-Regourd, Zodiacal light photopolarimetry. IV. Annual variations of brightness and the symmetry plane of the zodiacal cloud. Absence of solar-cycle variations, Astron. Astrophys., **64**, 9–16, 1978.

5.45 Dumont, R., A.C. Levasseur-Regourd, Zodiacal light gathered along the line of sight. Retrieval of the local scattering coefficient from photometric surveys of the ecliptic plane, Planet. Space Sci., **33**, 1–9, 1985.

5.46 Dumont, R., A.C. Levasseur-Regourd, The symmetry plane of the zodiacal cloud retrieved from IRAS data, in *Interplanetary Matter*, Proc. 10th European Regional Meeting of the IAU, Vol. **2**, ed. by Z. Ceplecha, P. Pecina, Publ. Astron. Inst. Czech. Acad. Sci., **67**, 281–284, 1987.

5.47 Dumont, R., A.C. Levasseur-Regourd, Properties of interplanetary dust from infrared and optical observations. I. Temperature, global volume intensity, albedo and their heliocentric gradients, Astron. Astrophys., **191**, 154–160, 1988.

5.48 Dumont, R., F. Sánchez, Zodiacal light photopolarimetry. II. Gradients along the ecliptic and the phase functions of interplanetary matter, Astron. Astrophys., **38**, 405–412, 1975.

5.49 Dumont, R., F. Sánchez, Zodiacal light photopolarimetry. III. All-sky survey from Teide 1964–1975 with emphasis on off-ecliptic features, Astron. Astrophys., **51**, 393–399, 1976.

5.50 East, I.R., N.K. Reay, The motion of interplanetary dust particles, I. Radial velocity measurements on Fraunhofer line profiles in the zodiacal light spectrum, Astron. Astrophys., **139**, 512–515, 1984.

5.51 Elsässer, H., Fraunhoferkorona und Zodiakallicht, Z. f. Astrophys., **37**, 114–124, 1955.

5.52   Fechtig, H., J.B. Hartung, K. Nagel, G. Neukum, D. Storzer, Lunar microcrater studies, derived meteoroid fluxes, and comparison with satellite-borne experiments, Proc. 5th Lunar Sci. Conf., 3, 2463–2474, 1974.

5.53   Fechtig, H., C. Leinert, E. Grün., Interplanetary dust and zodiacal light, Landolt-Börnstein, N.S., VI/2a, 228–243, 1981.

5.54   Fox, K., I.P. Williams, D.W. Hughes, The "Geminid" Asteroid (1983 TB) and its orbital evolution, Monthly Not. of the R.A.S., 208, 11P–15P, 1984.

5.55   Frey, A., W. Hofmann, D. Lemke, Spectrum of the zodiacal light in the middle UV, Astron. Astrophys., 54, 853–855, 1977.

5.56   Fulle, M., Meteoroids from comet Bennett 1970 II, Astron. Astrophys. 183, 392–396, 1987.

5.57   Fujiwara, A., G. Kamimoto, A. Tsukamoto, Destruction of basaltic bodies by high-velocity impact, Icarus, 31, 277–288, 1977.

5.58   Gault, D.E., J.A. Wedekind, The destruction of tektites by micrometeoroid impact, J. Geophys. Res., 74, 6780–6794, 1969.

5.59   Giese, R.H., Light scattering by small particles and models of interplanetary matter derived from the zodiacal light, Space Sci. Rev. 1, 589–611, 1963.

5.60   Giese, R.H., Optical properties of single-component zodiacal light models, Planet. Space Sci., 21, 513–521, 1973.

5.61   Giese, R.H., Zodiacal light and local interstellar dust: predictions for an out-of-ecliptic space-craft, Astron. Astrophys., 77, 223–226, 1979.

5.62   Giese, R.H., M.S. Hanner, C. Leinert, Colour dependence of zodiacal light models, Planet. Space Sci., 21, 2062–2072, 1973.

5.63   Giese, R.H., E. Grün, The compatibility of recent micrometeoroid flux curves with observations and models of the zodiacal light, in Ref. [5.2], pp. 135–139.

5.64   Giese, R.H., G. Kinateder, B. Kneissel, U. Rittich, Optical models of the three dimensional distribution of interplanetary dust, in Ref. [5.5], pp. 255–259.

5.65   Giese, R.H., B. Kneissel, U. Rittich, Three-dimensional zodiacal dust cloud, a comparative study, Icarus, 68, 395–411, 1986.

5.66   Gillett, F.C., Zodiacal light and interplanetary dust, thesis, University of Minnesota, 1966.

5.67   Göller, R.J., E. Grün, Calibration of the Galileo/ISPM dust detectors with iron particles, in Ref. [5.5], pp. 113–115.

5.68   Greenberg, J.M., Laboratory dust experiments – tracing the composition of cometary dust, in Cometary Exploration, ed. by T.I. Gombosi, Hungarian Academy of Sciences, Proc. Conf. Cometary Exploration Budapest, II, 23, 1983.

5.69   Greenberg, J.M., B.A. Gustafson, A comet fragment model for zodiacal light particles, Astron. Astrophys., 93, 35–42, 1981.

5.70   Grün, E., Physikalische und chemische Eigenschaften des interplanetaren Staubes – Messungen des Mikrometeoritenexperiments auf Helios, Bundesministerium für Forschung und Technologie, Forschungsbericht W 81-034, 1981.

5.71   Grün, E., N. Pailer, H. Fechtig, J. Kissel, Orbital and physical characteristics of micrometeoroids in the inner solar system as observed by Helios 1, Planet. Space Sci., 28, 333–349, 1980.

5.72   Grün, E., H. Fechtig, J. Kissel, Orbits of interplanetary dust particles inside 1 AU as observed by Helios, in Ref. [5.5], pp. 105–111.

5.73   Grün, E., H.A. Zook, H. Fechtig, R.H. Giese, Collisional balance of the meteoritic complex, Icarus, 62, 244–272, 1985.

5.74   Gustafson, B.Å.S., N.Y. Misconi, E.T. Rusk, Interplanetary dust dynamics. III. Dust released from P/Encke: distribution with respect to the zodiacal cloud, Icarus, 72, 582–592, 1987.

5.75   Halliday, I., A.T. Blackwell, A.A. Griffin, The frequency of meteorite falls on the earth, Science, 223, 1405–1407, 1984.

5.76   Hanner, M.S., On the albedo of interplanetary dust, Icarus, 43, 373–380, 1980.

5.77   Hanner, M.S., On the detectability of icy grains in the comae of comets, Icarus, 47, 342–350, 1981.

5.78   Hanner, M.S., J.G. Sparrow, J.L. Weinberg, D.E. Beeson, Pioneer 10 observations of zodiacal light brightness near the ecliptic: changes with heliocentric distance, in Ref. [5.2], pp. 29–35.

5.79   Hartung, J.B., D. Storzer, Lunar microcraters and their solar flare track record, Proc. 5th Lunar Sci. Conf., Geochim. Cosmochim. Acta, Suppl., 5 (3), 2527, 1974.

5.80   Haug, U., Über die Häufigkeitsverteilung der Bahnelemente bei den interplanetaren Staubteilchen, Z. f. Astrophys., 44, 71–97, 1958.

5.81   Hauser, M.G., F.C. Gillett, F.J. Low, T.N. Gautier, C.A. Beichman, G. Neugebauer, H.H. Aumann, B. Band, N. Boggess, J.P. Emerson, J.R. Houck, B.T. Soifer, R.G. Walker, IRAS observations of the diffuse infrared background, Astrophys. J., **278**, L15–L18, 1984.

5.82   Hauser, M.G., J.R. Houck, The zodiacal background in the IRAS data, in *Light on Dark Matter*, ed. by F.P. Israel, 39–44, D. Reidel, Dordrecht, 1986.

5.83   Hauser, M.G., T.N. Gautier, J. Good, F.J. Low, IRAS observations of the interplanetary dust emission, in Ref. [5.5], pp. 43–48.

5.84   Henry, R.C., R.C. Anderson, W.G. Fastie, Far-ultraviolet studies. VIII. Apollo 17 search for zodiacal light, in Ref. [5.4], pp. 41–44.

5.85   Hoffmann, H.J., H. Fechtig, E. Grün, J. Kissel, Temporal fluctuations and anisotropy of the micrometeoroid flux in the earth–moon system, Planet. Space Sci., **23**, 985–991, 1975.

5.86   Hong, S.S., I.K. Um, Inversion of the zodiacal infrared brightness integral, Astrophys. J., **330**, 928–935, 1987.

5.87   Hong, S.S., Henyey–Greenstein representation of the mean volume scattering phase function for zodiacal dust, Astron. Astrophys., **146**, 67–75, 1985.

5.88   Hörz, F., D.E. Brownlee, H. Fechtig, J.B. Hartung, D.A. Morrison, G. Neukum, E. Schneider, J.F. Vedder, D.E. Gault, Lunar microcraters: implications for the micrometeoroid complex, Planet. Space Sci., **23**, 151–172, 1975.

5.89   Hughes, D.W., Nodal regression of the quandrantid meteor stream, Observatory, **92**, 41–43, 1972.

5.90   Hughes, D.W., Meteors, in ref. [5.3], pp. 123–185.

5.91   van de Hulst, H.C., Zodiacal light in the solar corona, Astrophys. J., **105**, 471–488, 1947.

5.92   Humes, D.H., Results of Pioneer 10 and 11 meteoroid experiments: interplanetary and near-Saturn, J. Geophys. Res., **85**, 5841–5852, 1980.

5.93   Humes, D.H., J.M. Alvarez, R.L. O'Neal, W.H. Kinrad, The interplanetary and near Jupiter environments, J. Geophys. Res., **79**, 3677–3684, 1974.

5.94   Jacchia, L.G., F. Verniani, R.E. Briggs, An analysis of the atmospheric trajectories of 413 precisely reduced photographic meteors, Smithsonian Contr. to Astrophys., **10**, 1–139, 1967.

5.95   Jackson, A.A., H.A. Zook, A solar system dust ring with the earth as its shepherd, Nature, **337**, 629–631, 1989.

5.96   James, J.F., M.J. Smeethe, Motion of the interplanetary dust cloud, Nature, **227**, 588–589, 1970.

5.97   Jessberger, E.K., J. Kissel, H. Fechtig, F.R. Krüger, On the average chemical composition of cometary dust, ESA SP-249, 27–30, 1986.

5.98   Jessberger, E.K., A. Christoferidis, J. Kissel, Aspects of the major element composition of Halley's dust, Nature, **332**, 691–695, 1988.

5.99   Kirsten, T., Time and the solar system, in *The Origin of the Solar System*, ed. by S.F. Dermott, 267–346, John Wiley & Sons, New York, Chichester, Brisbane, Toronto, 1978.

5.100 Kissel, J., The Giotto particulate impact analyzer, ESA SP-1077, 67–83, 1986.

5.101 Kissel, J., F.R. Krüger, Ion formation by impact of fast dust particles and comparison with related techniques, Applied Physics A, **42**, 69–85, 1987.

5.102 Koutchmy, S., P.L. Lamy, The F-corona and the circum-solar dust evidences and properties, in Ref. [5.5], pp. 63–74.

5.103 Lamy, P.L., Interaction of interplanetary dust grains with the solar radiation field, Astron. Astrophys., **35**, 197–207, 1974.

5.104 Lamy, P.L., J. Lefèvre, J. Millet, J.P. Lafon, Electronic charge of interplanetary dust grains: new results, in Ref. [5.5], pp. 335–339.

5.105 Lamy, P.L., J.M. Perrin, Zodiacal light models with a bimodel population, in Ref. [5.4], pp. 75–80.

5.106 Lamy, P.L., J.M. Perrin, Volume scattering function and space distribution of the interplanetary dust cloud, Astron. Astrophys., **163**, 269–286, 1986.

5.107 Lanzerotti, L.J., W.L. Brown, J.M. Poate, W.M. Augustniak, Low energy cosmic ray erosion of ice grains in interplanetary and interstellar media, Nature, **272**, 431–433, 1978.

5.108 Leinert, C., Zodiacal light – a measure of the interplanetary environment, Space Sci. Rev., **18**, 281–339, 1975.

5.109 Leinert, C., Dynamics and spatial distribution of interplanetary dust, in Ref. [5.5], pp. 369–375.

5.110 Leinert, C., M. Hanner, I. Richter, E. Pitz, The plane of symmetry of interplanetary dust in the inner solar system, Astron. Astrophys. **82**, 328–336, 1980.

5.111 Leinert, C., H. Link, E. Pitz, Rocket photometry of the inner zodiacal light, Astron. Astrophys., 30, 411–422, 1974.

5.112 Leinert, C., H. Link, E. Pitz, R.H. Giese, Interpretation of a rocket photometry of the inner zodiacal light, Astron. Astrophys., 47, 221–230, 1976.

5.113 Leinert, C., I. Richter, E. Pitz, M. Hanner, Four years of zodiacal light observations from the Helios space probes: evidence for a smooth distribution of interplanetary dust, in Ref. [5.4], pp. 15–18.

5.114 Leinert, C., I. Richter, B. Planck, Stability of zodiacal light from minimum to maximum of solar cycle, (1974–1981), Astron. Astrophys., 110, 111–114, 1982.

5.115 Leinert, C., I. Richter, E. Pitz, B. Planck, The zodiacal light from 1.0 to 0.3 AU as observed by the Helios space probes, Astron. Astrophys., 103, 177–188, 1981.

5.116 Leinert, C., I. Richter, E. Pitz, M. Hanner, Helios zodiacal light measurements – a tabulated summary, Astron. Astrophys., 110, 355–357, 1982.

5.117 Leinert, C., S. Röser, J. Buitrago, How to maintain the spatial distribution of interplanetary dust, Astron. Astrophys., 118, 345–357, 1983.

5.118 Leinert, C., E. Pitz, Zodiacal light observed by Helios throughout solar cycle no. 21: stable dust and varying plasma, Astron. Astrophys. 210, 399–402, 1989.

5.119 LeSergeant d'Hendecourt, L.B., P.L. Lamy, On the size distribution and physical properties of interplanetary dust grains, Icarus, 43, 350–372, 1980.

5.120 LeSergeant d'Hendecourt, L.B., P.L. Lamy, Collisional processes among interplanetary dust grains, an unlikely origin for the meteoroids, Icarus, 47, 270–281, 1981.

5.121 Levasseur-Regourd, A.C., R. Dumont, Absolute photometry of zodiacal light, Astron. Astrophys., 84, 277–279, 1980.

5.122 Lillie, F.C., OAO-2 observations of the zodiacal light, in The Scientific Results from OAO-2, ed. by A.O. Code, NASA SP-310, 95–108, 1972.

5.123 Lumme, K., E. Bowell, Photometric properties of zodiacal light particles, Icarus, 62, 54–71, 1985.

5.124 MacQueen, R.M., Infrared observations of the outer solar corona, Astrophys. J., 154, 1059–1076, 1968.

5.125 Maihara, T., K. Misutani, N. Hiromoto, H. Takami, H. Hasegawa, A balloon observation of the thermal radiation from the circumstellar dust cloud in the 1983 total eclipse, in Ref. [5.5], pp. 55–58.

5.126 Maucherat-Joubert, M., P. Cruvellier, J.M. Deharveng, Ultraviolet observation of the zodiacal light from the D2B-Aura satellite, Astron. Astrophys., 74, 218–224, 1979.

5.127 McDonnell, J.A.M., Microparticle studies by space instrumentation, in Ref. [5.3], pp. 337–426.

5.128 McDonnell, J.A.M., W.C. Care, D.G. Dixon, Cosmic dust collection by the capture all technique on the space shuttle, Nature, 309, 237–240, 1984.

5.129 Millman, P., Meteors and interplanetary dust, in Ref. [5.2], pp. 359–372.

5.130 Misconi, N.Y., On the photometric axis of the zodiacal light, Astron. Astrophys., 61, 497–504, 1977.

5.131 Morfill, G.E., E. Grün, The motion of charged dust particles in interplanetary space. I. The zodiacal dust cloud, Planet. Space Sci., 27, 1269–1282, 1979.

5.132 Morfill, G.E., E. Grün, C. Leinert, The interaction of solid particles with the interplanetary medium, in The Sun and the Heliosphere in Three Dimensions, ed. by R.G. Marsden, D. Reidel, Dordrecht, Boston, Lancaster, Tokyo, 455–474, 1986.

5.133 Morgan, D.H., K. Nandy, G.I. Thompson, The ultraviolet galactic background from TD-1 satellite observations, Mon. Not. R. Astr. Soc., 177, 531–544, 1976.

5.134 Morrison, D.A., E. Zinner, 12054 and 76215: new measurements of interplanetary dust and solar flare fluxes, Proc. 8th Lunar Sci. Conf., 841–863, 1977.

5.135 Mukai, T., R.H. Giese, Modification of the spatial distribution of interplanetary dust grains by Lorentz forces, Astron. Astrophys., 131, 355–363, 1985.

5.136 Mukai, T., T. Yamamoto, A model of the circumsolar dust cloud, Publ. Astron. Soc. Japan, 31, 585–595, 1979.

5.137 Mukai, T., T. Yamamoto, Solar wind pressure on interplanetary dust, Astron. Astrophys., 107, 97–100, 1982.

5.138 Mukai, T., H. Fechtig, R.H. Giese, S. Mukai, Evolution of albedos of cometary dirty ice grains, Astron. Astrophys., 167, 364–370, 1986.

5.139 Murdock, T.L., D. Price, Infrared measurements of zodiacal light, Astron. J., 90, 375–386, 1985.

273

5.140 Nagel, K., G. Neukum, J.S. Dohnanyi, H. Fechtig, W. Gentner, Density and chemistry of interplanetary dust particles derived from measurements of lunar microcraters, Proc. 7th Lunar Sci. Conf., 1021–1029, 1976.

5.141 Naumann, R.J., The near earth meteoroid environment, NASA-TN-D-3717, 1966.

5.142 Naumann, R.J., D.W. Jex, C.L. Johnson, Calibraton of Pegasus and Explorer XXIII Detector Panels, NASA TR-R-321, 1969.

5.143 Neukum, G., E. Schneider, A. Mehl, D. Storzer, G.A. Wagner, H. Fechtig, H.R. Bloch, Lunar craters and exposure ages derived from crater statistics and solar flare tracks, Proc. 3rd Lunar Sci. Conf., 3, 2793–2810, 1972.

5.144 Nishimura, T., Infrared spectrum of zodiacal light, Publ. Astron. Soc. Japan, 25, 375–384, 1973.

5.145 Öpik, E.J., The stray bodies in the solar system. Part I. Survival of cometary nuclei and the asteroids, Advances Astron. Astrophys., 2, 219–262, 1963.

5.146 Pailer, N., E. Grün, The penetration limit of thin films, Planet. Space Sci., 28, 321–331, 1980.

5.147 Parker, E.N., The perturbation of interplanetary dust grains by the solar wind, Astrophys. J., 139, 951–958, 1964.

5.148 Peterson, A.W., Experimental detection of thermal radiation from interplanetary dust, Astrophys. J., 148, L37–L39, 1967.

5.149 Pitz, E., C. Leinert, A. Schulz, H. Link, Colour and polarization of the zodiacal light from the ultraviolet to the near infrared, Astron. Astrophys., 74, 15–20, 1979.

5.150 Rahe, J., "Comets", Landolt-Börnstein, N.S., VI/2a, 202–228, 1981.

5.151 Reach, W.T., Zodiacal emission. I. Dust near the earth's orbit, Astrophys. J., 335, 468–485, 1988.

5.152 Röser, S., Can short period comets maintain the zodiacal light?, in Ref. [5.2], pp. 319–322.

5.153 Röser, S., H.J. Staude, The zodiacal light from 1500 Å to 60 micron, Astron. Astrophys., 67, 381–394, 1978.

5.154 Salama, A., P. Andreani, G. Dall'Oglio, P. de Bernardis, S. Masi, B. Melchiorri, F. Melchiorri, G. Moreno, B. Nisini, K. Shivanandan, Measurements of near- and far-infrared zodiacal dust emission, Astron. J., 92, 467–473, 1987.

5.155 Shapiro, I.I., D.A. Lautman, G. Colombo, The earth's dust belt: fact or fiction?, J. Geophys. Res., 71, 5695–5704, 1966.

5.156 Schmidt, T., H. Elsässer, Dynamics of submicron particles in interplanetary space, in Ref. [5.1], pp. 287–289.

5.157 Schramm, L.S., D.S. McKay, H.A. Zook, G.A. Robinson, Analysis of micrometeorite material captured by the SOLAR MAX satellite, Lunar and Planet. Sci., XVI, 736–737, 1985.

5.158 Schuerman, D.W., Inverting the zodiacal light brightness integral, Planet. Space Sci., 27, 551–556, 1979.

5.159 Schuerman, D.W. (Ed.), Light Scattered by Irregularly Shaped Particles, Plenum Press, New York, London, 1980.

5.160 Schwehm, G., M. Rhode, Dynamical effects on circum solar dust grains, J. Geophys. 42, 727–735, 1977.

5.161 Sekanina, Z., H.E. Schuster, Meteoroids from periodic comet d'Arrest, Astron. Astrophys., 65, 29–35, 1978.

5.162 Sekanina, Z., H.E. Schuster, Dust from periodic comet Encke: large grains in short supply, Astron. Astrophys., 68, 429–435, 1978.

5.163 Smith, O., N.G. Adams, H.A. Kahn, Flux and composition of micrometeoroids in the diameter range 1–10 μm, Nature, 252, 101–106, 1974.

5.164 Southworth, R.B., Z. Sekanina, Physical and dynamical studies of meteors, NASA CR-2316, 1973.

5.165 Stohl, J., On the distribution of sporadic meteor orbits, in Asteroids Comets Meteors, ed. by C.I. Lagerkvist and H. Rickman, 419–424, Uppsala Universität, Uppsala, 1983.

5.166 Sykes, M.V., L.A. Lebofsky, D.M. Hunten, F. Low, The discovery of dust trails in the orbits of periodic comets, Science, 232, 1115–1117, 1986.

5.167 Sykes, M.V., IRAS observations of extended zodiacal structures, Astrophys. J., 334, L55–L58, 1988.

5.168 Temi, P., P. de Bernardis, S. Masi, G. Moreno, A. Salama, Infrared emission from interplanetary dust, Astrophys. J., 337, 528–535, 1989.

5.169 Toller, G.N., J.L. Weinberg, The change in near-ecliptic zodiacal light brightness with heliocentric distance, in Ref. [5.5], pp. 21–25.

5.170   Trulsen, J., A. Wikan, Numerical simulation of Poynting–Robertson and collisional effects in the interplanetary dust cloud, Astron. Astrophys., **91**, 155–160, 1980.

5.171   Verniani, F., An analysis of the physical parameters of 5759 faint radio meteors, J. Geophys. Res., **78**, 8429–8462, 1973.

5.172   Wallis, M.K., M.H.A. Hassan, Stochastic diffusion of interplanetary dust grains orbiting under Poynting–Robertson forces, Astron. Astrophys., **151**, 435–441, 1985.

5.173   Weinberg, J.L., Zodiacal light and interplanetary dust, in Ref. [5.5], pp. 1–6.

5.174   Weinberg, J.L., J.G. Sparrow, Zodiacal light as an indicator of interplanetary dust, in Ref. [5.3], pp. 75–122.

5.175   Weiss-Wrana, K., Optical properties of interplanetary dust: comparison with light scattering by larger meteoritic and terrestrial grains, Astron. Astrophys., **126**, 240–250, 1983.

5.176   Whipple, F.L., On maintaining the meteoriotic complex, in Ref. [5.1], pp. 409–426.

5.177   Wyatt, S.P., Jr., The electrostatic charge of interplanetary grains, Planet Space Sci., **17**, 155–171, 1969.

5.178   Wyatt, S.P., Jr., F.L. Whipple, The Poynting–Robertson effect on meteor orbits, Astrophys. J., **111**, 134–141, 1950.

5.179   Zellner, B., E. Bowell, Asteroid compositional types and their distributions, in *Comets Asteroids Meteorites, Interrelations, Evolution and Origin*, ed. by A.H. Delsemme, 185–197, Univ. of Toledo, 1977.

5.180   Zook, H.A., O.E. Berg, A source for hyperbolic cosmic dust particles, Planet Space Sci., **23**, 183–203, 1975.

5.181   Zook, H.A., Evidence for ice meteoroids beyond 2 AU, in Ref. [5.4], pp. 375–380.

5.182   Zook, H.A., R.E. Flaherty, D.J. Kessler, Meteoroid impacts on the Gemini windows, Planet Space Sci., **18**, 953–964, 1970.

5.183   Zook, H.A., G. Lange, E. Grün, H. Fechtig, The interplanetary micrometeoroid flux and lunar primary and secondary microcraters, in Ref. [5.5], pp. 89–96.

Subject Index

Abel transform 29
Absorption efficiency 216
Active regions: *see* solar active regions
Albedo 222, 223, 236, 241, 244, 245, 252, 253
Alfvén 127, 140, 194
– critical radius: *see* critical point
– speed 39, 48–49, 77, 113, 157
– waves: *see* waves
Allen–Baumbach model 28
Angular broadening 60, 63, 70
Angular momentum flux 11, 26, 149–152
Anisotropy 66, 70, 134, 135, 148, 154
Apollo telescope mount: *see* Skylab
Archimedean spiral: *see* Parker spiral
Asteroid 1, 222, 240, 247, 248, 249, 254, 259
– bands 255
– belt 244, 255, 262
– family 255
– particle 246
Astrophysical aspects 10, 11, 150

Ballerina model 127, 140–142, 161, 164, 194
Balloon experiment 217
Bartels 163
Beta-meteoroid 209, 235, 236, 250, 264
Bipolar magnetic regions 99, 164, 200
Brightness integral 216, 253

Capture cell 211
Calibrations 146, 150, 152, 158, 159, 221
Carbonaceous chondrite 220, 222
Carrington 25
Central pit 213, 215
Characteristic speed 118, 136–141
Charge on dust grains 227
Chondritic abundance 220
– material 222
– meteorites 8, 220, 221, 223, 256, 258
Chromosphere 11
Cluster mission 7
Collection of dust 210
Collisions 209, 231, 232, 244, 255, 256, 260, 261, 263

– catastrophic 231, 232
– erosive 231
– life times 231, 261, 264, 265
Columnar electron content: *see* electron content
Comet 1, 8, 209, 246, 247, 249, 255, 259
– d'Arrest 256
– Encke 246, 256
– Gunn 259
– Halley 222, 226, 256, 257
– Tempel–Tuttle 247
– particle 246, 254, 258
Composition 211, 220, 221, 223, 257, 258
Compound streams 106, 166, 170
Compression of streams 4, 101, 106, 114, 118, 120, 126, 129, 131, 133, 135, 141, 169
Constant speed approximations 103, 106, 114, 128
Convection zone 11
Corona(l)
– density 27–39
– density distribution 28–34
– density fluctuations 37, 39, 53, 61–67, 69–70, 82
– density models 28–34
– density variations 34–39
– ellipticity: *see* flattening index
– emission 19–24
– expansion models: *see* solar wind
– heating 8, 11, 131, 142, 168
– holes 6, 15, 19–20, 23, 25–26, 34–35, 45, 52–55, 57–59, 61, 76, 99, 101, 103, 118, 121, 125, 128, 140, 154, 163, 164, 169, 196, 203
– mass ejections(CME) 6, 8, 9, 48, 70–84, 133, 135, 147, 161, 164, 165, 166
– morphology 14–27
– rotation 25–27
– streamers 8, 15, 18, 20, 34–35, 41, 61, 70–71, 78, 130–132, 135, 141, 162, 169
– turbulence 52–70
Coronographs 6, 15-17, 27, 30, 35, 42, 55, 71, 74, 80, 135, 162, 166, 167
Corotating interaction regions (CIR) 106–120, 129, 130, 132, 135–141, 144, 166, 191
Corotating shocks: *see* shocks